T0328935

APPLIED FOOD SCIENCE

APPLIED FOOD SCIENCE

edited by:

Bart Wernaart

Bernd van der Meulen

Wageningen Academic
P u b l i s h e r s

EAN: 9789086863815
e-EAN: 9789086869336
ISBN: 978-90-8686-381-5
eISBN: 978-90-8686-933-6
DOI: 10.3920/978-90-8686-933-6

First published, 2022

© Wageningen Academic Publishers
The Netherlands, 2022

The individual contributions in this publication and any liabilities arising from them remain the responsibility of the authors.

The editors are grateful to Renate Smallegange for many years of cooperation

Table of contents

1. Introduction

Bernd M.J. van der Meulen[1,2,3,4]* and Bart F.W. Wernaart[5]*

[1]European Institute for Food Law, Zwanenwater 40, 1187 LC Amstelveen, the Netherlands; [2]Department of Food and Resource Economics, University of Copenhagen, Rolighedsvej 23, 1958 Frederiksberg C, Denmark; [3]Renmin University of China Law School, Huixian Rd, Beijing 100086, China P.R.; [4]Food Law Academy (www.Food-Law-Academy.com); [5]School of Business and Communication, Fontys University of applied sciences, P.O. Box 347, 5600 AH Eindhoven, the Netherlands; b.wernaart@fontys.nl; bernd.vandermeulen@food-law.nl

The food sector is a knowledge-intensive sector.

We live in times of transition. Climate change, demographic developments and war call for crops that are suitable for the new conditions, new sources of proteins, more sustainable methods of production and nutrition, increased attention to health, and a shift towards more plant-based sources of nutrition; in short, there is a need for innovation.

Businesses that aim to provide innovative foods to consumers face many challenges. The source materials they use should result in a safe product, and the production methods that are applied should achieve the intended result. Practically all foods are derived from plants, animals or microorganisms. Their properties are encoded in their DNA. Safety may be a matter of toxicology or microbiology. In processing, microbiology may play a role either in an instrumental way or as a hazard. Choices made in the selection and sourcing of food raw materials may raise ethical questions. It may take marketing efforts to bring the product to the attention of consumers. Psychology may be applied in nudging consumers towards more healthy consumption patterns. At every step of the way regulatory hurdles may need to be overcome. In each of these aspects another expert may take the lead in moving forwards from the idea to the consumer's plate: from vision to food.

The book you are holding in your hands right now is a response to the abovementioned challenges and is the result of the evolving conceptualisations and hard work of the authors. In the context of the European Institute for Food Law, the initial plan was to prepare a book on 'regulatory science'. This label is used for those sciences that are connected to the food legal system and

Bart Wernaart and Bernd van der Meulen (eds)
Applied food science
DOI: 10.3920/978-90-8686-933-6_1, © Bernd M.J. van der Meulen and Bart F.W. Wernaart 2022

that need to be applied by food businesses in order to be able to comply with their obligations. As (food) lawyers are not generally trained as scientists, they will not usually be able to provide all the necessary knowledge themselves, but will instead need to cooperate with the appropriate experts. So, the book initially intended to bridge the gap in background and culture between these food lawyers and the scientists with whom they work. At a very early stage we realised, however, that lawyers are not unique in their need to understand other food experts. This need applies to each expert in the food sector in more or less equal measure. As a consequence, the scope of the design was broadened to be all-inclusive. The purpose of this book is to introduce you to a wide range of sciences relevant to the food sector – i.e. food sciences. The aim is not to make you an expert in each and every field, but to provide you with sufficient knowledge and background concerning the content and methods of the different fields so that you can communicate and cooperate in a meaningful way with those who are the experts.

Some food sciences have a geographical scope. Other sciences are universal. Whether a substance is toxic does not depend on the country where the substance is present. But how to respond to the risk a toxic substance presents is a matter of choice. The way choices are made – on the basis of which law or policy, for example – may differ from country to country. Where geographical scope plays a role, we apply the EU perspective.

This book is organised around several themes (for a full overview see Figure 1.1). The first chapters mainly address normative issues and food governance aspects. These include a chapter on food legal requirements (Chapter 2; Van der Meulen and Wernaart), food ethics (Chapter 3; Wernaart), consumer protection (Chapter 4; Van Loon and Wernaart), food policy (Chapter 5; Gürsoy) and risk analysis (Chapter 6; De Boer). Food legal issues define what 'can and may and must' be done, i.e. what is allowed for food businesses, required or forbidden by law. Ethics is more about the morals of the food business. Distinguishing right from wrong is a business responsibility that is not limited to what legislation imposes. Many of the rules that apply to food businesses are intended to protect and sometimes also empower consumers. Governments design policies to pursue their objectives. In food policy in the EU the emphasis shifts between food security and food safety but other objectives also play a role. As of the new millennium food law and policy in the EU have the ambition to be science based. To this end, a methodology has been adopted known as risk analysis, which was initially developed in the USA. Being science-based means pledging not to follow the whims of successive political majorities but to limit interventions to what is necessary in scientific terms.

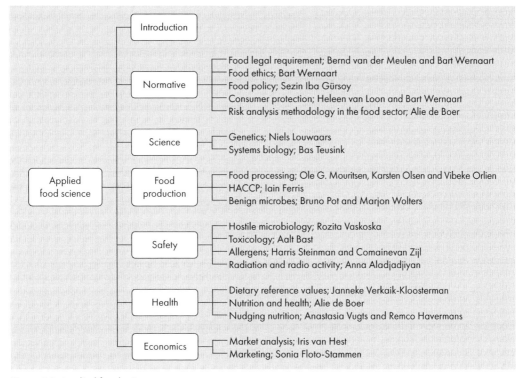

Figure 1.1. Applied food science.

The normative chapters are followed by two chapters on fundamental science. These chapters on genetics (Chapter 7; Louwaars) and systems biology (Chapter 8; Teusink) provide an understanding of developments in and among living organisms and apply both to potential food sources and to their consumers. These chapters provide a foundation for much of what follows.

Next, the book addresses food processing. First from the culinary perspective (Chapter 9; Mouritsen *et al.*). What do different forms of processing do to the food? Then from the safety perspective (Chapter 10; Ferris). What do businesses have to do to ensure food is safe (HACCP)? Thereafter, there is a discussion about the role that microorganisms may have in supporting humans (Chapter 11; Pot and Wolters). Microorganisms may contribute to food production as agents of fermentation. The same or other microorganisms may live with humans in symbiosis and support our health. When supplied through food we call the latter 'probiotics'.

Subsequent chapters focus on food safety hazards. Chapter 12 (Vaskoska, 2022) also looks at microorganisms, this time not the friendly ones addressed in Chapter 11 but the pathogens, the ones that make us ill. Other hazards that may be foodborne include chemicals, allergens and radioactivity. The first are discussed in the chapter on toxicology (Chapter 13; Bast, 2022), allergens are discussed in Chapter 14 (Steinman and Van Zeijl, 2022, and radioactivity in Chapter 15 (Aladjadjiyan, 2022). Like microorganisms, radioactivity is not only a threat to human health, but it can also play a role in protecting us.

It goes without saying that food is not just a potential threat to our health. Food is what sustains us and makes us what we are. To achieve optimal health we need to consume the right quantities of nutrients. Chapter 16 (Verkaik-Kloosterman, 2022) explains the science assessing what the right quantities are. The next chapter (Chapter 17; De Boer, 2022b) broadens the scope of the discussion on nutrition and health. It also addresses the claims that food businesses can make in this regard. Food-related health does not just depend on knowledge regarding and availability of healthy options. It depends most on the actual consumption decisions consumers make. Health is just one consideration informing such decisions. Many other factors including habit, pleasure, craving and convenience play a role as well, and often a dominant one. Chapter 18 (Vugts and Havermans, 2022) uses behavioural psychology to discuss possibilities to nudge consumers towards more healthy choices.

The book concludes with economic chapters. Market analysis (Chapter 19; Van Hest, 2022) helps businesses in deciding on their product portfolio, while marketing (Chapter 20; Floto-Stammen, 2022) provides instruments to actually reach customers with the products and services the business has to offer.

Although there is logic in the sequence of the chapters, they do not build on each other in the sense that it would be necessary to read one chapter first to be able to understand another chapter. Each chapter can be read independently and in the sequence preferred by the reader. The groupings of the chapters set out in the paragraphs above is purely indicative. Most chapters have a broader scope than this grouping might suggest.

Every chapter follows more or less rigidly a format beginning with a case illustrating how the topic discussed in the chapter may operate in practice. This case is followed by an introduction to some key concepts. In the remainder of the chapter, case and concepts are combined to provide an understanding of what the field of expertise discussed is all about and where it may be heading in the foreseeable future.

The team of authors consists of people who are acknowledged experts in their topic. They have been invited through the network of the editors. This may help explain why a relatively high percentage of authors is from the Netherlands. In inviting authors to join our team, we were primarily led by the expertise of the contributing authors. However, as is evidenced from the overview of authors, the team turned out to be nicely gender-balanced.

We have chosen to use gender-neutral language where appropriate. We follow – as much as possible – the recommendation in the English Style Guide of the European Commission[1] to phrase in plural where possible. In English, the plural is usually gender-neutral. The main purpose of gender-neutral language is to protect women from becoming invisible due to the use of the male form of words as a generic form. This purpose is not served where words are not used in a general meaning but to refer to actual women or man. In such situations we believe that the use of gendered language is justified.

There is no agenda – political or otherwise – in this book apart from knowledge dissemination. In so far as opinions are expressed, these are the opinions of the authors of the individual chapters. Between authors different views may be apparent. As will be the case between the readers.

Dear reader, in preparing this book we have pictured you as interested in the food sector with a more or less academic level of thinking. You may be a student, researcher or a practitioner. If you are a practitioner, you probably work in the food sector at a food business, NGO, research institute, public authority or consumers organisation. You may be an expert in one or more of the fields covered in this book. If this book opens some doors for you to additional fields of food sciences, our mission is accomplished.

References

Aladjadjiyan, A., 2022. Radiation and radioactivity in the food sector. In: Wernaart, B.F.W. and Van der Meulen, B.M.J. (eds) Applied Food Science. Wageningen Academic Publishers, Wageningen, the Netherlands, pp. 331-353.

Bast, A., 2022. Food toxicology. In: Wernaart, B.F.W. and Van der Meulen, B.M.J. (eds) Applied Food Science. Wageningen Academic Publishers, Wageningen, the Netherlands, pp. 267-288.

De Boer, A., 2022a. Risk analysis for foods. In: Wernaart, B.F.W. and Van der Meulen, B.M.J. (eds) Applied Food Science. Wageningen Academic Publishers, Wageningen, the Netherlands, pp. 99-123.

---

[1] See: https://ec.europa.eu/info/sites/info/files/styleguide_english_dgt_en.pdf.

De Boer, A., 2022b. Nutrition and health. In: Wernaart, B.F.W. and Van der Meulen, B.M.J. (eds) Applied Food Science. Wageningen Academic Publishers, Wageningen, the Netherlands, pp. 385-406.

Ferris, I.M., 2022. Hazard analysis and critical control points (HACCP). In: Wernaart, B.F.W. and Van der Meulen, B.M.J. (eds) Applied Food Science. Wageningen Academic Publishers, Wageningen, the Netherlands, pp. 187-213.

Floto-Stammen, S., 2022. Food marketing. In: Wernaart, B.F.W. and Van der Meulen, B.M.J. (eds) Applied Food Science. Wageningen Academic Publishers, Wageningen, the Netherlands, pp. 453-479.

Gürsoy, S.I., 2022. Food policy. In: Wernaart, B.F.W. and Van der Meulen, B.M.J. (eds) Applied Food Science. Wageningen Academic Publishers, Wageningen, the Netherlands, pp. 85-98.

Louwaars, N.P., 2022. Genetics. In: Wernaart, B.F.W. and Van der Meulen, B.M.J. (eds) Applied Food Science. Wageningen Academic Publishers, Wageningen, the Netherlands, pp. 125-140.

Mouritsen, O.G., Olsen, K. and Orlien, V., 2022. Food processing. In: Wernaart, B.F.W. and Van der Meulen, B.M.J. (eds) Applied Food Science. Wageningen Academic Publishers, Wageningen, the Netherlands, pp. 157-185.

Pot, B. and Wolters, M., 2022. Food microbiology. In: Wernaart, B.F.W. and Van der Meulen, B.M.J. (eds) Applied Food Science. Wageningen Academic Publishers, Wageningen, the Netherlands, pp. 215-245.

Steinman, H.A. and Van Zijl, C., 2022. Food allergies and food allergen control. In: Wernaart, B.F.W. and Van der Meulen, B.M.J. (eds) Applied Food Science. Wageningen Academic Publishers, Wageningen, the Netherlands, pp. 289-329.

Teusink, B., 2022. Systems biology. In: Wernaart, B.F.W. and Van der Meulen, B.M.J. (eds) Applied Food Science. Wageningen Academic Publishers, Wageningen, the Netherlands, pp. 141-155.

Van der Meulen, B.M.J. and Wernaart, B.F.W., 2022. Food law and regulatory affairs. In: Wernaart, B.F.W. and Van der Meulen, B.M.J. (eds) Applied Food Science. Wageningen Academic Publishers, Wageningen, the Netherlands, pp. 21-43.

Van Hest, I., 2022. Food market analysis. In: Wernaart, B.F.W. and Van der Meulen, B.M.J. (eds) Applied Food Science. Wageningen Academic Publishers, Wageningen, the Netherlands, pp. 429-452.

Van Loon, H. and Wernaart, B.F.W., 2022. Consumer protection. In: Wernaart, B.F.W. and Van der Meulen, B.M.J. (eds) Applied Food Science. Wageningen Academic Publishers, Wageningen, the Netherlands, pp. 65-83.

Vaskoska, R., 2022. Hostile microbiology. In: Wernaart, B.F.W. and Van der Meulen, B.M.J. (eds) Applied Food Science. Wageningen Academic Publishers, Wageningen, the Netherlands, pp. 247-266.

Verkaik-Kloosterman, J., 2022. Dietary reference values. In: Wernaart, B.F.W. and Van der Meulen, B.M.J. (eds) Applied Food Science. Wageningen Academic Publishers, Wageningen, the Netherlands, pp. 355-384.

Vugts, A. and Havermans, R., 2022. Nudging nutrition. In: Wernaart, B.F.W. and Van der Meulen, B.M.J. (eds) Applied Food Science. Wageningen Academic Publishers, Wageningen, the Netherlands, pp. 407-428.

Wernaart, B.F.W., 2022. Food ethics. In: Wernaart, B.F.W. and Van der Meulen, B.M.J. (eds) Applied Food Science. Wageningen Academic Publishers, Wageningen, the Netherlands, pp. 45-64.

2. Food law and regulatory affairs

This is what you need to know about EU food law

Bernd M.J. van der Meulen[1,2,3,4*] and Bart F.W. Wernaart[5]

[1]European Institute for Food Law, Zwanenwater 40, 1187 LC Amstelveen, the Netherlands; [2]Department of Food and Resource Economics, University of Copenhagen, Rolighedsvej 23, 1958 Frederiksberg C, Denmark; [3]Renmin University of China Law School, Huixian Rd, Beijing 100086, China P.R.; [4]Food Law Academy (www.Food-Law-Academy.com); [5]School of Business and Communication, Fontys University of applied sciences, P.O. Box 347, 5600 AH Eindhoven, the Netherlands; bernd.vandermeulen@food-law.nl

'No man is above the law and no man is below it:
nor do we ask any man's permission when we ask him to obey it.'
  – Theodore Roosevelt

Abstract

In the food business the position of legal experts is sometimes a lonely one: they are the ones who know the 'rules of the game' and speak truth to power. And this expertise is not always received with great enthusiasm by other professionals. Sometimes, frankly, law is perceived to 'stand in the way' of business developments. Nevertheless, proper knowledge of food law can also prevent a lot of harm, and safeguard against damage or unnecessary expenses. In this chapter we offer a brief introduction to law in general, and European food law in particular. After all, in the European Union the most important legislation on food is produced by the European legislature, and to a lesser extent the lawmakers of the Member States. We will introduce the framework regulation that is at the core of all this: the General Food Law and its most important features, definitions and procedures, as well as the European Food Safety Authority (EFSA) that provides the scientific substantiation for EU food law and policy. In this context, we reflect on the legal meaning of key concepts as 'food', 'unsafe food', 'informed consumer choice' and 'food business responsibilities'. Furthermore, we will discuss important issues that relate to food production and sales from a legal perspective: among other things, we will address the questions 'what ingredients can be used?'; 'what product benefits can be claimed?'; 'how should business processes be organised?'; 'what is to be done when food does not meet food safety requirements?'; and 'how can food law be enforced?' Then, we will focus on legal methodology: what are the most important sources of EU food law, and how can they be used? Finally, we will look ahead and discuss future challenges in the field of EU food law, and draw conclusions. Among other things, climate change is considered to be one of the most complex challenges that lie ahead, especially in the context of food security, and it needs to be addressed as urgently as possible. These developments will probably lead to dramatic changes in food law as well.

Key concepts

- ► Law organises just behaviour.
- ► Substantive law defines the content of just behaviour.
- ► Procedural law defines how to maintain the just behaviour.
- ► Public law governs the relationship between states and between the state and its citizens.
- ► Private law governs the relationship between citizens.
- ► Food law is a term used to address all the rules (mostly public but also private), legal and administrative requirements that apply to food in general and to food safety in particular.
- ► Food (or foodstuff) means any substance or product, whether processed, partially processed or unprocessed, intended to be, or reasonably expected to be ingested by humans.
- ► Food is considered unsafe when it is deemed to be injurious to health or unfit for human consumption.
- ► The European Food Safety Authority (EFSA) provides the scientific substantiation for EU food law and policy.
- ► The informed choice principle means that a consumer is able to make an autonomous food choice as a result of transparent communication by food businesses, without any misleading information.
- ► Food business are any undertaking, whether for profit or not and whether public or private, carrying out any of the activities related to any stage of production, processing and distribution of food.
- ► Food business operator, the person(s) responsible in the food business for complying with food law.
- ► The HACCP principles require businesses to analyse their processes to determine where hazards may occur and how they can be kept under control.
- ► Hygiene is the measures and conditions necessary to control hazards and to ensure fitness for human consumption of a foodstuff taking into account its intended use.

▶ Foods covered by EU authorisation requirements include food additives, GMOs and novel foods.
▶ Food additives are substances not normally consumed as a food that are added for a technological purpose.
▶ GMOs are foods that consist of or are made from genetically modified organisms.
▶ Novel foods are (other) foods that have not been consumed to a significant degree in the EU prior to May 1997.
▶ Food withdrawal takes place when a food business operator has reason to believe that a food which it has imported, produced, processed, manufactured or distributed is not in compliance with the food safety requirements.
▶ GRAS means Generally Recognized As Safe by experts qualified to make such an assessment, and is used in the USA Federal Drug and Cosmetic Act.

Case 2.1. We are what we eat: about contraceptive pills, sugar water, pigs and pork.

On 19 December 2017 the Court of Appeals in the Hague (the Netherlands) gave an intermediate ruling (ECLI:NL:GHDA:2017:3936) in a case between a pig farm (in the Netherlands) and a pharmaceutical business – AHP Manufacturing B.V. using the trade name Wyeth Medica Ireland ('Wyeth'). It all started around the turn of the millennium. Wyeth produced the lady-pill in Ireland. The pill as sold to consumers is sugar-coated. A waste product from the coating process was residue water with sugar content. Via a broker, Cara Environmental Technology Ltd. in Ireland, Wyeth shipped this waste material to Bioland, a waste processing business in Belgium. Although pharmaceutical waste is classified as hazardous, Wyeth failed to check whether Bioland had all the permits required to process pharmaceutical waste. Which, in fact, it had not. Moreover, Wyeth had not notified the authorities as would have been its statutory duty. Instead of processing it, Bioland sold the sugar-water to be used in animal feed. This included water contaminated with MPA. MPA (Medroxyprogesterone acetate) is an artificial hormone used to prevent pregnancy. The contamination of feed with MPA was discovered when breeding sows stopped producing piglets. Isn't it said that we are what we eat? From the feed, the contamination was transferred to the animals and through the consumption of their meat to humans. Once the presence of MPA in the food chain was discovered, the animals and their products were considered unfit for human consumption and had to be taken off the market. Along with BSE (mad cow disease), dioxin and several more, the MPA crisis was one of the big food scares around the turn of the millennium that sparked the reform of food law in the European Union (Van der Meulen and Wernaart (eds), 2020: Chapter 6).

Bioland went bankrupt. So, the pig farm sued Wyeth for damages it had suffered. Wyeth defended itself stating, among other things, that the pig farm had itself to blame as it had not checked the supplier's credentials including GMP+ certification. GMP+ (Good Manufacturing Practices plus; https://www.gmpplus.org/en) is a private standard for the feed sector. Due to the private nature of the scheme, participation is not compulsory (Van der Meulen (ed.), 2011).

The Court of Appeals ruled that Wyeth had acted negligently by discarding its waste without ascertaining that the receiving business was entitled and able to process it in the proper manner, and without notifying the authorities. Therefore, it was liable for the damages suffered by the pig farm. The pig farm had also acted negligently by not checking the origin of the sugar-water it used for feeding its pigs. Therefore, it had to bear part of the damages itself. On 25 September 2018 the Court of Appeals gave its final ruling (ECLI:NL:GHDHA:2018:2468) indicating that the damages should be divided. 30% should be borne by Wyeth and 70% by the pig farm. On 16 October 2020 the Supreme Court of the Kingdom of the Netherlands (ECLI:NL:HR:2020:1628) upheld the decisions of the Court of Appeals. At the time of writing the parties are still discussing the amount of the total damages suffered (to be shared 30-70).

2.1 Introduction and contemporary issues

When in 2009 the Food Law Academy (www.food-law-academy.com) sent a team of teachers to Sierra Leone with the purpose of helping to empower food exporters from this country to enter the EU market, the teachers were welcomed by the Vice President of the republic, Mr Samuel Sam-Sumana. In his welcome speech, the Vice President stressed the importance of understanding EU food law in order to realise the ambition of benefiting from exports of food products to the EU. 'If you want to play cricket, you have to know the rules.'

This metaphor fully applies to the food sector. If you want to be successful in the food business, regardless of whether you are an exporter in Sierra Leone, a multinational food business or an SME in the EU, you need to know the rules of the game. In the food sector the rules of the game are known as 'food law'.

2.1.1 But first we kill all the lawyers

Within food businesses, ensuring compliance with food legal requirements – i.e. 'knowing the rules of the game' – is usually the responsibility of a regulatory affairs (RA) manager or department. Sometimes this function is combined with other functions such as quality assurance (QA). The regulatory affairs people find their counterparts in public authorities, in particular in regulators at EU level and national level and in national food safety inspectors in the Member States where the business is located and where the business trades. People come from different backgrounds to regulatory affairs. Sometimes they are lawyers, sometimes they have a background in management or one of the food sciences. Most of them need to learn an important part of the tricks of the trade on the job.

People in legal positions are not always looked upon favourably. Although its meaning is subject to debate, Shakespeare's second most famous[2] quote is usually greeted with a smile: 'The first thing we do, let's kill all the lawyers' (Henry VI part 2, Act 4 scene 2). Lawyers are subject to many jokes. In the punchline they are destined for a bad end. Rarely does one find similar joy at the demise of cooks, bakers, or product developers. What is it about legal people that sparks such responses? Many explanations have been proposed. One element without a doubt is their adherence to the law. They often stand in the way of the realisation of people's dreams. Not only your opponent's lawyer may block your way, even your own may tell you that what you want is not in accordance with the law. To a lawyer, the law presents a higher authority than the business manager. Their insistence on hearing both sides of the argument may further contribute to suspicion about their loyalties. Also, resolving legal issues may take time. In the MPA case set out above, litigation was still ongoing twenty years after the damage occurred. In the context of a product launch, in particular, time is a great asset – or a great cost.

Regulatory affairs people are often perceived as individuals whose vocabulary is missing the word 'yes'. As a consequence, they are kept out of product development and marketing strategy as long as possible. This in turn brings frustration when the new product or the new strategy runs foul of the law. Life would be so much nicer in the food business if regulatory input were sought on questions such as 'can I use this ingredient?' or 'can I claim these benefits?' as early in product development as possible.

This chapter addresses these questions from the food legal/regulatory perspective. Law differs from country to country. In this chapter we focus on the EU, which we will explain further in Section 2.2.

---

[2] The first, of course, is: 'To be or not to be' (Hamlet); the third 'What's in a name? That which we call a rose By any other name would smell as sweet' (Romeo and Juliet).

2.2 What is European food law, and why is it important?

In this section, we will discuss the most important concepts that are needed to understand food law. We will briefly explore some basics of law and how law differs from related disciplines; we will furthermore discuss why and how food law is mostly dealt with in the context of the European Union. Then, we will briefly touch upon the definition of food and unsafe food, the principle of an informed consumer choice, and the main legal responsibilities of a food business operator.

2.2.1 Law

The purpose of law is to organise just behaviour in society (Wernaart, 2017: 11). This is done through substantive rules that define the content of just behaviour, and procedural rules that are adopted to maintain the just behaviour. Considering Case 2.1, the duty to inform authorities when shipping hazardous waste materials is an example of substantive law, where the procedure on how and when to inform the authorities, as well as rules regarding consequences when the obligation to inform is not met, are examples of procedural law. Law organises three specific relationships: the relationship between states; the relationship between the state and citizens, and the relationship between citizens. The first two legal relationships are governed by public law, and the latter by private law (Figure 2.1). Most EU law is therefore considered to be public law (it mostly addresses Member States). However, the law that governs the legal liability of the food business can be considered private law, since it addresses the legal remedies of consumers against business, or the legal remedies between businesses. Please note that in law a business is considered equal to a natural person. The distinction between public and private law is important, because the formal rules in both areas are significantly different.

national level   Law is adopted by the competent legislator. At the national level, the highest legislature is usually the elected – mostly bicameral – parliament in cooperation with the government. For instance, in Germany this is the Bundestag (House of Representatives) and the Bundesrat (Federal Council). The first is directly elected by the German people, whereas the latter is composed of representatives of the various States within the German Federation. Please note that laws are also produced by other legislators, for instance at the level of the federated areas or municipalities.

European level   At the European level, the most important laws are produced in procedures involving the European Commission, which initiates the law-making, and the Council of Ministers together with the European Parliament. Their laws are

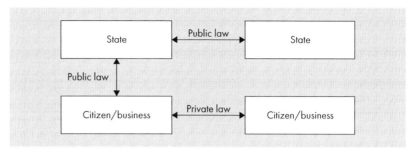

Figure 2.1. Public and private law.

mostly labelled as a regulation or a directive when they regulate generally binding issues, and when laws govern individual cases, they are called a decision. For instance, the General Food Law (as will be discussed in Section 2.2.2) is a regulation (Regulation (EC) 178/2002) (EC, 2002), the most important EU law on food waste is a directive (Directive 2018/851) (EC, 2018), and the permission of the European Commission to proceed with a proposed merger between two large food business operators is a decision.

policy

Law differs from policy insofar as policy is usually adopted by the executive branch (for instance, a national government or the European Commission) to make use of its margin of discretion offered by legislation in a coherent manner or to propose new legislation (see Chapter 5; Gürsoy, 2022). For instance, at the European level, the Directorate-General SANTE develops and carries out the EU Commission's policies in the field of health and food safety. Such policies are always within the margin of EU law and are aimed at executing EU laws.

ethics

While the question of what is right and wrong is important in both disciplines, law differs from ethics. Ethics is about values and virtues that define morality at the individual or organisational level (Chapter 3; Wernaart, 2022), and contribute to ethical decision-making. Law focuses on the formulation of societal norms that are put in a legally binding form so that they can be enforced in a court. Something can be considered morally praiseworthy – such as a healthy eating habit – but it is not a legal norm that can be enforced. Also, something can be legal – such as the production and sales of alcoholic products – and at the same time considered immoral by some.

2.2.2 Food law of the European Union

First and foremost, the European Union is one of the most advanced forms of economic integration worldwide with unique characteristics. Amongst other things, the EU has its own institutions that are able to produce laws that are of a supranational nature. When joining the Union, a Member State transfers some of its sovereignty to the Union and accepts that Union law can prevail over national legislation. Most EU legislation aims at facilitating free trade, and levels the playing field in competition. This results in harmonisation legislation, to make sure competitors are obliged to compete with one another based on the same product rules. The Court of Justice of the European Union has always played an important role in the process of economic integration. Landmark rulings that helped develop free trade within the Union have dealt with food-related characteristics or sales modalities. Famous examples are the Dassonville ruling (case C-8/74), the Cassis de Dijon ruling (Case C-120/78) and the Keck and Mithouard ruling (C-267/91 and C-268/91). Furthermore, a harmonised approach in food safety was prompted after various food crises in the nineties, shifting the focus of EU food law from the market to safety (Van der Meulen, 2020).

The label 'food law' is used for all the rules, legal and administrative requirements that apply to food in general and to food safety in particular. Food law can be considered a (relatively new) functional field of law (Urazbaeva *et al.*, 2019). In the 21st century most food law originates from the EU and to a lesser extent from its member states. It can be found in directives, but mostly in regulations of the EU. The basic regulation is Regulation (EC) 178/2002 of the European Parliament and of the Council laying down the general principles and requirements of food law, establishing the European Food Safety Authority and laying down procedures in matters of food safety (EC, 2002). This regulation is often referred to as the General Food Law Regulation, or simply the General Food Law (GFL). The definition of food law appears in Article 3(1) GFL.

According to Article 5 GFL the purpose of food law is to protect life and health and other interests of consumers and to ensure that consumers can make informed choices about the food they consume.

Food law aims to be science based (Article 6 GFL). This means, among other things, that the assessment of the safety of foods uses sciences, such as toxicology, microbiology and genetics. Such assessment is used, for example, for the authorisation of innovative foods and for the setting of limits to the presence of chemicals or microorganisms in food products. The scientific advice needed by the decision makers is provided by the European Food Safety Authority (EFSA) which is located in Parma (Italy). EFSA reviews the scientific evidence provided by applicants for authorisation or collects data regarding substances and organisms of concern. Decisions are usually taken by the European Commission – not by EFSA.

2.2.3 Food

It goes without saying that the key concept par excellence in food law is 'food'. If something is a food within the meaning of the legal definition, then it is subject to the rules and requirements of food law. There are many different definitions of food. The type of definition needed depends on the purpose for which it is used. The main purpose of food law is to protect consumers from harm. Therefore, the definition must cover anything that cannot cause harm when consumed. We eat to sustain our body. Consumption of macro- and micronutrients enables us to lead an active life. We eat to enjoy the pleasure of taste and texture. None of these elements are represented in the legal definition of food. Food safety law primarily addresses food as a potential pathway of harm into the human body. Therefore, a core element in the definition of food is 'intended to be or reasonably expected to be ingested'. Thus, the legal definition covers everything – with some limitations – that is put on the market with the purpose of consumption and even what – by its nature – may be expected to be ingested (Figure 2.2). The legislator chose the word 'ingested' instead of 'digested'. Digestion relates to what happens to the food in the body. Ingestion is about oral intake into the body regardless of the purpose. The definition further specifies that it covers products and substances, whether raw, processed or semi-processed. Therefore, the definition not only covers the final product as it appears on the consumer's plate but also its ingredients and raw materials. Food includes drink as well. Such a wide definition begs the question when in its lifecycle something begins to be a food. The definition says for plants that they are covered after harvest and animals after slaughter. Given that some animals – such as oysters – are eaten alive, an exception is included stating that living animals are also covered that are prepared to be placed on the market for human consumption.

Article 2
Definition of 'food'
For the purposes of this Regulation, 'food' (or 'foodstuff') means any substance or product, whether processed, partially processed or unprocessed, intended to be, or reasonably expected to be ingested by humans.
'Food' includes drink, chewing gum and any substance, including water, intentionally incorporated into the food during its manufacture, preparation or treatment. It includes water after the point of compliance as defined in Article 6 of Directive 98/83/EC and without prejudice to the requirements of Directives 80/778/EEC and 98/83/EC.
'Food' shall not include:
a) feed;
b) live animals unless they are prepared for placing on the market for human consumption;
c) plants prior to harvesting;
d) medicinal products within the meaning of Council Directives 65/65/EEC and 92/73/EEC;
e) cosmetics within the meaning of Council Directive 76/768/EEC;
f) tobacco and tobacco products within the meaning of Council Directive 89/622/EEC;
g) narcotic or psychotropic substances within the meaning of the United Nations Single Convention on Narcotic Drugs, 1961, and the United Nations Convention on Psychotropic Substances, 1971;
h) residues and contaminants.

Figure 2.2. Legal definition of food in Article 2 of Regulation (EC) 178/2002 (EC, 2002).

Products that are covered by their own legislation such as medicines and cosmetics are excluded from the definition. A synthetic hormone, such as MPA as discussed in Case 2.1, would not classify in its own right as a food.

unsafe food

Food shall not be placed on the market if it is unsafe (Article 14(1) GFL). Food is considered unsafe when it is deemed to be injurious to health or unfit for human consumption (Article 14(2) GFL). In assessing whether a food is unsafe, normal conditions of use are taken into account as well as the information available to consumers.

The concept 'unsafe' covers injury to future generations. Case 2.1 provides an example of what this could mean in practice. When food is contaminated with a synthetic hormone preventing conception, the future generation may suffer the ultimate harm of not existing at all.

In principle, it is the responsibility of food business operators to decide whether a food product is safe or unsafe. For certain categories of products the evidence that they are safe must be provided in an authorisation procedure before a product belonging to this category can be placed on the market as such or as an ingredient in another product.

2.2.4 Informed choice

> It's easier to fool people
> than to convince them that they have been fooled.
> – Mark Twain

The purpose of food law is to protect consumers' life and health and other consumer interests, in particular the right of consumers to be able to make informed choices about the food they consume. The right to informed choice is severely undercut when consumers are misled into believing that a food is something which in reality it is not. We find the prohibition to mislead consumers in several places in the legislation, including the General Food Law and the Food Information Regulation. Also, the Court of Justice of the EU has contributed to shaping this notion. In a landmark case (CJEU 4 June 2015, C-195/14, ECLI:EU:C:2015:361 'Teekanne') it ruled that a food package may still be misleading even if the list of ingredients provides all the relevant facts.

misleading information

Certain types of information are considered misleading by legal definition. It is never allowed to claim that a food has the property of preventing, treating or curing a human disease (Article 7(3) Regulation on the Provision of Food

Information to Consumers – FIC). This is the co-called prohibition of medicinal claims (Van der Meulen and Bremmers 2015). In Chapter 4 (Van Loon and Wernaart, 2022) we will further discuss the matter of consumer protection from a broader perspective.

2.2.5 Food business operators

Food law protects the interests of consumers. It provides powers to public authorities, but the starring role is reserved for food businesses. Food businesses place food on the market, therefore they are best placed to ensure that the food is safe and that consumers are well-informed. Food businesses are defined as 'any undertaking, whether for profit or not and whether public or private, carrying out any of the activities related to any stage of production, processing and distribution of food' (Article 3(2) GFL). Within the food business, the persons in charge are specifically addressed. These are the so-called food business operators (FBOs) (Article 3(3) GFL).

Most of food law is crafted by the EU institutions. It is the responsibility of food business operators to comply with food law. That is to say, they have to ensure within the businesses under their control that food legal requirements are met (Article 17(1) GFL). It is the responsibility of the Member States to enforce food law by organising official controls and providing for measures and sanctions in case of non-compliances (Article 17(2) GFL).

2.3 Key issues

Developing food products for the market in the EU (and anywhere else) usually consists of a combination of designing – or simply choosing – a recipe, developing a method to produce the recipe in the intended way at a viable scale and formulating the message with which to present the product to the consumer. From a legal perspective, this leads to various key questions that are relevant in food production: 'what ingredients can be used?'; 'what product benefits can be claimed?'; 'how should business processes be organised?'; 'what is to be done when food does not meet food safety requirements?'; and 'how can food law be enforced?'

2.3.1 Can I use this ingredient?

Food products consist of one or more ingredients. It may depend on the nature of the developed or chosen ingredients to what extent and under which conditions they may be used in the EU.

Generally speaking, conventional ingredients may be freely used in the EU both in traditional recipes and in new combinations. 'Conventional ingredient' is not a technical term. We use it here to indicate all foods that are not covered by authorisation requirements. As a rule of thumb, one may think of foods that were known to our grandparents.

food authorisation (EU)

Increasingly, the law defines categories of products that may only be used if they have been authorised and only in accordance with the terms of the authorisation. The categories include food additives, genetically modified organisms (GMOs) and (other) novel foods. Food additives are substances not normally consumed as a food that are added for a technological purpose (Article 3(2)(a) Regulation (EC) 1333/2008 on food additives) (EC, 2008). GMOs are food that consist of or are made from genetically modified organisms (Regulation (EC) 1829/2003) (EC, 2003a). Novel foods are (other) foods that have not been consumed to a significant degree in the EU prior to May 1997 (Regulation (EU) 2015/2283) (EC, 2015).

All authorised food additives, GMOs and novel foods are included in so-called positive lists. These are lists that indicate which products are allowed. All products within the same category that are not included in the list are prohibited. These positive lists have been made accessible through online databases (see below in the section 'important websites'), including the additives database, the GMO register and the EU list of novel foods. Furthermore, there is a database of products and substances that have come to the attention of the European Commission where the Commission provides a preliminary position regarding their potential novel food status. This is the so-called novel food catalogue.

Most businesses will use these databases to ensure they only use ingredients that are not prohibited in the EU. However, innovating businesses may also undertake procedures to apply for the authorisation of products that classify as food additives, GM food or novel food but have not as yet been authorised. Normally, regulatory affairs would lead the team preparing the file to be submitted with the application for authorisation.

GRAS (USA)

Authorisation procedures are time consuming, costly and their outcome is uncertain. On the one hand EFSA applies very high standards to the scientific evidence that the applicant must provide. On the other hand, Member State politics is involved through the committee procedure. Several Member States do not look favourably upon food additives, GMOs and novel foods. For this reason, many businesses choose to bring innovative products first to market in the USA. The USA is seen as more open to innovations than the EU. The USA does not distinguish different categories for products with a technological purpose, GMOs and otherwise new foods. Anything that is added to food is considered a 'food additive' (in a much wider sense of the word). Food additives are subject to authorisation in the USA. However, products that are 'GRAS' escape from the classification as additive and thus from the authorisation requirement. GRAS means Generally Recognized As Safe by experts qualified to make such an assessment (Federal Food Drug and Cosmetic Act[3]). Food businesses can come to an agreement with the Food and Drug Administration (FDA) in an exchange of letters that this requirement is met. If a product does not present reason for concern based on its chemical composition, nutritional properties and the like, there is no need to perform safety testing. This makes the road to the market considerably shorter than in the EU. If consumers in the USA receive the product favourably, investment in market access in the EU may be worth considering. It will be interesting to see developments in this regard in the UK now that Brexit is a fact.

In Case 2.1, MPA was not intended to be an ingredient. If it had been, the business placing it on the market as a food would have had to consider its status. The purpose of MPA is to prevent pregnancy. This is not a technological purpose within the meaning of the food additives definition. Therefore, MPA would not be an additive. As it has not been consumed as a food prior to 1997, it would classify as a novel food. As a consequence, its use as an ingredient would be prohibited except when specifically authorised. MPA would not stand a chance of being authorised as it is not safe for human consumption.

2.3.2 Can I claim these benefits for my product?

Much to the annoyance of many marketing people, the communication by businesses to consumers about their food products is strictly regulated. Regulation 1169/2011 on food information to consumers (also known as the Food information to consumers regulation or FIC regulation) provides for information that must appear on the label, information that is prohibited both in labelling and in other commercial communications and requirements that apply to information that is neither mandatory nor prohibited: i.e. voluntary information (EC, 2011).

mandatory food information The FIC Regulation lists 12 pieces of mandatory information that must appear on every food label in the EU. For very small labels a few exceptions apply. The mandatory particulars are: the name of the product, the list of ingredients, the presence of allergens, the quantity of highlighted ingredients, the net quantity of the product, the date of durability, conditions of storage and use, contact

---

[3] 21 USC § 321(s): 'generally recognized, among experts qualified by scientific training and experience to evaluate its safety, as having been adequately shown through scientific procedures (...) to be safe under the conditions of its intended use'.

details of the responsible food business, the origin of certain ingredients, instructions of use, percentage of alcohol, and nutrition information (Article 9 FIC Regulation).

The FIC regulation uses the word 'name' as a technical term. It does not refer to the brand name or fantasy name of the product (such as Golden Wonder, or Raider). Regardless of the presence of such names there must be an indication making it clear to a consumer who is not familiar with the product what kind of product it actually is. Ingredients must be listed in descending order of weight. In the list of ingredients, allergens must be highlighted. Unfortunately, allergens are not a closed group of substances. Moreover, the prevalence of allergies varies across the globe. The FIC regulation lists the 14 most prevalent allergens in the EU. As to alcoholic strength, this only needs to be mentioned on alcoholic beverages.

In addition to the requirements of the FIC it is also mandatory to mention on the label if an ingredient is used that consists of or is produced from a GMO (Regulations (EC) 1829 and 1830/2003) (EC, 2003a,b).

prohibited food information

It is generally prohibited to mislead consumers. Therefore, any message or form of presentation that may make the consumer believe that the product is something that in reality it is not or that the product has properties that in reality it does not have, is prohibited. Moreover, any claim that states, suggests or implies that a food product can be used for the diagnosis, treatment or cure of a human disease is considered misleading by legal definition (Article 7 FIC) even if there were scientific evidence that the claim was true. This is the so-called prohibition on medicinal claims (also known as therapeutic claims).

voluntary food information

If space on the label allows and also in other commercial communications such as advertisements, non-mandatory information may be added, provided it is not misleading. Certain types of non-mandatory information are strictly regulated. Claims that a product has certain health benefits because of what it contains (nutrition claims) or because of what it does (health claims) may only be made in accordance with Regulation (EC) 1924/2006 on Nutrition and Health Claims ('the Claims Regulation') (EC, 2006b). The Claims Regulation has an annex with allows nutrition claims such as 'light' (meaning 33% less energy or less fat). Authorised health claims are included in an EU register of health claims database. Figure 2.3 shows a screenshot of this EU register.

The procedure to get a new claim authorised is burdensome. The applicant has to submit a scientific dossier to EFSA providing evidence of the link between a food component and certain health benefits. After EFSA has given its opinion, the European Commission decides whether to authorise the claim. In principle,

Claim type [i]	Nutrient, substance, food or food category [i] ▲	Claim [i]	Conditions of use of the claim / Restrictions of use / Reasons for non-authorisation [i] ⇕	Health relationship [i] ⇕	EFSA opinion reference / Journal reference [i] ⇕	Commission Regulation [i] ⇕	Status [i]	Entry ID [i] ⇕
Art.13(1)	Zinc	Zinc contributes to the normal function of the immune system	The claim may be used only for food which is at least a source of zinc as referred to in the claim SOURCE OF [NAME OF VITAMIN/S] AND/OR [NAME OF MINERAL/S] as listed in the Annex to Regulation (EC) No 1924/2006.	function of the immune system	2009;7(9):1229	Commission Regulation (EU) 432/2012 of 16/05/2012	Authorised	291, 1757

Figure 2.3. Screen shot from the EU Claims register.

authorised claims are generic. This means that they may be used by everyone. If the authorisation is based on proprietary data owned by the applicant, the applicant may be granted a period of exclusivity of five years.

2.3.3 How to organise my business processes?

HACCP

Food safety is not something that you have; food safety is something that you do. Processes of food businesses should be organised in such a way that hazards are under control and risks are minimised. Regulation (EC) 852/2004 on food hygiene requires food businesses to have systems based on HACCP principles (EC, 2004a). HACCP stands for *Hazard Analysis and Critical Control Points*. The principles require businesses to analyse their processes to determine where hazards may occur and how they can be kept under control. The system should be made explicit in writing. Its application must be documented. In Chapter 10 (Ferris, 2022) the HACCP principle is discussed further. Annexes to the food hygiene regulation as well as Regulation (EC) 853/2004 on hygiene for foods of animal origin provide many additional requirements (EC, 2004b).

The word 'hygiene' is defined in this context as 'the measures and conditions necessary to control hazards and to ensure fitness for human consumption of a foodstuff taking into account its intended use' (Article 2(1) Regulation (EC) 852/2004).

Part of hygiene is to ensure that legal limits for pathogenic microorganisms or chemicals are respected.

Several pieces of legislation set limits on certain contaminants including Regulation 2073/2005, which sets limits on the presence of certain micro-organisms in certain foods (EC, 2005a); Regulation 1881/2006 which sets limits on certain chemicals (EC, 2006a); and Regulation (EURATOM) 2016/52 which sets limits on radioactive contamination (EC, 2016). Other regulations address residues of pesticides (Regulation 396/2005) and veterinary drugs (Regulation 37/2010) (EC, 2005b, 2010). The use of hormones is prohibited (Directive 96/22) (EC, 1996).

Case 2.1 provides an example in which the hazard analysis was flawed. The pig farm should have realised that the feed fed to the pigs is a potential source of contamination of the pork. Therefore, they should have paid more attention to the source and the composition of the sugar-water it purchased to feed their pigs.

2.3.4 What to do when something goes wrong?

A situation may occur in which a food business operator 'has reason to believe that a food which it has imported, produced, processed, manufactured or distributed is not in compliance with the food safety requirements'. If this is the case, immediate action should be undertaken. Products still present in the business should be blocked. Products supplied to customers should be withdrawn from the market. Products that may already have reached consumers must be recalled through the media if other measures are insufficient to restore food safety (Article 19 GFL) (EC, 2002).

Again, the case set out in Section 2.1 provides an example of what this means in practice. Much of the damage suffered by the pig farm was caused by the fact that its products were considered unfit for consumption and, therefore, could not be placed on the market and had to be withdrawn from the market and recalled where they had already left the business.

2.3.5 Food law enforcement

When food businesses ignore the rules of ethics, discussed in Chapter 3 (Wernaart, 2022), this may damage their reputation. They may face consequences in their sales or on social media. If the rules of law are ignored, this may lead to legal consequences. A business causing harm by placing an unsafe food on the market may be held liable to pay for the damages suffered by the victims. Furthermore, they may face enforcement actions by public authorities.

The Official Controls Regulation (Regulation (EU) 2017/625) provides inspection agencies with legal instruments to check compliance by food business operators (EC, 2017). Furthermore, it provides for measures to be taken in case of non-compliance, such as imposing a recall. In addition to the measures to remedy the harm, Member States should have sanctions available to punish wrongdoers. In the case of deceptive and fraudulent actions these sanctions should either be high enough to ensure that they are higher than the illegal benefits gained, or they should be based on the annual turnover of the business concerned.

2.4 Methods

In this section we will reflect on the sources of law, and how they can be used.

2.4.1 Sources of law and regulatory intelligence

Legislation is built in a hierarchical structure. The EU Treaties grant power – always within limits – to the European legislature consisting of the European Parliament (representing the people), the Council of Ministers (representing the Member States) and the European Commission (the day-to-day administration, i.e. 'the government' of the EU) to make legislation on certain issues including food. The legislation consists of Regulations, which are directly binding on the people and businesses in the EU, and Directives, which require the legislatures in the Member States to harmonise their national requirements. The EU legislature can delegate part of the task of making legislation to the European Commission. In making its decisions – including legislation – the European Commission needs the approval of a committee representing the Member States. The Directorate-General (DG: the EU equivalent of a ministry) of the European Commission in charge of food law is the DG for Health and Food Safety (DG Sante). The committee in charge of food law is the Standing Committee on Plants, Animals, Food and Feed (SCoPAFF). All legislation can be found in the EU database of official documents Eur-Lex. All proposals for legislation submitted to the European Parliament can be found (and tracked) in the online legal observatory (see reference list). The European Commission invites feedback on proposed delegated legislation (see reference list).

To facilitate access to case law – i.e. rulings of the courts – the EU Council of Ministers has devised a system of European Case Law Identifiers (ECLIs). To every court ruling from the EU, from one of the Member States or from the European Court of Human Rights[4] that is published online, a unique ECLI is conferred. This makes them easy to trace.

---

[4] Please note that the European Court of Human Rights is not an EU institution, but a body of the Council of Europe.

As part of their policy the European Commission and EFSA publish, among other things, guidance documents. To distinguish them from legislation, these are known as 'soft-law'.

Regulatory affairs people work their way up from the bottom. The legislation is an important part of the 'regulatory intelligence' but usually guidance documents and information from business associations provide more detail for application in individual cases. Part of the regulatory intelligence is also the scientific data affecting procedures such as safety studies.

2.4.2 Selection and interpretation of legal sources

Lawyers and regulatory affairs people possess the skill, based on their knowledge and training, of being able to select the rules that may apply to a given case or question. These rules are the starting point of the analysis of what may and may not be done or omitted in a given situation.

To any situation, rules apply that have been formulated before the situation occurred. The rules are general and rule makers cannot have taken into account all aspects of the given situation. This is one of the reasons why the rules need to be understood – interpreted – in light of the specific situation. If the rule is 'food shall not be placed on the market if it is unsafe', how do I give meaning to all the figures and graphs on the datasheet and laboratory analysis?

In the case of a litigated conflict, it is the work of the courts to decide what the rules mean in the situation for which they have to cast their verdict. For this reason, lawyers and regulatory affairs people pay close attention to the rulings of the Court of Justice of the EU and other courts dealing with food legal issues to see if they have already ruled on similar issues. Also, they will use the same methods of interpretation as courts. They will break down rules into conditions (the food is unsafe) and consequences (it shall not be placed on the market). They apply the condition to the situation and the consequence to the action that was, will or should be taken. A first step in making sense of the rule is reading the words in their normal meaning – except when the law itself provides definitions of the words it uses, such as the word 'food' discussed above. The EU has no less than 24 official languages. Therefore, to understand the normal meaning of a word it may be helpful to look at other language versions of the same text. In addition to the wording, other clues may be important such as the context of the rule or its purpose.

2.5 Looking ahead

The world is changing fast. Around 2013 the European Commission commissioned a foresight study. A report was issued by the EU's Joint Research Council (JRC) 'Delivering on EU Food Safety and Nutrition in 2050 – Future challenges and policy preparedness' (Mylona *et al.*, 2016). The report sets out four possible future scenarios. The future is unpredictable. Events such as Brexit and Covid-19 were not foreseen in the foresight study. The game-changing factor it does address is climate change. Climate change is expected to lead to desertification in the south of the EU. This in turn will lead to loss of agricultural production and hence to loss of food security in the EU. How will we cope? International trade is an option. Trade agreements have been concluded with Canada, Mercosur[5] and China. The trade agreement with the USA, TTIP (Transatlantic Trade and Investment Partnership), launched in 2013, met with huge resistance and was abandoned. After Trump, it may take years if not decades before trust in the USA as a reliable partner is restored. Fortunately, trade dependence is not the only scenario. Another scenario is scientific progress. New technologies and new food sources may contribute to restoring EU's food security. Know-how from the pharmaceutical sector may come to the aid of the food sector and merge into something the report labels as 'Phood' – i.e. food with enhanced health properties. However, as we have seen above, the legal infrastructure places enormous obstacles in the way of such innovation. Yet another possible scenario is the emergence of so-called 'urban agriculture'. Many individuals may start producing food in their gardens, on their balconies and rooftops and in their kitchens. These food products may be traded online from consumer to consumer. Under the Covid pandemic we have seen parts of this scenario emerge, e.g. 'home cooks' offering their products online. The report raises the question of whether food law is not too complicated if consumers become food businesses. Experience under Covid has confirmed that few home cooks have the slightest idea about food safety legislation and good hygiene practices.

The European Commission under Ursula von der Leyen presented a green agenda in 2019 which was elaborated in a green deal and a farm-to-fork strategy (see websites in reference list). The food sector must become circular and climate neutral. The overall ambition is for the entire EU to be climate neutral by 2050.

---

[5] Mercosur: 'Mercado Común del Sur' (Spanish) means 'Southern Common Market'.

2.6 Conclusions

Food law, as it stands at the moment, has food safety as its main objective. Products that do not (fully) comply with food safety requirements are discarded. Innovative products that have not gone through an authorisation procedure cannot be placed on the market. This current paradigm of food law contributes to considerable waste streams and still food is not 100% safe. To become future proof we may need to find a different balance between food safety and an acceptance that risk is a part of life, innovation and circularity. Future crises will play an important role in shaping the new balance.

References

European Commission (EC), 1996. Council Directive 96/22/EC of 29 April 1996 concerning the prohibition on the use in stockfarming of certain substances having a hormonal or thyrostatic action and of β-agonists, and repealing Directives 81/602/EEC, 88/146/EEC and 88/299/EEC. Official Journal of the European Union L 125, 23.5.1996: 3-9.

European Commission (EC), 2002. Regulation (EC) No 178/2002 of the European Parliament and of the Council of 28 January 2002 laying down the general principles and requirements of food law, establishing the European Food Safety Authority and laying down procedures in matters of food safety. Official Journal of the European Union L 31, 1.2.2002: 1-24.

European Commission (EC), 2003a. Regulation (EC) No 1829/2003 of the European Parliament and of the Council of 22 September 2003 on genetically modified food and feed. Official Journal of the European Union L 268, 18.10.2003: 1-23.

European Commission (EC), 2003b. Regulation (EC) No 1830/2003 of the European Parliament and of the Council of 22 September 2003 concerning the traceability and labelling of genetically modified organisms and the traceability of food and feed products produced from genetically modified organisms and amending Directive 2001/18/EC. Official Journal of the European Union L 268, 18.10.2003: 24-28.

European Commission (EC), 2004a. Regulation (EC) No 852/2004 of the European Parliament and of the Council of 29 April 2004 on the hygiene of foodstuffs. Official Journal of the European Union L 139, 30.4.2004: 1-54.

European Commission (EC), 2004b. Regulation (EC) No 853/2004 of the European Parliament and of the Council of 29 April 2004 laying down specific hygiene rules for food of animal origin. Official Journal of the European Union L 139, 30.4.2004: 55-205.

European Commission (EC), 2005a. Commission Regulation (EC) No 2073/2005 of 15 November 2005 on microbiological criteria for foodstuffs. Official Journal of the European Union L 338, 22.12.2005: 1-26.

European Commission (EC), 2005b. Regulation (EC) No 396/2005 of the European Parliament and of the Council of 23 February 2005 on maximum residue levels of pesticides in or on food and feed of plant and animal origin and amending Council Directive 91/414/EEC. Official Journal of the European Union L 70, 16.3.2005: 1-16.

European Commission (EC), 2006a. Commission Regulation (EC) No 1881/2006 of 19 December 2006 setting maximum levels for certain contaminants in foodstuffs. Official Journal of the European Union L 364, 20.12.2006: 5-24.

European Commission (EC), 2006b. Regulation (EC) No 1924/2006 of the European Parliament and of the Council of 20 December 2006 on nutrition and health claims made on foods. Official Journal of the European Union L 404, 30.12.2006: 9-25.

European Commission (EC), 2008. Regulation (EC) No 1333/2008 of the European Parliament and of the Council of 16 December 2008 on food additives. Official Journal of the European Union L 354, 31.12.2008: 16-33.

European Commission (EC), 2010. Commission Regulation (EU) No 37/2010 of 22 December 2009 on pharmacologically active substances and their classification regarding maximum residue limits in foodstuffs of animal origin. Official Journal of the European Union L 15, 20.1.2010: 1-72.

European Commission (EC), 2011. Regulation (EU) No 1169/2011 of the European Parliament and of the Council of 25 October 2011 on the provision of food information to consumers, amending Regulations (EC) No 1924/2006 and (EC) No 1925/2006 of the European Parliament and of the Council, and repealing Commission Directive 87/250/EEC, Council Directive 90/496/EEC, Commission Directive 1999/10/EC, Directive 2000/13/EC of the European Parliament and of the Council, Commission Directives 2002/67/EC and 2008/5/EC and Commission Regulation (EC) No 608/2004. Official Journal of the European Union L 304, 22.11.2011: 18-63.

European Commission (EC), 2015. Regulation (EU) 2015/2283 of the European Parliament and of the Council on novel foods, amending Regulation (EU) No 1169/2011 of the European Parliament and of the Council and repealing Regulation (EC) No 258/97 of the European Parliament and of the Council and Commission Regulation (EC) No 1852/2001. Official Journal of the European Union L 327, 11.12.2015: 1-22.

European Commission (EC), 2016. Council Regulation (Euratom) 2016/52 of 15 January 2016 laying down maximum permitted levels of radioactive contamination of food and feed following a nuclear accident or any other case of radiological emergency, and repealing Regulation (Euratom) No 3954/87 and Commission Regulations (Euratom) No 944/89 and (Euratom) No 770/90. Official Journal of the European Union L 13, 20.1.2016: 2-11.

European Commission (EC), 2017. Regulation (EU) 2017/625 of the European Parliament and of the Council of 15 March 2017 on official controls and other official activities performed to ensure the application of food and feed law, rules on animal health and welfare, plant health and plant protection products, amending Regulations (EC) No 999/2001, (EC) No 396/2005, (EC) No 1069/2009, (EC) No

1107/2009, (EU) No 1151/2012, (EU) No 652/2014, (EU) 2016/429 and (EU) 2016/2031 of the European Parliament and of the Council, Council Regulations (EC) No 1/2005 and (EC) No 1099/2009 and Council Directives 98/58/EC, 1999/74/EC, 2007/43/EC, 2008/119/EC and 2008/120/EC, and repealing Regulations (EC) No 854/2004 and (EC) No 882/2004 of the European Parliament and of the Council, Council Directives 89/608/EEC, 89/662/EEC, 90/425/EEC, 91/496/EEC, 96/23/EC, 96/93/EC and 97/78/EC and Council Decision 92/438/EEC (Official Controls Regulation). Official Journal of the European Union L 95, 7.4.2017: 1-142.

European Commission (EC), 2018. Directive (EU) 2018/851 of the European Parliament and of the Council of 30 May 2018 amending Directive 2008/98/EC on waste. Official Journal of the European Union L 150, 14.6.2018: 109-140.

Ferris, I.M., 2022. Hazard analysis and critical control points (HACCP). In: Wernaart, B.F.W. and Van der Meulen, B.M.J. (eds) Applied Food Science. Wageningen Academic Publishers, Wageningen, the Netherlands, pp. 187-213.

Gürsoy, S.I., 2022. Food policy. In: Wernaart, B.F.W. and Van der Meulen, B.M.J. (eds) Applied Food Science. Wageningen Academic Publishers, Wageningen, the Netherlands, pp. 85-98.

Mylona, K., Maragkoudakis, P., Bock, A.K., Wollgast, J., Caldeira, S. and Ulberth, F., 2016. Delivering on EU food safety and nutrition in 2050 – future challenges and policy preparedness. JRC science for policy report. Publication Office of the European Union, Luxembourg, Luxembourg. Available at: https://publications.jrc.ec.europa.eu/repository/handle/JRC101971.

Urazbaeva, A., Szajkowska, A., Wernaart, B., Franssens, N.T. and Vaskoska, R.S., 2019. The functional field of food law, reconciling the market and human rights. Wageningen Academic Publishers, Wageningen, the Netherlands.

Van der Meulen, B.M.J. (ed.), 2011. Private food law. Governing food chains through contract law, self-regulation, private standards, audits and certification schemes. European Institute for Food Law series, Volume 6. Wageningen Academic Publishers, Wageningen, the Netherlands. https://doi.org/10.3920/978-90-8686-730-1

Van der Meulen, B.M.J. and Bremmers, H., 2015. The prohibition of medicinal claims. Food in fact but medicinal product in law? Wageningen Working Paper in Law and Governance 2015/03. http://dx.doi.org/10.2139/ssrn.2605881

Van der Meulen, B.M.J. and Wernaart, B. (eds), 2020. EU Food law handbook. Wageningen Academic Publishers, Wageningen, the Netherlands.

Van der Meulen, B.M.J., 2020. Food law: development, crisis and transition. In: Van der Meulen, B.M.J. and Wernaart, B. (eds) EU food law handbook. Wageningen Academic Publishers, Wageningen, the Netherlands, pp. 137-158.

Van Loon, H. and Wernaart, B.F.W., 2022. Consumer protection. In: Wernaart, B.F.W. and Van der Meulen, B.M.J. (eds) Applied Food Science. Wageningen Academic Publishers, Wageningen, the Netherlands, pp. 65-83.

Wernaart, B., 2017. International law and business, a global introduction. Noordhoff, Groningen, the Netherlands.

Wernaart, B.F.W., 2022. Food ethics. In: Wernaart, B.F.W. and Van der Meulen, B.M.J. (eds) Applied Food Science. Wageningen Academic Publishers, Wageningen, the Netherlands, pp. 45-64.

Important websites

Eur-Lex: https://eur-lex.europa.eu/index.html

The EU additives database: https://ec.europa.eu/food/safety/food_improvement_agents/additives/database_en

EU Commission and EFSA guidance documents: https://europa.eu/european-union/topics_en; https://ec.europa.eu/food/overview_en and https://www.efsa.europa.eu/

The EU farm to fork strategy: https://ec.europa.eu/food/farm2fork_en

The EU Green Deal: https://ec.europa.eu/info/strategy/priorities-2019-2024/european-green-deal_en

The EU GMO register: https://webgate.ec.europa.eu/dyna/gm_register/index_en.cfm

The EU legal observatory: https://oeil.secure.europarl.europa.eu/oeil/home/home.do

Have your say (page to provide feedback on proposed delegated legislation): https://ec.europa.eu/info/law/better-regulation/have-your-say

The EU novel food list: https://ec.europa.eu/food/safety/novel_food/authorisations/union-list-novel-foods_en

The EU novel food catalogue: https://ec.europa.eu/food/safety/novel_food/catalogue_en.

The EU register of health claims: https://ec.europa.eu/food/safety/labelling_nutrition/claims/register/public/.

The website of Directorate General SANTE: https://ec.europa.eu/food/overview_en

Further reading

European Food and Feed Law Review (Lexxion Berlin).

Van der Meulen, B. and Wernaart, B. (eds), 2020. EU food law handbook. Wageningen Academic Publishers, Wageningen, the Netherlands.

Van der Meulen, B., 2018. Legal method and theory. Orc-Grid and Cobra-C Matrix: tools for research and teaching law. European Institute for Food Law working paper 2018/03. http://dx.doi.org/10.2139/ssrn.3135873

Van der Meulen, B., 2018. The safe food principle, a critical reflection on the key concept of EU food safety law. European Institute for Food Law working paper 2018/08. http://dx.doi.org/10.2139/ssrn.3271027

Polinski, D. and Van der Meulen, B., 2019. Unfit for human consumption, the elusive element in the EU food safety concept of Article 14 GFL. European Institute for Food Law working paper 2019/01. http://dx.doi.org/10.2139/ssrn.3371444

3. Food ethics

This is what you need to know about ethical decision-making in the food sector

Bart F.W. Wernaart

School of Business and Communication, Fontys University of applied sciences, P.O. Box 347, 5600 AH Eindhoven, the Netherlands; b.wernaart@fontys.nl

Abstract

Ethics is the philosophy of doing the right thing – morally. And it is an important discipline in the food sector which may cause multiple societal and environmental challenges in relation to food and food production. Dealing with these challenges may affect important values for stakeholders of the food industry, such as autonomy, animal welfare, transparency, social justice (equality), health (food safety) and environmental prosperity. This leads to ethical dilemmas when possible decisions lead to opposing – and mutually exclusive – ethical solutions. To deal with such dilemmas we need to understand how people take ethical decisions, and how normative theories can help justify these decisions. For instance, consumers may want to balance different values when they decide on what food they purchase and may want to see such moral considerations reflected in the strategies of the food business; business leaders may want to reconcile profits or societal and ecological values in their business strategies; governmental institutions may want to adopt balanced food policies respecting these involved values; and non-governmental organisations may want to address ethical issues in the food sector that may not be fully recognised by other stakeholders. The aim of this chapter is to offer the reader an introduction to the most important theories in applied food ethics, with a focus on ethical decision-making. To this end, we will explore contemporary issues in food ethics, discuss the basics of empirical and normative ethics, introduce business ethics, and then conclude with some expectations for the near future.

Bart Wernaart and Bernd van der Meulen (eds)
Applied food science
DOI: 10.3920/978-90-8686-933-6_3, © Bart F.W. Wernaart 2022

Key concepts

- ▶ In ethics, we explore value judgements in what is right and wrong.
- ▶ Morality is the combined norms and values of an individual, group or organisation.
- ▶ Norms are rules or actions that aim at the fulfilment of a value.
- ▶ A value is an abstract notion of something we find valuable and seek to realise by moral actions.
- ▶ In descriptive ethics we empirically assess how people take ethical decisions.
- ▶ In normative ethics we explore how people ought to take ethical decisions.
- ▶ In food ethics, ethical issues that relate to food substances (what to eat and what not to eat) or the societal and environmental consequences of food production are explored.
- ▶ The food ethics matrix is a way to structurally analyse how the values of food business stakeholders relate to one another.
- ▶ In bio-ethics, the ethical issues in life sciences are explored, including biology and medicine.
- ▶ In business ethics we explore how commercial organisations can organise their corporate social responsibility.

Case 3.1. Hang the elephant, feed the dog and eat the cow!

Topsy was the first baby elephant that was held in captivity in the United States. Initially, she was adored by people but then seemed to develop a bad temper and when she matured, she became a 'dangerous elephant'. During her circus years she killed three people in unexpected outbursts of rage. The circus had no other option than to sell her to a zoo (Luna Park, USA). Unfortunately, her bad temper did not improve, and her owners decided to execute her. This was not done quietly or privately, but by means of a public execution in front of a crowd of paying visitors. Topsy was brought to a scaffold that was built for the occasion. Her executioners attached electrodes to her feet and her large body was wrapped in wires and iron chains. To ensure her death, she was given a last meal of carrots and cyanide. Eventually, the latter proved unnecessary, since the electric shock of 6,600 volts did its work, and Topsy was no more (New York Times, 5 January 1903).

The execution of Topsy the Elephant was not an exceptional case. It is estimated that at least 36 (circus)-elephants met a similar fate in the United States between 1880-1930 (Wood, 2012). Among them, 'Murderous Mary' must have been the most infamous. In 1916 – after crushing one of her trainers – she was hanged from a

---

railroad crane in a public execution. It took two attempts to actually kill the elephant. In the first attempt the iron wire snapped and Mary broke one of her hips (Burton, 2009: 217-227).

Animal welfare and food ethics

We have a complicated relationship with animals. Some we keep for company, some to work for us, some we keep in a zoo or theme park for our enjoyment, some we protect because they are endangered species, and others we factory farm for food processing and consumption. Occasionally we apply a combination of the above which leads to challenging moral issues. Why do some of us eat beef and at the same time host a funeral for our deceased dog (Wernaart, 2020)? As we can see in the cases of Topsy and Mary the Elephant, we assume that some animals have a character or personality and can act – to a certain degree – autonomously and sometimes we even hold these animals accountable for their behaviour.

What does this mean for our ethical relationship with animals? To what degree do we have a responsibility to balance the interest of animals with our own? To what level do we need to respect the autonomy of the animal species? And to what extent do we have the right to decide on their fate? Such moral questions are important for food businesses when they produce and sell animal-based products, for consumers when they decide on their eating habits, and for governments when they have to balance their policies in this field. It is perhaps not surprising that animal ethics is a firmly established theme in food ethics (Mepham, 2000).

3.1 Food ethics

Ethics and food go back a long way; the ethics of food handling, processing and consuming has for ever been part of moral discourse (Zwart, 2000). For instance, the ancient Greek philosophers focused on the golden mean in food consumption and temperance was the key word as it was considered an important virtue when preparing and eating food. The Jews, through their Hebrew Bible, focused on strict rules on what they were or weren't allowed to eat: a more result-oriented approach. In the 18th century, especially in the post-industrial era, food ethics was mostly discussed in a social context: food production and consumption affecting public health and the environment. In today's society we still recognise traces of the old food ethics discourse.

In a nutshell, moral issues in food ethics either relate to food substances (what to eat and what not to eat) or the societal and environmental consequences of food production and consumption. The important values that are at stake in such discourses on ethics are consumer autonomy, animal welfare, transparency, social justice (equality), health (food safety) and environmental prosperity, as we can see in Figure 3.1. These values are not 'stand-alone' issues; they correspond to and interrelate with one another in the context of particular moral issues. In this section, we will further discuss some of the most important contemporary moral issues in food ethics and show how these core values are involved.

Food ethics is a very old concept that is not part of the core of the contemporary ethics debate, which is mostly associated with and debated in the broader context of bioethics or business ethics (Costa, 2018; see also Section 3.2). This may clarify why most methods in ethics applied to the food sector are 'borrowed' from bioethics (for instance, the food ethics matrix) or business ethics (for instance, EDM theories). In bioethics, we explore the ethical issues in life sciences, including biology and medicine; while in business ethics, we explore how commercial organisations can organise their corporate social responsibility.

3.2 Contemporary issues in food ethics

In this section, we will explore some of the most important contemporary issues in food ethics. These are moral issues that relate to GMO technology, animal welfare, healthy eating habits and the environmental impact of food production.

GMO technology GMO technology has always led to an intense debate and has many layers (Jasanoff, 2016: 87-115). A useful distinction is made by Comstock (2010), who observed that ethical concerns about GMO production may be led by extrinsic and intrinsic objections.

In the case of intrinsic objections, the very nature of GMO technology is questioned, and considered an unnatural process which is unethical for that reason (Comstock, 2000: 182-195). These intrinsic motivations sometimes

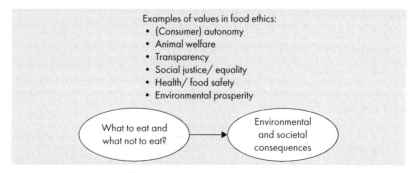

Figure 3.1. Moral issues in food ethics.

Applied food science

have a religious charge, and the main concern is that it is not for humanity to play God by altering His species. Other arguments are not directly religious but relate to the nature of things in their own right. It is considered immoral to try to change the essence of things because it may then lose its meaning. Some reject the idea of crossing species boundaries, or engineering species as if they are machines. Perhaps the popular novel 'The Island of Dr Moreau' written by the famous author H.G. Wells (1896) is a colourful example in which the crossing of species boundaries is firmly criticised.

In case of extrinsic objections, moral issues arise when the potential benefits do not outweigh the potential harm that may result from GMO technology. For instance, GMO technology may damage existing ecosystems when they are not carefully assessed before they are used (e.g. Fitting, 2006). Other concerns are more societal: there may be moral issues in case of a lack of GMO products, but also in case of an abundance of GMO products. In case of the first, the absence of GMO technology could affect people's food security. In case of the last, the autonomy of consumers may be harmed.

In theory, GMO technology could contribute to improving food security and food quality which, in poor regions in particular, could solve major problems. For instance, crops could – through genetic engineering – become better resistant to drought which would be ideal in dry regions, such as sub-Saharan countries (Komparic, 2015), and would improve people's access to adequate food. However, in practice, most GMO technologies are owned by large private multinational organisations that are not primarily interested in contributing to food access in poor countries. They strongly protect their innovations with intellectual property (IP) rights (Bird and Cahoy, 2015), or are unwilling to invest in poor countries due to a lack of local IP protection (Adenle, 2011). There is also a geopolitical aspect to it; some of the developing countries which could benefit from GMO technology are reluctant to embrace such technologies due to their close political connections with European countries. EU countries are traditionally less optimistic about the benefits of GMO technology compared to the USA (Paalberg, 2010). This leads to a situation in which the potential of GMO contributing to food security is not necessarily put to good use.

When GMOs are widely used in food products, other ethical issues may present themselves. Consumer autonomy is seriously affected when consumers only have a limited choice of non-GMO food products to purchase or consume due to the wide use of GMO products. This is especially problematic when supermarkets offer a very limited number of alternatives to GMO products or offer them as niche products for a much higher price (Roff, 2007). This leads to a 'take it or leave it' situation for those who cannot afford non-GMO products.

animal well-being   As we can see in Case 3.1 of this chapter, the matter of animal well-being leads to moral issues in food production. To what extent do we need to take the well-being of animals into consideration in food production? Ever since we industrialised food production and were able to factory farm animals, the well-being of these animals has been a continuous cause for debate (Anomaly, 2015). In essence, there are three leading arguments in this discourse. The first is that humans do not have a moral responsibility towards animals because animals cannot be moral agents. After all, animals are generally incapable of having a moral consciousness or taking ethical decisions. Consequentially, we do not need to take into consideration the interest of animals in our ethical decision-making. This means that we can use animals in the way we see fit. Please note that this does not necessarily mean we have to treat animals in a cruel or disrespectful manner. It primarily means that we can treat animals in such a way that it matches our own needs and interests, without the need to worry about their well-being in our moral considerations (Carruthers, 1992; Hsiao, 2017). This also means that if we want to keep one animal as a companion, and kill the other for food, it would be morally unproblematic. An alternative view is offered by Tom Regan (1983) who argues that some animals are – just like humans – self-aware and have a psychological identity. This means that they are able to experience pain and pleasure, and they are able to remember where they came from and where they intend to go to. In such cases, fundamental rights should not be reserved for humans only; they should also include these animal species with a psychological identity. In other words: we can kill and eat insects, but not a cow or a pig. Yet another view is offered by the famous utilitarianist Peter Singer (1975). He states that when animal species have similar interests to human beings, 'the greater happiness for the greatest number' principle should also apply to them regarding these interests. So, if animals are able to experience pain, they have an interest in the absence of pain, the same way humans have. These approaches in ethics inspired to notable changes in legislation. For instance, traces of Regan's approach can be found in Article 13 of the Treaty on the Functioning of the European Union. This article recognises that animals are sentient beings, and therefore Member States should 'pay full regard to the welfare requirements of animals' in formulating and implementing EU policies[6]. Another example of Singer's viewpoints expressed in law comes from Switzerland, where the

---

[6] The full article reads as follows: 'In formulating and implementing the Union's agriculture, fisheries, transport, internal market, research and technological development and space policies, the Union and the Member States shall, since animals are sentient beings, pay full regard to the welfare requirements of animals, while respecting the legislative or administrative provisions and customs of the Member States relating in particular to religious rites, cultural traditions and regional heritage.' See for a further and more thorough discussion: Simonin and Gavinelli, 2019.

culinary practice of boiling lobsters alive was banned by law in 2018. This was a response to research that established that lobsters are able to experience pain (Magee and Elwood, 2013).

healthy eating habits

As already discussed in the first paragraph of this section, the Greek already emphasised the importance of healthy eating habits in their virtue ethics. The virtue 'temperance' was considered to be a golden standard in between self-abnegation and gluttony. Where the Greek focused on good character to define a healthy lifestyle, this was later determined by scientific measurements. The scientific assessment of what is healthy and what not evolves each time new insights are gained. For instance, the official UK advice regarding the consumption of alcohol during pregnancy changed from no advice at all, to the advice to limit alcohol consumption during pregnancy, to the advice not to consume alcohol at all in a timespan of less than ten years (Inskip *et al.*, 2009). A central theme in this context is the autonomy of the consumer and the transparency of the food business. A consumer can make autonomous food choices if these decisions are independent, authentic, and the result of balancing various options in a well-informed way (Christman, 1989; Valdman, 2010). In the case of food, consumer autonomy can be encouraged by the proper labelling of products, as well as food education and the open access to reliable sources on the definitions of a healthy lifestyle. Especially in the latter case, a new ethical issue presents itself regarding the transparency of food communication. The information about the exact composition or the effects of food products are mostly provided by the companies who produce and sell them. There may sometimes be a fine line between informing consumers and marketing: a fast-food undertaking will probably not admit that some of their products do not contribute to a healthy lifestyle, while their advertising contributes to obesity (Fister, 2005; Owen, 2018). Below, in Section 3.2, we will discuss the tension between ethics and business in more detail.

environmental impact

An important theme in food ethics relates to the environmental impact of food production. Matters such as chain responsibility, the role of the consumer and circular economy, are recurring themes in this context. The most important issue here is the discourse on which stakeholder bears what responsibility in protecting environmental prosperity.

3.3 Methods in ethical decision-making

As we have seen in the previous section, there are complicated moral issues in food ethics that need to be dealt with. Stakeholders involved will need to make ethical decisions to realise the values they pursue. In this section, we will explore how models in ethics can help decision-making processes. We will first discuss the nature of an ethical dilemma, then we will highlight useful models in descriptive ethics (ethical decision-making) and normative ethics. Finally, we will explore the tension between ethics and business.

ethics

In ethics, we explore value judgements about what is right and what is wrong. In pursuit of what is right, we seek to realise values. A value is an abstract notion of something we consider valuable and seek to realise by moral actions. Such an action is driven by moral norms; rules or actions that help us realising a value. For instance, if we want to realise the value animal welfare, the norm could be not to purchase animal-based products. The interplay of norms and values then defines someone's morality. Morality therefore is the combined norms and values of an individual. However, combined norms and values can also be shared amongst people, and so we can recognise the particular morality of a group, culture or even an organisation.

ethical dilemma

An ethical dilemma (or moral problem) arises when different value judgements can be applied to a given case, and these judgements mutually exclude one another (Statman, 1996). For instance, if we would want to reduce the number of people who suffer from obesity, it would be helpful to ban certain unhealthy substances that contribute to it. At the same time, such a ban could interfere with the autonomy of the consumer, who wants to make independent food choices. From a utilitarian perspective, banning unhealthy substances will probably be justifiable, or even a requirement, to contribute to the greatest happiness of the greatest number principle. However, from a more Kantian perspective, where the autonomy of the individual is emphasised, this is probably a no-go.

We distinguish between descriptive (or empirical) ethics, in which we describe how people take ethical decisions, and normative ethics, in which we explore how people ought to take ethical decisions. For example, in case of the first, we *describe* the factors that influence an employee's decision to come forward when they observe a violation of the company's ethics code (see also Chapter 4; Van Loon and Wernaart, 2022). In case of the latter, we *prescribe* how the employee ought to act when observing a violation of the company's ethics code.

descriptive ethics

An impressive number of works have been published on the Ethical Decision Making (EDM) theory in which the human ethical behaviour is observed, analysed and explained from various perspectives and disciplines. There are mainly two central questions:

1. What are the characteristics of the different stages in ethical decision-making?
2. Which factors influence ethical behaviour during these stages?

While there are some variations to the theme (Jones, 1991), James Rest (1986) offers the most commonly used model for characterising EDM stages. He states that the process of ethical decisions is composed of four steps (Figure 3.2):

1. recognising the moral issue;
2. making a moral judgment;
3. establishing moral intent; and
4. moral acting.

In other words: an individual first becomes aware of a moral problem, then evaluates the issue by taking an ethical decision, after that becomes determined to act accordingly, and finally behaves by implementing the moral decision (Crane *et al.*, 2019: 139-140).

Various factors influence how individuals go through these stages, and how they make ethical decisions. In this case too, we find an abundance of literature describing these factors (Craft, 2013; Schwartz, 2016). A useful distinction is made by Jones (1991), who subdivides influencing factors into individual, situational and issue-related factors (Figure 3.2).

Individual factors relate to the particularities of the person that is confronted with an ethical issue. Examples are demographics, psychological features and ethical experiences which, together, shape the moral capacity of an individual. For instance, it is established that gender is a demographic factor that may influence the way individuals take ethical decisions (Gilligan, 1982). When we explore this further in the context of vegetarianism, we see that men and woman appear to have different relationships with meat consumption. They also see vegetarianism through different eyes (Ruby, 2012). To illustrate one of the aspects, Rozin *et al.* (2012) established a (metaphoric) relationship between masculinity and meat. In other words, meat consumption is a metaphor for masculinity, which may influence moral viewpoints on vegetarianism amongst men.

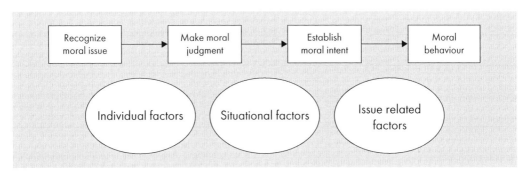

Figure 3.2. Ethical decision-making and factors of influence.

Situational factors refer to the organisational and personal context of the individual that is confronted with an ethical issue (Trevino *et al.*, 2006). The ethical infrastructure of an organisation, including the reward system, code of conduct or business culture, strongly affects the way individuals take ethical decisions when acting in their professional capacity. Someone's personal context is also important, such as their living conditions or their financial situation. For instance, low-income consumers make different value judgements about eating habits compared to high-income consumers – the first group is more likely to be associated with unhealthy eating habits than the latter (French *et al.*, 2019).

Issue-related factors refer to the moral issue in itself. Jones (1991) observes that the moral intensity of a situation affects the way moral actors take ethical decisions. In his issue-contingent model, Jones introduces various dimensions that together define moral intensity. By way of example, one of the dimensions is the magnitude of the consequences: the total sum of harms or benefits for the moral stakeholder influences the way they take an ethical decision. Another dimension is the proximity of a moral issue: How close to the moral issue is the moral stakeholder for them to be affected by it? When applied to consumer behaviour, these two dimensions may explain why some consumers feel affection for a chicken held in a local petting zoo (and would never consider eating that particular chicken), and also have chicken for dinner. The particular death of one chicken will probably not make a difference in how chickens are factory farmed, and the factory farm isn't right next to your home. The sense of urgency to take the interests of the chicken into consideration is lacking due to the perceived low intensity of the moral issue.

normative ethics

Normative ethics does not focus on how people make ethical decisions; it focuses on how people ought to take ethical decisions, or what is morally right. We distinguish between normative ethics theories that assume that moral truth exists and those that assume that morality is a relative concept. The theories in the first case start from the assumption that certain ethical rules are always applicable (universalism) or always right (absolutism). The theories in the second case are theories of moral relativism, in which we assume there is no moral truth. Instead, morality depends on individual or societal features, such as moral intentions, moral characteristics, moral impulses and emotions, moral relations and moral discourse (Wernaart, 2018: 69-96).

Good examples of universalist normative theories are egoism and utilitarianism. In both theories, something is morally right when it leads to the best possible consequences for either the moral actor (egoism) or society in general (utilitarianism). Therefore, we sometimes label such approaches as consequentialist theories. The famous work 'Wealth of Nations' by Adam

Smith[7] starts from the position that the so-called 'invisible hand' theory works (and only works) because consumers use their buying power only for their own short-term self-interest. This is how supply and demand is balanced in the most optimal way. Egoism is considered to be the corner stone of neoliberal macroeconomic theory. In Utilitarianism, founded by British Enlightened thinkers like Jeremy Bentham and John Stuart Mill, something is morally right when it leads to the greatest happiness for the greatest number. We referred to Peter Springer and his animal ethics earlier: a fine example of applied utilitarianism.

Examples of universalist theories that are also absolutist are the so-called deontological theories. These theories are universally applicable principles that encompass morality as an absolute truth. Examples are duty ethics, moral rights and the principle of equality. Duty ethics mostly refers to the Kantian categorical imperative, where Kant states that human beings have a duty to make sure that moral actions are in line with the universal expectations of morality. Moral rights stem from the idea that a society is based on a social contract that governs the freedoms of the individual and protects against violations of fundamental rights. The principle of equality mostly relates to the Rawlsian concept of justice, centred around various (and cumulative) interpretations of equality principles.

Examples of relativism theories are virtue ethics, post-modernist ethics, feminist ethics and discourse ethics (Crane *et al.* 2019: 114-125; Wernaart, 2018: 84-91). Something is morally right when this is done by good people (virtue ethics), morality can be found in moral impulses and emotions (post-modernist ethics), morality contributes to maintaining healthy relations (feminist ethics) or is a result of a fair discourse (discourse ethics). What these approaches have in common is that neither consequences nor principles determine the morality of an ethical decision. Instead, there is a focus on the intention of the moral actor. This may differ per person, situation or (sub)-culture.

In most Western literature on ethics, there is a strong focus on universalist and absolutist theories that originally stem from the 18th-century Enlightenment movement (Crane *et al.*, 2019: 91). In food ethics we also see an emphasis on these theories. For instance, the well-known ethical matrix for assessing moral issues in the context of food (Mepham, 1996; 2000; 2004; 2013) exclusively focuses on utilitarianism, duty ethics and the principle of equality. This matrix –

---

[7] Please note that although egoism is an important feature of 'Wealth of Nations', Adam Smith was not an egoist per se. In his broader works on ethics, his work can best be labelled as a version of virtue ethics. See Huhn (2017).

which has been adopted and improved – is still the core method that is used by leading organisations in food ethics, most notably by the Food Ethics Council (www.foodethicscouncil.org), situated in the UK. Using this matrix results in an overview of core values per stakeholder in a given food-related moral issue. For instance, when applied to the matter of obesity, Mepham analyses the values of food producers, food marketers, consumers and society members in the context of well-being, autonomy and fairness (Mepham, 2010, Figure 3.3). Wellbeing loosely represents a utilitarianism approach and is result-driven; autonomy refers to Kant's categorical imperative; and fairness is an interpretation of the principle of equality. By way of example, when applied to the issue of obesity, the core values of consumers are 'reduction of risks of obesity and associated diseases' (well-being); informed food choice (autonomy) and equal access to healthy food (fairness). This way, values of various stakeholders can be mapped in light of different approaches in normative ethics and can serve as input for ethical decision-making.

The food matrix can be a helpful tool for policy makers and business leaders in the food sector. However, there is an exclusive focus on consequentialism and deontology while relativistic theories are not represented. A way to overcome this gap is to make use of the agent-deed-consequence (ADC) model (Dubljević and Racine, 2014). In each moral action, the intention of the agent, the deed itself, as well as the consequences of the action, are assessed all together. These elements represent the moral approaches we find in universalist and absolutist approaches (consequentialism and deontology) and moral relativism. The model can be used for a more balanced application of normative ethics, without a merely one-sided focus on consequences or principles. This means that, in the context of anti-obesity policies for instance, it is not only the adoption of the policy (deed) or the results (consequence) that matter, but also the motivation of the policymaker. If the policy makers are mainly driven by the desire to reduce healthcare costs, or aim at higher tax returns due to a higher turnover in the food industry, it would reduce the moral quality of the policy.

Respect for	Wellbeing	Autonomy	Fairness
Food producers			
Food marketers			
Consumers			
Society			

Figure 3.3. The food ethics matrix, according to Mepham (2010).

ethics and business The models in descriptive and normative ethics discussed above are primarily designed for individual use, and not for organisational use per se. While these models can be used to analyse individual ethical decision-making or explore how individuals ought to behave in an organisation, they do not always offer a solution when regarding the ethics of the organisation. After all, the ethics of a food company is not necessarily the same as the total sum of individual ethics of those who work on behalf of that company. At the same time, a shared vision on morality within a commercial organisation may help in unifying individual norms and values – at least when they are used in a professional capacity on behalf of the company. One way or another, there is, unmistakably, a certain tension between individual and business ethics, because in the latter, profit is always a core value that cannot be ignored. There are several models in business ethics that may help understand how ethics can be channelled in the organisational structure of a company. Generally, we can say that there is a fine line between business ethics and a business model. There is a reason why Milton Friedman (1962; 1970) famously stated: 'the business of business is business'. He is sometimes mistaken for a radical believer in profit. Instead, it would do his theory more justice if we accepted his concerns about the motivation of a business to 'do' ethics, which is to increase its profits in the end. In his view, the models we discuss in this section are no more than window dressing and distract from the true aim of a private organisation. In other words, there is a certain contradiction in the phrase 'business ethics' (Collins, 1994). It is unlikely that a company wants to contribute to solving societal problems if this does not at the same time lead to profit optimisation. However, this viewpoint is not as widely embraced as the models it criticises.

John Elkington's famous triple bottom line (1997) is the starting point of many models in business ethics. Elkington argued that business should not only focus on a single bottom line (profit) but instead direct its business decisions from a triple perspective: people, planet and profit. The idea is simple; when businesses invest in people and the environment, they will sustain a healthier business model in the long run. This way, business not only takes its moral responsibility towards society and the environment, but also ensures a sustainable profit. The idea that businesses should take responsibility for their societal and ecological impact is also referred to as 'corporate social responsibility' (CSR). Those involved in the decision-making of a business should not act solely in the interest of shareholders but move from a shareholder approach to a stakeholder approach (Freeman, 2010). In a stakeholder approach the interests of all relevant stakeholders are mapped, measured and used as input for balanced decision-making. Stakeholders could be defined as those individuals or groups that can affect or are affected by the company's decision-making (see also Crane et al., 2019: 59-65). In earlier models these groups were narrowed down to shareholders, suppliers, customers and employees. Later, governments, competitors and civil society were added to the list (Freeman, 1984). In a more contemporary approach, the company uses a network model that not only maps out the various stakeholders; it also shows how these stakeholders in turn are related to their own stakeholders, giving a full and more complete impression of mutual and conflicting interests (Rowley, 1997). An overview of business ethics management can be found in Figure 3.4.

In business ethics management, transparency and accountability are important values that can help reduce the risk of window dressing. There are plenty of examples of companies that have been involved in massive scandals, fraud and other unethical practices, while claiming professional stakeholder management goals, a sustainable mission and vision statement and a triple bottom line strategy. When ethics is taken seriously, it should move beyond the sphere of good intentions. Therefore, it is important to be transparent about the impact of a business on society and the environment, and – where needed – give an account of matters when the impact appears to be negative (Wernaart, 2018: 257-289). To this end, companies are increasingly giving account of their environmental and societal impact through 'external cost accounting'. Next to their usual financial statement that focuses on economic achievements, companies also report on their achievements in other fields. This is called 'integrated reporting'. Thorough integrated reporting works, to a certain extent. A link can be established between companies that thoroughly monitor their energy submissions and report about this and their financial success – see for instance, the data that is collected by the Corporate Disclosure Project (www.CDP.net). However, it needs to be noted here that a sound methodology

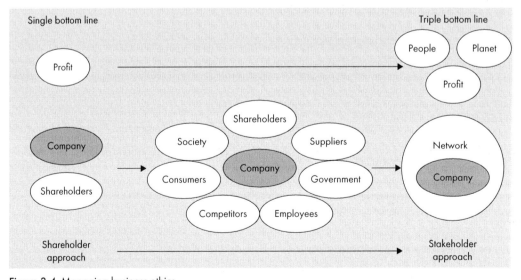

Figure 3.4. Managing business ethics.

in reporting is extremely important in order to avoid, yet again, the risk of window dressing (Murtha and Hamilton, 2012). As it seems, the pursuit of profit maximisation and integrated reporting is not always a constructive combination.

3.4 Looking ahead

As we have noted in this chapter, food ethics may be one of the oldest forms of applied ethics. However, in current ethics discourse, it is not a stand-alone issue but mostly discussed in the wider context of bioethics or business ethics. Considering the seriousness of the involved values and interests in food ethics, this applied field perhaps deserves some more explicit attention amongst those who work in the food branch.

Some major challenges can be observed that will require intense moral leadership in ethical decision-making in the near future. Ethics and business are not necessarily a natural combination. It is a challenge to use individual ethics in an institutional context, and it is a challenge to reconcile the interest of profit maximisation with social and environmental values. This means that business leaders will not only have to show inspirational or effective leadership; they also need to show moral leadership. In particular when food production processes are heavily automated and in a global production chain (Wernaart and Van Halderen, 2020).

Case 3.2. Influencers and food ethics.

Influencers unmistakably affect the lifestyle of young people especially. After all, this is what they are primarily paid to do. The success of food bloggers, vloggers and influencers should not be underestimated, and influencer marketing is becoming increasingly important in the food business sector (Hartman Group, 2012). Influencers have a mixed relationship with food science and food policy and seem to struggle between informing their followers and pursuing their financial goals as an influencer. Transparency regarding their agenda and sources is a key issue that is at the core of moral challenges.

Clay for cleansing

In 2016, Dutch food influencer Rens Kroes advised her followers to add clean clay to their diets for cleansing the body. While various food scientists questioned this approach (Koen Venema in RTL Boulevard, 8 March 2016; Edith Feskens in NPO Radio 1, 30 December 2019), Kroes argued that she preferred to approach food from an intuitive, rather than a scientific perspective.

Body positivity

In 2020, Dutch TV Chef Miljuschka Witzenhausen aired a four-part documentary in which she explores the story behind influencer posts (Witzenhausen, 2020). She explores the mismatch between fit and healthy bodies that are shown on many Instagram accounts, and the unrealistic body perception and expectation this causes, especially among young women. Her key message was that a group of influencers has a dangerous effect on the lifestyle and eating habits of its followers; the (sponsored) messages they send do not reflect a healthy or realistic lifestyle. Instead, influencers should be honest about their bodies and lifestyles, and more transparent about the messages they send. This should contribute to body positivity among followers.

Let's eat!

In 2020, 18-year-old TikToker Sara Sadok went viral for uploading clips in which she invited people to eat together. Especially during the pandemic lockdowns, people with an eating disorder would have found it difficult to get themselves to eat. In order to help them, she uploaded her videos and was widely praised for her initiative (Dogson, 2020).

Another issue that requires special attention can be found on the consumer's side. Amplified by massive communication platforms, there is an increasing desire to consider scientific data as fluent concepts that can be used in discussions, but they do not automatically have exclusive academic authority. The 'influencers and food ethics' case illustrates this phenomenon. This is a real challenge for food scientists, whose natural input in moral discourse on values such as healthy lifestyle or social impact is not always taken seriously.

3.5 Conclusions

Morality is an innate part of being human. Ethics is the attempt to understand and develop this morality in a structured way. What is considered right and wrong may develop dramatically over time. Not too long ago it was believed right to hold an elephant to account for its misbehaviour through public execution. Today only a few would not reject such a spectacle as inherently evil. At present we seem to be in the middle of a reassessment of our relationship with the creatures – animals and plants – with whom we share the planet and indeed with the planet itself. One factor, however, is constant: people generally aim at doing the right thing.

People not only apply morality to themselves, but it is also a part of their expectation and valuation of others including the businesses to which they are related as stakeholders. In organisational decision-making, taking account of the expectations of internal and external stakeholders may contribute to developing the ethics of individual businesses or even societies. As a result, ethics may contribute to constituting the rules that the business or society adheres to or imposes upon others.

References

Adenle, A.A., 2011. Response to issues on GM agriculture in Africa: are transgenic crops safe? BMC Research Notes 4: 388. https://doi.org/10.1186/1756-0500-4-388

Anomaly, J., 2015. What's wrong with factory farming? Public Health Ethics 8: 246-254. https://doi.org/10.1093/phe/phu001

Bird, R.J. and Cahoy, D., 2015. Human rights, technology, and food: coordinating access and innovation for 2050 and beyond. American Business Law Journal 52:435-500. https://doi.org/10.1111/ablj.12050

Burton, T.G., 2009. The hanging of a circus elephant. In: Olson, T. and Cavender, A.P. (eds) A Tennessee folklore sampler, selected readings from the Tennessee Folklore Society Bulletin. University Of Tennessee Press, Knoxville, TN, USA.

Carruthers, P., 1992. The animals issue, moral theory in practice. Cambridge University Press, Cambridge, UK. https://doi.org/10.1017/CBO9780511597961

Christman, J.C., 1989. Introduction. In: Christman, J.C. (ed.). The inner citadel: essays on individual autonomy. Oxford University Press, Oxford, UK, pp. 3-23.

Collins, J.W., 1994. Is business ethics an oxymoron? Business Horizons 37: 1-8. https://doi.org/10.1016/0007-6813(94)90013-2

Comstock, G., 2000. Vexing nature? On the ethical case against agricultural biotechnology. Kluwer Academic Publishers, Boston, MA, USA.

Comstock, G., 2010. Ethics and genetically modified foods. In: Gottwald, F.T., Ingensiep, H.W. and Meinhardt, M. (eds) Food ethics. Springer, New York, NY, USA, pp. 48-66.

Costa, R., 2018. Teaching food ethics. In: Costa, R. and Pittia, P. (eds) Food ethics education. Springer, Cham, Switzerland, pp. 3-24.

Craft, J.L., 2013. A review of the empirical ethical decision-making literature: 2004-2011. Journal of Business Ethics 117: 221-259. https://doi.org/10.1007/s10551-012-1518-9

Crane, A., Matten, A., Glozer. and Spence, L., 2019. Business ethics (5th ed.). Oxford University Press, Oxford, UK.

Dogson, L., 30 October 2020. A TikToker is going viral for helping people who struggle to eat sit down for a meal. Business Insider. Available at: https://www.businessinsider.nl/a-tiktoker-is-going-viral-for-helping-people-who-struggle-to-eat-sit-down-for-a-meal/.

Dubljević, V. and Racine, E., 2014. The ADC of moral judgment: opening the black box of moral intuitions with heuristics about agents, deeds and consequences. AJOB Neuroscience 5: 3-20. https://doi.org/10.1080/21507740.2014.939381

Elkington, J., 1997. Cannibals with forks. Capstone Publishers, Oxford, UK.

Fister, K., 2005. Junk food advertising contributes to young Americans' obesity. British Medical Journal 331: 1426. http://doi.org/10.1136/bmj.331.7530.1426-c

Fitting, E., 2006. Importing corn, exporting labor: the neoliberal corn regime, GMOs, and the erosion of Mexican biodiversity. Agriculture and Human Values 23: 5-26. https://doi.org/10.1007/s10460-004-5862-y

Freeman, E., 1984. Strategic management. Cambridge University Press, Cambridge, UK.

Freeman, E., 2010. Stakeholder theory. Cambridge University Press, Cambridge, UK.

French, S.A., Tangney, C.C., Crane, M.M., Wang, Y. and Appelhans, B.M., 2019. Nutrition quality of food purchases varies by household income: the SHoPPER study. BMC Public Health 19: 231. https://doi.org/10.1186/s12889-019-6546-2

Friedman, M., 13 September 1970. The social responsibility of business is to increase its profit. New York Times Magazine. Available at: https://www.nytimes.com/1970/09/13/archives/a-friedman-doctrine-the-social-responsibility-of-business-is-to.html.

Friedman, M., 1962. Capitalism and freedom. University of Chicago Press, Chicago, IL, USA.

Gilligan, C., 1982. In a different voice. Harvard University Press, Cambridge, MA, USA.

Hartman group., 2012. Clicks and cravings, the impact of social technology on food culture. Hartman Group, Bellevue, WA, USA. Available at: http://store.hartman-group.com/clicks-cravings/

Hsiao, T., 2017. Industrial farming is not cruel to animals. Journal of Agriculture and Environmental Ethics 30: 37-54. https://doi.org/10.1007/s10806-017-9652-0

Huhn, M.P., 2017. Adam Smith's philosophy of science: economics as moral imagination. Journal of Business Ethics 155: 1-15. https://doi.org/10.1007/s10551-017-3548-9

Inskip, H.M., Crozier, S.R., Godfrey, K.M., Borland, S.E., Cooper, C. and Robinson, S.M., 2009. Women's compliance with nutrition and lifestyle recommendations before pregnancy: general population cohort study. British Medical Journal 2009: 338. https://doi.org/10.1136/bmj.b481

Jasanoff, S., 2016. The ethics of invention. Norton and Company Ltd., New York, NY, USA.

Jones, T.M., 1991. Ethical decision making by individuals in organizations: an issue-contingent model. Academy of Management Review 16: 366-395.

Komparic, A., 2015. The ethics of introducing GMOs into sub-Saharan Africa: considerations from the sub-Saharan African Theory of Ubuntu. Articles from international congress of bioethics 2014. http://doi.org/10.1111/bioe.12191

Magee, B., and Elwood, R., 2013. Shock avoidance by discrimination learning in the shore crab (*Carcinus maenas*) is consistent with a key criterion for pain. Journal of Experimental Biology 216: 353-358. http://doi.org/10.1242/jeb.072041

Mepham, B., 1996. Ethical analysis of food biotechnologies: an evaluative framework. In: Mepham, B. (ed.) Food ethics. Routledge, London, UK, pp. 101-119.

Mepham, B., 2000. A framework for the ethical analysis of novel foods: the ethical matrix. Journal of Agricultural and Environmental Ethics 12: 165-176. https://doi.org/10.1023/A:1009542714497

Mepham, B., 2004. A decade of the ethical matrix: a response to criticisms. In: De Tavernier, J. and Aerts, S. (eds) Science, ethics and society, preprints of 5th EURSAFE congress, Leuven, Belgium, pp. 271-274.

Mepham, B., 2010. The ethical matrix as a tool in policy interventions: the obesity crisis. In: Gottwald, F.T., Ingensiep, H.W. and Meinhardt, M. (eds) Food ethics. Springer, New York, NY, USA, pp. 16-30.

Mepham, B., 2013. Ethical principles and the ethical matrix. In: Clark, J.P. and Ritson, C. (eds) Practical ethics for food professionals: ethics in research, education and the workplace. John Wiley and Sons, Ltd., Hoboken, NJ, USA, pp. 39-56.

Murtha, T.O. and Hamilton, A., 2012. Sustainable asset management. In: Krosinsky, C. (ed.) Evolutions in sustainable investing. John Wiley and Sons Inc., Hoboken, NJ, USA, pp. 53-77.

New York Times, 5 January 1903. Coney elephant killed; Topsy overcome with cyanide of potassium and electricity. She was Adam Forepaugh's 'original baby elephant' twenty-eight years ago – her keeper, 'Whitey,' would not see her die.

NPO Radio 1, 30 December 2019. Misverstanden over voeding: 'Detoxen met klei, waar haal je het vandaan?' Available at: https://www.nporadio1.nl/wetenschap-techniek/20766-misverstanden-over-voeding-Detoxen-met-klei-waar-haal-je-het-vandaan.

Owen, J., 2018. Childhood obesity: government's plan targets energy drinks and junk food advertising. British Medical Journal 2018: 361. https://doi.org/10.1136/bmj.k2775

Paarlberg, R., 2010. GMO foods and crops: Africa's choice. New Biotechnology 27: 609-613. http://doi.org/10.1016/j.nbt.2010.07.005

Regan, T., 1983. The case for animal rights. University of California Press, Berkeley, CA, USA.

Rest, J.R., 1986. Moral development: advances in research and theory. Praeger, New York, NY, USA.

Roff, R.J., 2007. Shopping for change? Neoliberalizing activism and the limits to eating non-GMO. Agriculture and Human Values 24: 511-522. https://doi.org/10.1007/s10460-007-9083-z

Rowley, T., 1997. Moving beyond dyadic ties: a network theory of stakeholder influences. Academy of Management Review 22: 887-910. https://doi.org/10.2307/259248

Rozin, P., Hormes, J.M., Faith, M.S. and Wansink, B., 2012. Is meat male? A quantitative multimethod framework to establish metaphoric relationships. Journal of Consumer Research 39: 629-643. https://doi.org/10.1086/664970

RTL Boulevard, 8 March 2016. Rens Kroes: Kleidieet is mijn persoonlijke ervaring. Available at: https://www.rtlboulevard.nl/lifestyle/artikel/682846/rens-kroes-kleidieet-mijn-persoonlijke-ervaring

Ruby, M.B., 2012. Vegetarianism. A blossoming field of study. Appetite 58: 141-150. https://doi.org/10.1016/j.appet.2011.09.019

Schwartz, M.S., 2016. Ethical decision-making theory: an integrated approach. Journal of Business Ethics 139: 755-776. https://doi.org/10.1007/s10551-015-2886-8

Simonin, D. and Gavinelli, A., 2019. The European Union legislation on animal welfare: state of play, enforcement and future activities. In: Hild, S. and Schweitzer, L. (eds) Animal welfare: from science to law. La Fondation Droit Animal, Paris, France, pp. 59-70. Available at: http://www.fondation-droit-animal.org/documents/AnimalWelfare2019.v1.pdf

Singer, P., 1975. Animal liberation: a new ethics for our treatment of animals. HarperCollins, New York, NY, USA.

Statman, D., 1996. Hard cases and moral dilemmas. Law and Philosophy 15: 117-148. Available at: http://www.jstor.org/stable/3504826

Trevino, L.K., Weaver, G.R. and Reynolds, S.J., 2006. Behavioural ethics in organizations. Journal of Management 32: 951-990. https://doi.org/10.1177/0149206306294258

Valdman, M., 2010. Outsourcing self-government. Ethics 120: 761-790.

Van Loon, H. and Wernaart, B.F.W., 2022. Consumer protection. In: Wernaart, B.F.W. and Van der Meulen, B.M.J. (eds) Applied Food Science. Wageningen Academic Publishers, Wageningen, the Netherlands, pp. 65-83.

Wernaart, B. and Van Halderen, M., 2020. Digital transformation requires moral leadership. Innovation Origins. Available at: https://innovationorigins.com/digital-transformation-requires-moral-leadership/

Wernaart, B., 2018. Ethics and business, a global introduction. Noordhoff, Groningen, the Netherlands.

Wernaart, B., 2020. Whitepaper – dieren en ethiek: voedsel of vriend? Noordhoff, Groningen, the Netherlands.

Witzenhausen, M., 2020. #Eerlijke foto. Videoland. Available at: https://www.videoland.com/series/500494/eerlijkefoto/3779

Wood, A.L., 2012. 'Killing the elephant': murderous beasts and the thrill of retribution, 1885-1930. Journal of the Gilded Age and Progressive Era 11: 405-444. Available at: https://www.jstor.org/stable/23249163

Zwart, H., 2000. A short history of food ethics. Journal of Agricultural and Environmental Ethics 12: 113-126. https://doi.org/10.1023/A:1009530412679

4. Consumer protection

This is what you need to know about legal standards and industry self-regulation that contribute to consumer protection

Heleen van Loon[1]* and Bart F.W. Wernaart[2]

[1]Business Administration & Agribusiness, HAS University of Applied Sciences, P.O. Box 90108, 5200 MA 's-Hertogenbosch, the Netherlands; [2]School of Business and Communication, Fontys University of applied sciences, P.O. Box 347, 5600 AH Eindhoven, the Netherlands; h.vanloon@has.nl

Abstract

In the mid-nineteen-nineties the European food industry faced considerable negative media attention. Major food scandals had caused food safety and public health concerns amongst consumers. This resulted in significant damage to the reputation of the food industry and a serious loss of consumer trust. These incidents showed that the then applicable European legislation had fallen short in protecting the consumer. The European Legislature felt the urge to adjust the legislation on this matter. At the core of these legal reforms was the General Food Law (in force as of 2002). Furthermore, in the food business industry, various forms of self-regulation where introduced to improve consumer protection. The aim of this chapter is to offer the reader an introduction to the regulatory methods of consumer protection, with a focus on the specific legal requirements to food business as well as leading forms of industry self-regulation. We will also explore contemporary issues and recent developments on these topics. Finally, we will draw conclusions and look ahead, and briefly reflect on the potential of using the Nutri-score as a uniform European food label.

Key concepts

- ► Food scandals are important drivers for legislative and non-legislative improvements in regulating food safety and consumer protection.
- ► Consumer protection is a value in the food business industry that governs the relationship between food business and consumers.

Bart Wernaart and Bernd van der Meulen (eds)
Applied food science
DOI: 10.3920/978-90-8686-933-6_4, © Heleen van Loon and Bart F.W. Wernaart 2022

▶ Food information is a value in the food business industry that aims to facilitate informed consumer choice.
▶ Food advertising standards aim to facilitate informed consumer choice by regulating advertisement quality standards.
▶ Industry self-regulation is a system of normative standards voluntarily set by individual companies or collectives.

Case 4.1. Food scandals and consumer trust.

During the 1990s the European food industry faced considerable negative media attention due to various food scandals that were at the forefront of public debate in the European Union. The most notorious food scandal in this series concerned the Bovine Spongiform Encephalopathy (BSE) crisis, also known as Mad Cow disease that broke out in the UK in the 1980s. BSE literally means spongiform encephalopathy in cows and develops due to eating contaminated feed. Several hundred thousand cattle were infected with BSE. A link was discovered between the consumption of BSE-infected meat products and the development of a fatal variant of Creutzfeldt Jacob disease in humans. In 1996, ten years after the first BSE infection was detected, the first person died of this fatal variant. Currently, more than 200 people have died due to Creutzfeldt Jacob worldwide. A few hundred thousand contaminated cattle were slaughtered (WUR, n.d).

Another notorious food scandal occurred in Belgium in 1999. The Belgian authorities discovered a case of poultry fodder contaminated with industrial oil containing the harmful chemical dioxin. Huge quantities of food products such as eggs, mayonnaise and even biscuits were taken off the shelves in shops because of dioxin contamination. During the dioxin crisis, 7 million chickens and 60,000 pigs were slaughtered and almost 2,000 farms were shut down for months (Het Belang van Limburg, 2015).

4.1 Consumer trust and European food law

The series of food scandals in the nineteen-nineties (see also Case 4.1) demonstrated that the food law rules in force at the time, which had been drawn up from an economic perspective in order to stimulate the internal market and thus promote trade in food products between the Member States, could not guarantee the safety of the products traded between them (Vogel, 2001; van der Meulen, 2009). Furthermore, these major events highlighted the fact that food safety concerns not only the final product, but also the whole production process (De Jonge *et al.*, 2004).

The food scandals, under the spotlight of enormous media attention, not only caused consumer concerns about food safety (Pennings *et al.*, 2002), but also led to a decline in consumer confidence in the ability of both European and national authorities to protect public health (Vogel, 2001). A reform of European food law that puts consumer protection first was therefore necessary and indispensable (Section 4.2). Next to that, the business sector itself adopted various forms of self-regulatory standards to take further responsibility in the field of consumer protection (Section 4.3). Food law and food ethics have been covered in Chapters 2 and 3 (Van der Meulen and Wernaart, 2022; Wernaart, 2022), discussing legal and self-regulatory issues in a broader sense. However, regulating consumer protection and food safety is a topic in its own right and deserves a separate chapter in the context of applied food science where we reflect in more detail on how the instrument of law and self-regulation is used to contribute to the protection of the consumer. Also, in university curricula, we find that consumer protection is increasingly becoming a subject on its own, where law and non-legal regulatory methods are jointly discussed.

4.2 Consumer protection and EU legislation

In order to restore consumer confidence in food safety, the instrument of law has been used to regulate food law more specifically, well beyond the 'general' legal protection offered to consumers through e.g. contract law or product liability law. As we have seen in Section 2.2.2, Regulation (EC) No 178/2002, which laid down the general principles and requirements of food law establishing the European Food Safety Authority as well as procedures in matters of food safety, also known as the General Food Law (GFL), was adopted in 2002 (EC, 2002). The general objectives of the General Food Law reads as follows: 'food law shall pursue one or more of the general objectives of a high level of protection of human life and health and the protection of consumers' interests, including fair practices in food trade, taking account of, where appropriate, the protection of animal health and welfare, plant health and the environment' (Article 5 GFL). The GFL framework is the umbrella-law for EU-wide food legislation, and consumer protection is a cornerstone of EU food law. In particular, adding more specific legal standards on communication about food in a business to consumer relationship is an important way to help consumers make an informed choice, and prevent them from being misled.

Consumers have a right to expect that the food they purchase and consume will be safe and of good quality. Communication between food business and the consumer plays an important role. On the one hand, communication aims at informing the consumer about the characteristics of the food product. On the other hand, communication aims at convincing the consumer to buy certain products through advertising. Informing and convincing are two different goals that may conflict. To this end, laws are produced that on the one hand guarantee particular information through e.g. labelling of prepacked food products (food information standards), and laws that restrict painting an overly rosy picture of the advantages of consuming certain food products (advertising standards).

4.2.1 Food information standards

food information

One of the general principles of the General Food Law is laid down in Article 8 which deals specifically with the protection of consumer interests. The aim of a high level of consumer protection is further elaborated in Regulation 1169/2011 on the provision of food information to consumers (also known as FIC) (EC, 2011). On this basis, food companies[8] must provide consumers with honest and unambiguous information about the products they intend to consume in order to enable them to make an informed and balanced food choice that is appropriate to their individual dietary needs. Food information means not only information about a product by means of a label, but also 'other accompanying material or means, including modern technology tools or oral communication' (Article 2(2)a FIC). This means that, for example, advertising and leaflets, product packages and online food sales must also comply. The basic principle here is that consumers must not be misled, particularly about the characteristics, effects or properties of food (FIC, preamble, recital 20). This principle is enshrined in Article 7 of FIC, which stipulates that food information must not be misleading and must be accurate, clear and easy to understand for the consumer.

One of the most important means of informing consumers about a product is labelling (Holle, 2020). The FIC therefore stipulates that it is compulsory to provide certain information on the label of pre-packaged food products. This concerns the following 12 mandatory particulars on pre-packed food products:

1. the name of the food;
2. the list of ingredients;
3. allergen declaration;
4. the quantity of certain ingredients or categories of ingredients;
5. the net quantity of the food;
6. the date of minimum durability or the 'use by' date;

---

[8] Those rules on the provision of food information to consumers apply to foods intended for consumption by the final consumer, the consumer, and available as such, for example, in supermarkets or through delivery services or web shops, but also to foods provided to or by mass caterers, such as restaurants, canteens and schools.

7. any special storage conditions and/or conditions of use;
8. the name or business name and address of the food business operator;
9. the country of origin or place of provenance;
10. instructions for use;
11. the actual alcoholic strength by volume;
12. a nutrition declaration.[9]

name of the food

Each food product must have a name. This is not the fantasy name or brand name, like Tony's Chocolonely or Snickers, used by food companies to describe the product, but the name of the product that makes it clear to the consumer which product is being bought or consumed. In principle, the name of the product is the legal name of the product. This is a reserved designation that must be used if the product has a particular composition. These legal designations can be found in further EU directives and regulations.

For example, pursuant to Directive 2000/36/EC of the European Parliament and of the Council of 23 June 2000 relating to cocoa and chocolate products intended for human consumption, the reserved designation 'white chocolate' must be used on the label if a 'product obtained from cocoa butter, milk or milk products and sugars which contains not less than 20% cocoa butter and not less than 14% dry milk solids obtained by partly or wholly dehydrating whole milk, semi- or full-skimmed milk, cream, or from partly or wholly dehydrated cream, butter or milk fat, of which not less than 3,5% is milk fat.'[10]. The same applies, for example, to vodka,[11] milk[12] and fruit juice[13].

For food products for which there is no EU legal designation, a legal designation may have been formulated in national regulations and must then be used in that country. There may be differences in regulation and interpretation between Member States in this regard, such as ice cream in the neighbouring countries of the Netherlands and Belgium (Dommels, 2020). In Belgium, the reserved designation for ice cream applies to ice cream with a minimum milk fat content of 8% and milk protein content of 2.5%. In the Netherlands, the reserved designation may only be used for ice cream intended to be consumed frozen and with, among other things, a milk fat content of at least 5%. This implies that Belgian ice cream must contain 3% more milk fat than Dutch ice cream.

---

[9] This is also required for non-prepacked food products (article 44 (1) FIC).
[10] Article 4 Annex I Directive 2000/36/EC.
[11] Regulation 110/2008.
[12] Regulation 1308/2013.
[13] Directive 2001/112/EC.

If no EU or national legal name exists for a food product, a food business may opt for a so-called customary name. A customary name is a name that makes it immediately and without further explanation clear to the consumer which product it concerns, such as pesto or sushi. Again, these names can differ per country or per region. For instance, in the Netherlands, crème fraîche is thick soured cream, while in Belgium crème fraîche refers to fresh cream.

In the absence of an EU or national statutory name or customary name, a descriptive name must be given to a product. With this descriptive name, a consumer who does not know the product must understand what kind of product she or he is dealing with, such as 'Airy toasts with whole wheat flour' (Liga Cracotte crackers whole wheat)[14].

A thorny issue in the EU regarding the name of a food product is the use of meat-related and dairy-related names for products of plant origin. In 2017, the Court of Justice of the EU issued an important ruling on this issue. This judgment was prompted by a preliminary question from a German court asking for an interpretation of the EU legal designations 'milk' and 'dairy products' and their use in new products. The case concerned a dispute between the Verband Sozialer Wettbewerb (VSW), a German association fighting unfair competition, and TofuTown, a German company manufacturing and marketing vegetarian and vegan products.[15] According to VSW, TofuTown infringed EU rules laid down in the Common Market Regulation for Agricultural Products, which states that names of certain regulated products shall not be used for other products, by promoting vegetable products with names such as 'Soyatoo Tofubutter', 'Pflanzenkäse' and 'Veggiecheese'. In this Regulation, milk is exclusively defined as:

> the product normally secreted by the mammary glands and obtained from one or more milkings without either addition thereto or extraction therefrom

The names used by TofuTown refer to dairy products and thus suggest that the products are made from milk, which is not the case, according to VSW. The Court concluded that 'milk' is reserved exclusively for dairy products, i.e. products made from milk, and that names such as 'butter' and 'cheese' may

---

[14] Annexes III and VI to Regulation 1169/2011 list various indications that must accompany a name where it applies, such as the indication of the physical state of the product or the specific treatment it has undergone, such as powder, concentrate or smoked.

[15] CJEU 14 June 2017.Case C-422/16, Verband Sozialer Wettbewerb e.V.v. TofuTown.com GMBH. ECLI:EU:C:2017:458.

not be used to designate a purely vegetable product. Not even if a clarifying or descriptive addition, such as non-animal or vegetable, is added to indicate the vegetable origin of the product.

There has been an initiative to further expand the protection of dairy names, as happened when the European Parliament voted on 23 October 2020 in favour of an amendment further limiting the use of dairy designations for plant-based dairy substitutes (Irving, 2020; Van Dinther, 2020). This ban would have gone beyond the existing restrictions resulting from the TofuTown case. With this ban, descriptions such as 'contains no milk', 'cheese alternative', 'alternative to dairy products', and climate footprint comparisons between vegetable and dairy products like 75% less CO_2 and 'creamy' would also no longer be allowed. The reason for this proposal was that such products might cause confusion among consumers. The proposal led to massive protests from the plant industry, NGOs and action groups. The proposal would, in their view, amount to censorship of vegetable dairy products. In addition, the proposal was not in line with the EU goal to achieve a more sustainable food system. In mid-2021, the proposal was off the table in the final vote on European agricultural policy. This led the European Parliament to withdraw the proposal (Van Dinther, 2021). On 23 October 2020, the European Parliament also voted on an amendment to ban the use of meat related names for plant-based alternatives. The marketing of non-meat products under meat names would be misleading and the use of these meat names would suggest that these vegetable alternatives are equivalent to 'original' meat products. In contrast to the ban on the use of dairy product names for vegetable dairy substitutes, the European Parliament voted directly against this amendment (Irving, 2020). By adding the words 'vegetarian' or 'vegan', it is sufficiently clear to the consumer what kind of product it is (Van Dinther, 2020).

list of ingredients

Back to the mandatory labelling particulars. Most of the particulars concern the list of ingredients. This list of ingredients used in the preparation of the product must have an appropriate title, such as 'list of ingredients', and must name all the ingredients in descending order by weight. The ingredients shall be designated by the name as described above for the name of the food.[16]

---

[16] Annex III of the FIC contains additional statements for certain categories of food products that must be mentioned on the label, such as the additional statement 'High caffeine content', 'Not recommended for children and pregnant or breast-feeding women' in the case of beverages with a high caffeine content or foods with added caffeine.

In the Teekanne judgment, the European Court of Justice decided that a complete and correct list of ingredients is not always sufficient to remove the ambiguous impression created by the package of a food product.[17] In this case, the German company Teekanne marketed a fruit tea under the name 'Felix Himbeer-Vanille Abenteuer' (Felix raspberry-vanilla adventure). The package of this fruit tea included pictures of raspberries and vanilla blossom, the words 'naturally flavoured fruit tea' and 'naturally flavoured fruit tea – vanilla raspberry flavour', and a graphically designed seal with the words 'only natural ingredients' in a gold-coloured circle. However, the list of ingredients of this fruit tea indicated that it did not contain any vanilla or raspberry flavouring. In the judgement, the Court of Justice put an end to the so-called Labelling doctrine (Van der Wal, 2019), according to which a communication cannot be misleading if the list of ingredients is correct, and stated that in addition to the list of ingredients, the rest of the package must also be taken into account when determining whether the consumer is under the impression that the food product contains a certain ingredient when, in fact, it does not.

allergen declaration In order to protect consumers with intolerances or allergies, so that they can make an informed and safe choice, the presence of ingredients liable to cause intolerances or allergens shall be clearly and markedly indicated on the label of a prepacked food.[18]

QUID Where the presence of an ingredient in a product is particularly emphasised in any way, for example by its name or by the way it is depicted on the package of the food product, the percentage of the ingredient must be stated on the label. Furthermore, the net quantity of liquid products shall be indicated on the label in litres, centilitres, millilitres, and for other products in (kilo)grams.

minimum durability date The label must also state the shelf life, the 'best before' indication for products that do not deteriorate quickly and the 'use by' indication for products that 'are highly perishable and are therefore likely after a short period to constitute an immediate danger to human health' (Regulation 1169/2011, Article 24) (EC, 2011). If special storage or use conditions apply to a product, these must also **storage or use conditions** be stated, such as 'keep refrigerated at a maximum of 4 °C and consume within 2 days of opening'. In view of the fact that a food company is responsible for the safety of the product, the (trade) name and address of the food business must be stated on the label.

---

[17] CJEU 4 June 2015. Case C-195/14, Bundesverband der Verbraucherzentralen und Verbraucherverbände v. Teekanne GmbH& Co.KG. ECLI:EU:C:2015:361.
[18] Annex II of the FIC contains the 14 main sources of intolerance and allergy in Europe, such as crustaceans and eggs.

origin of the product

Because of the extensive trade in food within the Member States, but also outside them, food products come from all corners of the world; Machego cheese from Spain, kiwis from New Zealand and Limoncello from Italy. How does a consumer know where a product comes from? By reading the origin of the product on the label. For unprocessed meat, eggs, fish, vegetables, fruit, olive oil and honey, it is compulsory to indicate the country of origin. In addition, it is compulsory to indicate the origin of the product on the label if the failure could mislead the consumer about the real origin of the product. For example, if a yoghurt is presented on the packaging as Greek yoghurt while it was prepared in the Netherlands, it must be explicitly stated on the label that the yoghurt was produced in the Netherlands. The origin indication should not be confused with the commercial name of a product that refers to a tradition or recipe, as is the case with Irish coffee and French fries, where it is not compulsory to state the origin as long as this does not misleads the consumer (Rog, 2020).

instructions for use alcohol percentage

In addition, the label must include instructions for use, and for drinks with an alcohol content above 1,2%, the alcohol percentage must be stated and finally the label must include a nutritional statement (a summary of the amount of energy (kilocalories and/or kilojoules), carbohydrates, sugars, protein, fat, saturated fatty acids and salt per 100 grams or 100 millilitres).

novel foods

For 'novel foods', i.e. foods and ingredients which were not marketed as foods within the European Union before 15 May 1997 and which are currently authorised within the European Union, additional rules apply in the context of food safety. Novel foods may include phosphatidylserine from soy phospholipids, which is used for example in chocolate-based candy, cereal bars, yoghurt-based foods and drinks,[19] or algal oil from the micro-algae *Schizochytrium* sp, which is used for example in dressings, dairy products, bakery products and breakfast cereals.[20] If the European Food Safety Authority (EFSA) has positively assessed the safety of a novel food and the European Commission has approved the authorisation of the novel food, this novel food will be included in the so-called Union list of authorised novel foods, or 'Union list' for short.[21] In order to protect consumers' interests, the Union list contains

---

[19] Commission implementing decision of 19 August 2011 authorising the placing on the market of Phosphatidylserine from soya phospholipids as a novel food ingredient under Regulation (EC) No 258/97 of the European Parliament and of the Council (2011/513/EU).

[20] Commission implementing regulation (EU) 2018/1032 of 20 July 2018 authorising the extension of use of oil from the micro algae *Schizochytrium* sp. as a novel food of the European Parliament and of the Council, and amending Commission Implementing Regulation (EU) 2017/2470.

[21] Commission Implementing Regulation (EU) 2017/2470 of 20 December 2017, establishing the Union list of novel foods in accordance with Regulation (EU) 2015/2283 of the European Parliament and of the Council on novel foods.

strict conditions under which the 'novel food' may be used and additional specific rules, in addition to the general labelling requirements described above, for labelling. This is to ensure that consumers are adequately informed about the nature and safety of the product. Next to the term 'novel foods', there is the legal concept of 'genetically modified foods'. These are foods that have been genetically manipulated by certain scientific techniques in order to obtain new or improved qualities. Bread or bakery products containing flour, protein, oil, fat or lecithin derived from genetically modified soybeans or sugar from genetically modified sugar beet are examples of genetically modified food. If

genetically modified organisms (GMOs) foodstuffs contain or consist of genetically modified organisms (GMOs), are produced from GMOs or contain ingredients produced from GMOs, this must be explicitly indicated on the label.[22] The aim is to inform the consumer as accurately as possible about the origin of the food and to give the consumer the freedom of choice to buy products with or without genetically modified ingredients.

4.2.2 Food advertising standards

Food companies try to entice consumers to buy their food products and use various marketing techniques (see also Chapter 20; Floto-Stammen, 2022) to do so. For example, food companies often try to present products as healthier than they actually are. Food packages often includes terms such as 'low sugar', 'fat free', 'source of fibre', 'fresh', 'light', 'pure' and 'natural'. This is voluntary food information to entice consumers to buy these products, but to what extent is the use of these terms allowed? When is a consumer being misled?

nutrition and health claims In order to ensure that consumers are fairly informed about the function and effects of food and are not misled, EU rules apply that govern the use of such claims, also known as nutrition and health claims. A nutrition claim says something positive about the composition of a product and a health claim indicates that the product has a positive effect on health. Regulation (EC) no 1924/2006, also known as the Claims Regulation, contains the regulations regarding food and health claims (EC, 2006). Food companies may only use nutrition claims, such as light and high-fibre, if it appears on the European list of permitted food claims and if the composition of the product complies with the conditions set for it.[23] This list and its conditions are included as an annex to the Claims Regulation. For example, the following is stated about the claim 'high-fibre':

---

[22] Regulation (EC) No 1829/2003.
[23] Regulation (EC) Nr. 1924/2006.

A claim that a food is high in fibre, and any claim likely to have the same meaning for the consumer, may only be made where the product contains at least 6 g of fibre per 100 g or at least 3 g of fibre per 100 kcal.[24]

The use of a health claims, such as 'Calcium is needed for the normal growth and development of the bones', is permitted if this health claim is included in the annex to Regulation (EC) No 432/2012 and complies with the conditions set therein (EC, 2012). The approved nutrition and health claims have been assessed by the EFSA. The European Claims Register is a tool to find out whether a food or health claim is permitted or not and which regulation is the legal basis for it.

Currently, the use of the claim 'natural' on products is under discussion as no legal definition of this claim exists (Ligtermoet, 2021), except for Regulation (EC) 1334/2008 on the use of flavourings in food (EC, 2008). Many food companies use claims such as '100% natural' or 'only natural ingredients' on their products; however, in practice the composition of these products are not so natural at all. A report published in November 2020 shows that consumers associate the word 'natural' with the origin of ingredients, the minimal processing of the product or the absence of additives and that these expectations do not match the actual composition of the product. This is why MEPs are now calling for a legal definition of the term 'natural' and stricter rules on its use. This is also in line with the European Union's commitment to a sustainable food system in which consumers are encouraged to eat healthily and sustainably. Therefore, in order to help consumers to adopt healthier eating patterns and make the right choices, the information provided on the actual composition of the product should not be misleading.

4.3 Consumer protection and self-regulation

Consumers are not protected by the instrument of law alone. Sometimes, the legislature leaves regulating elements of consumer protection to the private sector, and sometimes the private sector initiates protective measures on its own initiative.

First, some thoughts on the word self-regulation itself. In this chapter, we use this jargon, but should also underline there are different ways to refer to the idea that regulating food safety is not done by a legislator, such as private food law (Van der Meulen, 2011) or voluntary accountability (Wernaart, 2021: 273-277). It needs to be noted here that the word 'self' should not be confused

---

[24] Annex of Regulation (EC) Nr. 1924/2006.

with 'autonomous'. In case of self-regulation, the regulator could be, but is not *per se* the same organisation as the regulatee, or the entity that certifies/oversees implementation of the self-regulation. When we consider the Unilever example below, the company at the end of the product chain imposes its self-regulation on the full product chain. In taking so called 'chain responsibility', supplying companies are far from free in determining their own standards on food security.

Next, there are different levels of self-regulation, and various methods to self-regulate. Considering the level of self-regulation, we could make a distinction between individual and collective self-regulation. Individual self-regulation is usually a way of establishing ethical standards and other norms within an organisation (see also Chapter 3 (Wernaart, 2022) on business ethics). Most Multinational Organisations (MNOs) have their own code of conduct, which includes standards to protect consumers.

By way of example, Unilever states in its code of business principles:

> Unilever is committed to providing purposeful branded products and services which consistently offer value in terms of price and quality, and which are safe for their intended use. Products and services will be accurately and properly labelled, advertised and communicated. (Unilever, n.d.)

In the same code, we find a balancing of interest between consumer safety and animal wellbeing:

> Uphold Unilever's commitment to eliminate animal testing without compromising on consumer safety (Unilever, n.d.13)

This is further elaborated in 'Unilever's position on: Alternative approaches to animal testing':

> We do not test on animals and believe that animal testing is not needed to make sure that our products are safe for people to use and safe for our planet. We use leading edge human-relevant safety science, not animals, to evaluate the safety of our products and ingredients for consumers, our workers and the environment. We pro-actively share our non-animal safety approaches with others, collaborating with partners across the world to help bring about an end to animal testing for consumer products, now and in the future. That's why we also develop and advance the use of 'next generation' safety assessment approaches, based on modern science, that do not rely on new animal data' (Unilever, 2021).

individual self-regulation

Such individual self-regulation initiatives are companywide standards that are imposed on a full product chain. Where the effect of law is usually restricted to the boundaries of a jurisdiction, self-regulation is not hindered by these restrictions. Instead, the scope of the company's activities limits the effects of the standards. This means that the individual self-regulation of a multinational can reach many jurisdictions at the same time, and – if so desired and organised – can cover an entire product chain.

industry self-regulation

Self-regulation can also be done collectively, mostly sector-wide, or in collaboration with other stakeholders. For instance, various stakeholders are involved with the Dutch Advertising Code,[25] including advertisers, media, consumer organisations and content creators. The stakes of these groups are brought together in the commercial code. Industry self-regulation is mostly supported by self-regulation bodies that are installed to implement and protect the standards. In case of the Advertising Code, this is the Advertising Code Committee. Such self-regulating organisations may contribute to consumer

food quality labels

protection in different ways. As we can see in Case 18.1, food quality labels are created to guarantee a certain quality aspect in sustainable food production. In some cases, collective self-regulation is not only about guaranteeing food quality, but also contributes to effectuating these standards by offering complaint procedures to consumers. A good example is the Advertising Code Committee, as further elaborated below.

Case 4.2. Vanilla custard without vanilla.

On 16 May 2019, the Board of Appeal of the Dutch Advertising Code Committee (ACCP), a self-regulatory body which means that companies voluntary submit to the competence of such body, decided that the name of a popular Dutch desert – *vanillevla* (vanilla custard) was misleading because *vanillevla* does not contain natural vanilla.[26] The consumer organisation Foodwatch flagged this case against FrieslandCampina and claimed that the *vanillevla* was not labelled according to the strict EU requirements. The *vanillevla* contained the name 'Optimel Vanillevla', with the description 'soft vanilla taste' on the front of the package. The list of ingredients indicated it contained vanilla flavouring without specifying if the vanilla was of natural or chemical origin. In fact, the characteristic taste of vanilla was obtained by the use of a chemical flavouring substance. As both the name of the product and the packaging specifically refer to the ingredient 'vanilla', the average consumer may assume that the product actually contains 'vanilla', according to the Committee. Final decision: the consumer is misled.

---

[25] See https://www.reclamecode.nl/nrc_taxonomy/general/?lang=en
[26] ACCP 16 May 2019, 2018/00701 – CVB.

Although decisions by the Advertising Code Committee are not legally binding, most food companies conform to them. Friesland Campina was one of them and the text on the package was adjusted. Other food companies, such as Albert Heijn, Jumbo, Lidl and Aldi, also decided to follow the ruling and changed the packaging of their vanilla products (Foodwatch, 19 June 2019).

advertising codes

As the decisions in Case 4.2 show, the question of whether or not the consumer is misled by the information presented about the product always arises. More often, consumers, consumer organisations or anyone else who wishes to do so, turn to national independent self-regulatory bodies – such as the Dutch Advertising Code Committee – which examine whether advertising by companies complies with the rules on advertising that the company has imposed on itself. The added value of using self-regulated procedures instead of formal court procedures are numerous. For the consumers, such procedures are usually better accessible and affordable compared to legal proceedings. Also, self-regulated procedures are not as lengthy as court cases. For legislators, the added value is that they do not directly interfere in matters of conflicting rights. Setting strict legal standards for advertising might interfere with the freedom of speech, while at the same time the right of consumers not to be harmed or misled is also firmly embedded in law. To avoid balancing on a very thin line, self-regulation that results from a cooperation of advertisers, media, consumers and content-creators can be a feasible solution. For business, the motivation to initiate self-regulation could be that it is driven by an intrinsic motivation of companies to contribute to consumer protection. Depending on the required protection, this can be more effective than legislation that imposes particular ways of protecting the consumer in the private sector. Another argument can be that the private sector feels comfortable being in the lead and avoids unexpected legal interference by lawmakers when they adequately contribute to consumer protection. A strong incentive to comply with self-regulation is the possible image damage in case of non-compliance (Wernaart, 2021: 281-282). Preventing the sector from being publicly shamed is a motive for companies to voluntarily submit to rules set by the sector itself (Baarsma *et al.,* 2003).

The various national independent complaint bodies for advertising in the Member States fall under the umbrella of the European Advertising Standards Alliance (EASA). The EASA not only deals with complaints about cross-border advertising, but also coordinates and supports the various independent national complaint bodies within the Member States in order to create as much uniformity as possible. On an international level, a similar institute exists, namely the International Council for Ad Self-Regulation (ICAS). ICAS is a global platform which promotes – in cooperation with the International Chamber of Commerce – effective advertising self-regulation to make sure that marketing communications are legal, honest, truthful and decent.

Compliance with legal food standards and self-regulation are typically part of a corporate social responsibility strategy (see also Section 3.2). Critical society and consumers demand that companies consider the social and environmental consequences of their actions. To what extent do companies in the food sector feel the ethical responsibility to comply with the rules set and to what extent are they prepared to focus their strategy on value creation in order to win the trust and respect of the consumer (Nguyen *et al.*, 2020)? For the food sector, there are considerable challenges in this respect, given the critical opinion of consumers on what is eaten. This leads to a complex set of ensuing requirements for the production chain and the quality, health and safety of products (Hartmann, 2011). This responsibility is not limited to an individual food **chain responsibility** company, but extends to chain responsibility in which the entire chain must feel responsible for fulfilling obligations, as we can see in Case 18.3.

Case 4.3. The Fipronil crisis: who is responsible for consumer protection?

A distressing chain responsibility case was the Fipronil crisis that came to light in the summer of 2017. It became known that the Dutch company Chickfriend mixed the blood louse pesticide Dega16, authorised for chicken houses, with the toxic substance Fipronil. The mixture was used at hundreds of poultry farms in the Netherlands, Belgium and Germany. The poultry ingested the toxic substance and as a result the eggs were contaminated. In other words, a potential danger to public health. Result: around 200 farms were shut down for months, hundreds of millions of eggs were destroyed and over 1.5 million chickens were culled. The costs for the poultry farmers: between €65 and €75 million. The costs for supermarkets: €7 to €8 million.

Who was responsible? The parties involved pointed at each other. The poultry farmers thought that the food safety supervisor, the NVWA, had failed, and the NVWA in turn thought that the poultry farmers themselves should have taken responsibility (De Jager, 2019).

The outcome of an independent investigation into this food crisis was harsh: all the main players in the egg chain were responsible. The egg chain gave insufficient priority to food safety which resulted in failure in their legal responsibility. Food safety must be given top priority in the entire chain – both by the poultry farmers

---

and the NVWA – partly by means of self-regulation, and commercial or other interests must be subordinate to this. Politicians and government are also involved and are therefore partly responsible for food safety in the chain (Sorgdrager, 2018).

4.4 Conclusions and looking ahead

The above shows that in response to various food crises in Europe, a legal framework as well as industry self-regulation have been established to ensure food safety for consumers. Various statutory process and substantive requirements combined with private standards and controls place the responsibility for ensuring food safety on food companies operating throughout the food chain. The extent to which these standards are complied with depends on the extent to which food companies feel socially responsible for food.

The European Union continues with the farm-to-fork strategy, which is a core element of the Green Deal.[27] The 'farm to fork' strategy stands for a fair, healthy and environmentally friendly food system.[28] An environment in which healthy and sustainable choices are also the easiest when there are labels that help consumers eat healthily and sustainably. A promising initiative in this field is a new labelling system for nutritional value, established by the French Government,[29] and recommended in Germany, Belgium, the Netherlands, Luxembourg, Spain and Switzerland: the Nutri-Score. This algorithmic ranking system inform consumers in a uniform and easy way what a healthy choice is. A growing group of consumer organisations and companies would like to have this as a uniform European food label (WHO, 2021).

References

Baarsma, B., Felsö, F., Van Geffen, S., Mulder, J. and Oostdijk, A., 2003. Zelf doen? Inventarisatiestudie van zelfreguleringsinstrumenten. SEO, Amsterdam, the Netherlands.

De Jager, G.J.M., 2019. Fipronil in eieren en mest; never let a serious crisis go to waste? Tijdschrift voor Agrarisch Recht 2019: 213-222.

---

[27] By 2050, Europe should become the first climate-neutral continent.

[28] Communication from the Commission to the European Parliament, the Council, the European economic and social committee and the committee of the regions. A Farm to Fork Strategy for a fair, healthy and environmentally-friendly food system.

[29] Journal Officiel de la République Française, Arrêté du 31 octobre 2017 fixant la forme de présentation complémentaire à la déclaration nutritionnelle recommandée par l'Etat en application des articles L. 3232-8 et R. 3232-7 du code de la santé publique. https://www.legifrance.gouv.fr/eli/arrete/2017/10/31/SSAP1730474A/jo/texte

De Jonge, J., Frewer, L.J., van Trijp, J.C.M., Renes, R.J., De Wit, W. and Timmers, J.C.M., 2004. Monitoring consumer confidence in food safety: an exploratory study. British Food Journal 106: 837-849. https://doi.org/10.1108/00070700410561423

Dommels, Y., 5 June 2020. Etiketteren in België en Nederland blijft wereld van verschil. VMT Vakblad voor de voedingsmiddelenindustrie. https://www.vmt.nl/wetgeving-toezicht/artikel/2020/06/etiketteren-in-belgie-en-nederland-blijft-wereld-van-verschil-10141437?_login=1&_ga=2.104351491.2024782401.1619632960-1040793667.1619632960.

European Commission (EC), 2002. Regulation (EC) No 178/2002 of the European Parliament and of the Council of 28 January 2002 laying down the general principles and requirements of food law, establishing the European Food Safety Authority and laying down procedures in matters of food safety. Official Journal of the European Union L 31, 1.2.2002: 1-24.

European Commission (EC), 2006. Regulation (EC) No 1924/2006 of the European Parliament and of the Council of 20 December 2006 on nutrition and health claims made on foods. Official Journal of the European Union L 404, 30.12.2006: 9-25.

European Commission (EC), 2008. Regulation (EC) No 1334/2008 of the European Parliament and of the Council of 16 December 2008 on flavourings and certain food ingredients with flavouring properties for use in and on foods and amending Council Regulation (EEC) No 1601/91, Regulations (EC) No 2232/96 and (EC) No 110/2008 and Directive 2000/13/EC. Official Journal of the European Union L 354, 31.12.2008: 34-50.

European Commission (EC), 2011. Regulation (EU) No 1169/2011 of the European Parliament and of the Council of 25 October 2011 on the provision of food information to consumers, amending Regulations (EC) No 1924/2006 and (EC) No 1925/2006 of the European Parliament and of the Council, and repealing Commission Directive 87/250/EEC, Council Directive 90/496/EEC, Commission Directive 1999/10/EC, Directive 2000/13/EC of the European Parliament and of the Council, Commission Directives 2002/67/EC and 2008/5/EC and Commission Regulation (EC) No 608/2004. Official Journal of the European Union L 304, 22.11.2011: 18-63.

European Commission (EC), 2012. Commission Regulation (EU) No 432/2012 of 16 May 2012 establishing a list of permitted health claims made on foods, other than those referring to the reduction of disease risk and to children's development and health. Official Journal of the European Union L 136, 25.5.2012: 1-40.

Floto-Stammen, S., 2022. Food marketing. In: Wernaart, B.F.W. and Van der Meulen, B.M.J. (eds) Applied Food Science. Wageningen Academic Publishers, Wageningen, the Netherlands, pp. 453-479.

Foodwatch, 19 June 2019. Succes tegen misleiding! AH Jumbo e.a. passen hun misleidende vanillevla aan. https://www.foodwatch.org/nl/current-nieuws/2019/succes-tegen-misleiding-ah-jumbo-ea-passen-hun-misleidende-vanillevla-aan/.

Hartmann, M., 2011. Corporate social responsibility in the food sector. European Review of Agricultural Economics 38: 297-324. https://doi.org/10.1093/erae/jbr031

Het Belang van Limburg, 14 April 2015. Wat was dat ook weer, die dioxinecrisis? https://www.hbvl.be/cnt/dmf20150414_01629025.

Holle, M., 2020. Food information. In: Van der Meulen, B.M.J. and Wernaart, B. (eds) EU food law handbook. Wageningen Academic Publishers, Wageningen, the Netherlands, pp. 327-378. https://doi.org/10.3920/978-90-8686-903-9

Irving, D., 23 October 2020. Veggieburger mag zo blijven heten, kaasalternatief mag niet meer. VMT Vakblad voor de voedingsmiddelenindustrie. https://www.vmt.nl/wetgeving-toezicht/nieuws/2020/10/veggieburge-mag-zo-blijven-heten-10144143?_login=1.

Ligtermoet, 20 April 2021. Einde aan misleidend gebruik van de term 'natuurlijk' op levensmiddelen? VMT Vakblad voor de voedingsmiddelenindustrie. https://www.vmt.nl/wetgeving-toezicht/artikel/2021/02/einde-aan-misleidend-gebruik-van-de-term-natuurlijk-op-levensmiddelen-10149231.

Nguyen P-M.D., Vo, N., Phuc Nguyen, N. and Choo, Y., 2020. Corporate social responsibilities of food processing companies in Vietnam from consumer perspective. Sustainability 12: 71. https://doi.org/10.3390/su12010071

Pennings, J.M.E., Wansink, B. and Meulenberg, M.T.G., 2002. A note on modeling consumer reactions to a crisis: the case of the mad cow disease. International Journal of Research in Marketing 19: 91-100. https://doi.org/10.1016/S0167-8116(02)00050-2

Rog, J., 30 March 2020. Land van oorsprong en plaats van herkomst: wat betekenen de nieuwe regels? VMT Vakblad voor de voedingsmiddelenindustrie. https://www.vmt.nl/wetgeving-toezicht/nieuws/2020/03/land-van-oorsprong-en-plaats-van-herkomst-wat-betekenen-de-nieuwe-regels-10140844?_ga=2.261739085.171728361.1619683896-314281158.1619683896&_login=1.

Sorgdrager, W., 2018. Commissie onderzoek fipronil in eieren juni 2018. Available at: https://www.anevei.nl/dynamic/media/4/documents/Dossiers/Fipronil/Rapport%20Commissie%20onderzoek%20fipronil%20in%20eieren.pdf

Unilever, July 2021. Unilever's position on: alternative approaches to animal testing. Available at: https://assets.unilever.com/files/92ui5egz/production/5f08c41a40e03128d79e5a6161da28b5adb2c507.pdf/alternative-approaches-to-animal-testing.pdf.

Unilever, n.d. Code of business principles and code policies. Available at: https://www.unilever.com/Images/code-of-business-principles-and-code-policies_tcm244-409220_en.pdf.

Van der Meulen, B.M.J. and Wernaart, B.F.W., 2022. Food law and regulatory affairs. In: Wernaart, B.F.W. and Van der Meulen, B.M.J. (eds) Applied Food Science. Wageningen Academic Publishers, Wageningen, the Netherlands, pp. 21-43.

Van der Meulen, B.M.J., 2009. The system of food law in the European Union. Deakin Law Review 14: 305-339.

Van der Meulen, B.M.J., 2011. Private food law. Governing food chains through contract law, self-regulation, private standards, audits and certification schemes. Wageningen Academic Publishers, Wageningen, the Netherlands. https://doi.org/10.3920/978-90-8686-730-1

Van der Wal, I.E., 2019. Op de voorkant de leugen, op de achterkant de waarheid? Over de Nederlandse uitwerking van het *Teekanne*-arrest. Intellectuele Eigendom en Reclamerecht (IER). 63 pp. Available at: https://www.njb.nl/umbraco/uploads/2018/8/scriptie-privaatrecht-i.e.-van-der-wal.pdf.

Van Dinther, M., 23 October 2020. 'Vleesnamen' voor plantaardige alternatieven niet verboden in EU – 'zuivelvariaties' wel. De Volkskrant. https://www.volkskrant.nl/nieuws-achtergrond/vleesnamen-voor-plantaardige-alternatieven-niet-verboden-in-eu-zuivelvariaties-wel~b22617a9/.

Van Dinther, M., 26 May 2021. Europees Parlement: plantaardige variatie op yoghurt mag blijven. De Volkskrant. https://www.volkskrant.nl/nieuws-achtergrond/europees-parlement-plantaardige-variatie-op-yoghurt-mag-blijven~baa4dd81/.

Vogel, D., 2001. The new politics of risk regulation in Europe. Centre for Analysis of Risk and Regulation, London School of Economics and Political Science, London, UK.

Wageningen University & Research (WUR), n.d. BSE, aka mad cow disease. Available at: https://www.wur.nl/en/Research-Results/Research-Institutes/Bioveterinary-Research/Animal-diseases/Prion-diseases/BSE-aka-mad-cow-disease.htm.

Wernaart, B., 2021. Ethics and business: a global introduction (1st ed.). Routledge, Oxford, UK. https://doi.org/10.4324/9781003193951

Wernaart, B.F.W., 2022. Food ethics. In: Wernaart, B.F.W. and Van der Meulen, B.M.J. (eds) Applied Food Science. Wageningen Academic Publishers, Wageningen, the Netherlands, pp. 45-64.

World Health Organization (WHO), 2021. The nutri-score: a science-based front-of-pack nutrition label helping consumers make healthier food choice. IARC Evidence Summary Brief No. 2. Available at: https://www.iarc.who.int/wp-content/uploads/2021/09/IARC_Evidence_Summary_Brief_2.pdf

5. Food policy

This is what you need to know about the actions of actors in the food sector

Sezin İba Gürsoy

International Relations, Kırklareli University, Kayalı Kampüsü, Kırklareli, Turkey; seziniba@klu.edu.tr

> Who controls the food supply controls the people; who controls the energy can control whole continents; who controls money can control the world.
> – Henry Kissinger

Abstract

Food is a direct policy tool as well as an underlying condition of policy in international diplomacy. Food is a political issue as well, with questions about how food systems are established, how they have changed, and who has won or lost in power relationships between various actors. Around the world, various forms of intensive food policy are emerging. The term 'Food Policy' has several different definitions, and even within these definitions concepts like 'Food Security', 'Food System' and 'Food Sovereignty' are commonly used, especially in relation to the Food Policy. Food policy is a relatively new and developing area of international relations. This chapter introduces some fundamental food policy concepts and examines the evolution of food policy taking into account EU food policy. It starts with the Mansholt Plan as a case study.

Key concepts

- ▶ Food regime is a prominent analysis of the role of food and agriculture in global capitalism.
- ▶ Food security is a measure of the availability of food and the ability of individuals to access it.
- ▶ Food system refers to everything and everyone who influences and is influenced by the agricultural activities and input use, processing, transformation, distribution and consumption of food.

Bart Wernaart and Bernd van der Meulen (eds)
Applied food science
DOI: 10.3920/978-90-8686-933-6_5, © Sezin İba Gürsoy 2022

▶ Food sovereignty refers to people's rights to define their own food and agriculture policies, as well as to protect and regulate domestic agricultural production and trade.

5.1 Introduction

Food, which has always been used as a weapon and a tool for wielding power throughout history, is fundamental for humanity's survival. Food is regarded as a particularly political commodity because of its importance to society and states. Since the middle of the twentieth century debates on food policy have been the subject of studies regarding security, trade, and development, not just in relation to the expertise in agricultural and nutritional sciences. In this sense, the state is regarded as the most important actor in ensuring food policy and food production, and food is regarded as the most basic power element required for the state's survival. Food policy is defined as the influence of government policy on food production and distribution. As a result, food policy is an important component for a state, both economically and in terms of maximising state power.

threats

The threats to agricultural productivity and stability – such as climate change, rising energy prices, and dwindling agrobiological resources, changing patterns of production and consumption, the 2007 food price crisis, and the COVID-19 pandemic – have attracted the attention of academics, policymakers, activists, and businesses to the precarious state of global food production systems. These crises have revealed that food policies, developed in higher governance levels, have significant shortcomings in addressing the underlying issue of food security. Citizens and governments believe that the current global food system is weakening as a result of these problems.

food policy definition

Food policy encompasses a wide range of issues, including food safety and food quality, production, consumption, distribution, transportation, culture and traditions, among others (Chatzopolo, 2018). As a result, food policies emerge within multi-level governance, multi-sector, and multi-actor chains. Based on a broad definition of food policy is 'a wide range of actions and decisions concerning the production and processing of food, its impact on public health and well-being, the environment, and natural resources' (Lang *et al.*, 2009: 215). According to Chambolle (1988: 456) food policy is 'a balanced government strategy regarding the food economy that takes into account its interrelationships with both the national and international economies'. Timmer (1983: 787) notes that 'Food policy involves both consumers and producers. If prices are too high for consumers to afford nutritional food products, then it reduces the amount they can purchase. High food prices can cause lower income households to have a poorer quality diet. Producers rely on food prices for income and therefore cannot make the prices so low that they are not able to survive. There is a fine line between supply and demand which creates a challenge for food policy.' Also, Coff and Kemp (2014: 4) state that food policy is an ambiguous concept. Climate change, security policy, development and aid policy, agricultural policy, and health policy are all recognised as being part of or included in food policy. Lang *et al.* (2009:21) also emphasise the welcoming nature of food policy. They define food policy as 'ranging from how food is produced and grown, to how it is processed, distributed, and consumed; from the structures that shape food supply, to those that determine health and the environment; from the sciences and processes that unlock food's potential, to the formal governance and lobbies that seek to control it; from the impact of the food system's dynamics on society, to the way its demands are factored into policy-making'. Food policy, according to Neil D. Hamilton (2002), is 'any decision by a government that shapes the type and costs of food used or available or affects opportunities for farmers and workers in food choices available to consumers.'

Food literature is typically divided into sections based on key concepts such as food security, food safety, and food system. 'Food security' refers to the four dimensions such as availability, access, utilisation, and stability. This chapter focuses on these concepts in relation to food policy. The evolution of food security in food policy, as well as the separation of concepts over time, will be discussed. EU food policy will be reviewed.

5.2 Understanding the importance of food policy

Food policy is a type of public policy that deals with, shapes, or regulates the food system in order to ensure food and nutrition security. Food policy encompasses all government actions affecting food security, supply, price, production, distribution, food quality, and safety (Harper *et al.*, 2009: 9). With the concepts of food security, food safety, and the food sovereignty movement, the food system has in recent years begun to devise broad, comprehensive approaches to food policy that encompass all aspects of the food system. Food security efforts are linked to food safety, environmental sustainability, food quality, global health, the environment, and human safety. Therefore, according to Harper *et al.* (2009), food policy is multidisciplinary and multisectoral, addressing particularly social, political, economic, and environmental factors. In this sense, food policy gives information about what is produced and consumed and how fair it is, rather than being production-oriented (Lang *et al.*, 2009). Food policy is primarily concerned with the production and distribution of food.

According to Gupta and Sethi (1967), the national food policy ought to be, in order of priority: (1) maximisation of the quantities of foodgrains brought by the producers for sale in the market; (2) acquisition by the state of such proportions of the total available supply as would be adequate to give the state a commanding position in the market in order to influence the price level of foodgrains; (3) equitable distribution of the foodgrains that the state procures; and (4) maintenance of a price structure for foodgrains that is free from violent fluctuations and yet is flexible over time.

stakeholders

In addition to this broad range of issues, a comprehensive food policy will also address the interests of actors and consumers in the food value chain. According to Harper *et al.* (2009: 9-10) for a comprehensive food policy, all stakeholders must be involved in the policy-making process, from formulation and decision-making to implementation and evaluation. These stakeholders make up the food chain, which is influenced by legislation, political rules such as taxes and subsidies, and other policies. The state has certain tools with which to create a comprehensive food policy. Food policy instruments proposed by Lugo-Morin **food policy instruments** (2022: 11) are: (1) food control tools; (2) food promotion tools; (3) food taxes; and (4) feedback tools. In food control tools, ancestral knowledge from indigenous food systems must be preserved. Control instruments must be based on theoretical certainty. Incentives and cooperation should be developed in terms of food promotion tools. The third instrument is a tax, which must be defined for the food system. Food policy should be constantly monitored in the context of feedback tools. This will also make it possible to observe the effectiveness of the implemented policies. Zerbian and De Luis Romero (2021) stated that the main features of governance for food security are the inter-organisational dimension of policy-making and the early participation of social actors, stakeholders and civic groups in decision-making processes. Power dynamics are inherent in the organisations that create and enforce the rules, such as governments, international organisations, and corporations. In recent studies, cities have surfaced as new food policy players worldwide. They are becoming key players in addressing complex socioeconomic and sustainability challenges related to food security (FAO, 2019). According to Zerbian and De Luis Romero, (2021) in comparison with higher governance levels, cities have adopted a more collaborative approach using tools that promote joined-up food policies, the democratisation of policy-making and knowledge exchange.

5.3 Evolution of food policy

The concept of food policy has always existed, even if it is not known by that name. Although food policy dates back to ancient times, it only became one of the most important goals of state policies, both as a national security and a social need, during the famine years caused by World War II. After World War

II, one of the most important goals of efforts to optimise food and agricultural production was to ensure food security (Barrett and Maxwell, 2005; Shaw, 2007). Maxwell and Slater (2003), making a distinction between 'old' food policy (1970s) and 'new' food policy (2000s), suggest that food policy has changed in several ways. Lang and Barling (2012: 27-42) link changes in food policy over time to developments in the modern policy agenda and summarise the issue in four main phases (Lang *et al.*, 2009: 27-42): The first phase of food policy is the 1940-1950s, which focused on agricultural production. The second stage is the 1970s, prioritising food markets or development policy. The third period is the gradual emergence of environmental crisis and market failure (between 1980 and 2000). The last period is classified as the 21st century when food policy struggles with the public health problem.

5.3.1 The first phase (1940-1950s)

Food policy has primarily been designed to increase food production to meet the demands of a growing population. Food was a major political and social issue during World War II. Even in the post-war period, major famines occurred in Asia (e.g. Bangladesh, Cambodia, China, India) and Africa (e.g. Ethiopia and Nigeria), thus engendering the international dimension of food security. The institutionalisation of this internationalisation came about with the United Nations Food and Agriculture Organization (FAO, nd), the first specialised agency of the United Nations, which was established on 16 October 1945 (since then celebrated annually as World Food Day), immediately following the end of the war. Its goal is to achieve food security for all peoples and ensure that they have regular access to sufficient high-quality food in order to live active, healthy lives. From the 1950s, governments changed their political strategies and chose to establish a new food policy regime based on agricultural production, family farming, and smallholding (Lizzi and Righettini, 2018). Instead of national self-sufficiency targets, new agricultural development plans for promoting agricultural commodity production for global markets began to be adopted during this process. During the 1940s and 1950s, a food-policy framework based on protectionism was developed. In that era, the solution to hunger was usually to increase food production. This framework was based on agricultural production and agricultural reform in order to increase agricultural output, reduce waste, and adequately feed people. Farmers were heavily subsidised and paid based on quantity produced as an incentive to increase output (Hoop, 2015: 5). 1954 is an important date, being the year that the Food for Peace Program was initiated by Public Law 480 (PL480). Through the Public Law 480 Program, the US government provided funding to other countries such as Vietnam, Cambodia, Chile, Korea, and Syria from the 1950s to the 1970s. PL480 established a broad basis for US distribution of foreign food aid. During the same period, the US contributed significantly to

Europe's post-war reconstruction through the Marshall Plan, ensuring stability and capitalism in the Western world. Contradictions within the post-war international food order began to emerge in the late 1960s. The post-war production-oriented understanding gained a different dimension with the Green Revolution beginning in the 1950s. From the late 1960s to the 1980s, the Green Revolution, characterised by the development of high-yielding wheat and rice varieties, nearly tripled yields, particularly in Asia (Fan, 2020). The Green Revolution in Asia led to significant increases on returns as regards land and farmer incomes.

5.3.2 The second phase (1970s)

After 1972, food policy reforms became more national than international in nature. The majority of the rich food-importing countries changed their national agricultural policies. Many of these countries planned to reduce their reliance on food imports in the future, particularly American grain. National food regulations were enacted by the European Union, the Soviet Union, and Japan. Food prices skyrocketed, hunger was on the rise, and the world was in the grip of a global food crisis. The mid-1970s global food crisis sparked debate in the international food regime. The main focus of attention was on food supply. The main goal of this supply-side concept was to ensure adequate food production while also maximising stability. The 1973-1974 food crisis exposed serious flaws in the food regime. According to Emma Rothschild (1976: 285), the political consequences of the 1973 food crisis extend far beyond food policy as a power factor. The 1970s were years when the world's agricultural product supply was insufficient, demand was high, and agricultural product prices rose as a result. When the oil shock was added to these developments, agricultural policies had to be re-evaluated and new measures implemented.

In light of the international food crisis, several countries, both developed and developing, asked the United Nations to convene an international conference to assess the situation and agree on potential solutions. The United Nations World Food Conference was convened in November 1974 in response to this demand. This conference was a catalysing moment in food problems among the states. The conference not only promoted food security as a new concept, but it also established institutional arrangements, such as new information, resources for improving food security, and forums for political dialogue. The 1974 Conference called for international cooperation to increase reserves and promote agricultural development in developing countries. After the conference, in 1975 the United States more than doubled its food aid to the world's poorest countries. According to Rothschild (1976: 295) it was because these were the countries 'most severely affected' by the economic crisis.

In addition to all these developments, food policy also developed as a scientific field in this period. One of the most obvious indicators of this is the 1975 publication of the academic journal Food Policy. The establishment of the International Food Policy Research Institute (IFPRI) in the same year, with the goal of eradicating hunger and malnutrition, paved the way for the development of food policy. FAO, the World Food Program (WFP), and the International Fund for Agricultural Development (IFAD) were institutionally consolidated in order for food policy to become international.

5.3.3 Third phase (1980-2000s)

With the end of the Cold War, there was an increase in studies on global food problems, and issues such as the environment, water, and climate began to be recognised as a major threat to international security. Thus, food security became a topic of discussion in the 1970s and 1980s, primarily in relation to the Third World, hunger, and poverty reduction (Timmer, 2001). There have been numerous approaches to define food security as a concept in the literature. When it was accepted that the problems of poverty and hunger were the result of a lack of effective demand in the 1980s, the food problem became the focus of serious research. Amartya Sen, for example, developed a different understanding with the entitlement approach in 1981, apart from the basic assumption of Thomas Malthus and his supporters, on the availability of food (Sen, 1981). Even the World Bank's 1986 report on 'Poverty and Hunger' highlighted the dynamic (variable) nature of food security over time. Concerns about food safety, food security, nutritional balance, food system and nutritional requirements for an active and healthy life have been included in food policy since the 1980s. At this point, social and cultural eating habits also started to gain importance.

Several international conferences on environment, food, human rights, social development, women, were held between 1974 and 1996, all of which were directly or indirectly related to food policy. The World Food Summit in Rome in 1996 was the first official sign of a shift in food policy that was universally accepted by the international community. This summit produced a widely accepted definition of food security, emphasising the concept's multidimensionality. Accordingly, FAO, food security exists 'when all people, at all times, have physical and economic access to sufficient safe and nutritious food that meets their dietary needs and food preferences for an active and healthy life' (FAO, 1996).

Otherwise, food sovereignty, was discussed at the same FAO summit as an alternative approach to neoliberal policies, focusing on community control over how food is produced, traded, and consumed. Since the beginning of the

2000s, producer groups that did not see neo-liberal economic policies as an opportunity for their own development organised and inflamed the discussions on food sovereignty or food democracy. In this context, the concept of food sovereignty emerged from an international movement called La Via Campesina as an open objection to neo-liberal rural development policies and constituted one of the main discussion topics of the 1996 World Food Summit.

From 1980 to 2000, environmental issues, food quality and food safety were increasingly incorporated into food policy, with the emphasis shifting to ecological public health in 2000 and later. In the late 1980s and early 1990s, environmental perspectives gained traction, and environmental sustainability discourse gradually began to permeate food policy. Food safety was the defining feature of food policy from the 1980s to the 2000s. However, an outbreak of BSE (bovine spongiform encephalopathy), also known as Mad Cow Disease, put food safety to the top of the agenda in food policy (Lang *et al.*, 2009; Oosterveer, 2005; Pennington, 2003). Ultimately, the tools and strategies used to ensure food security need to be compatible with sustainability as well as food safety and public health.

5.3.4 Last phase (the 21st century)

Current food policy has been shaped and moved to a different stage since the 2000s. The year 2000 was a watershed moment in the fight against global poverty. In the early twenty-first century, in September 2000, the United Nations Millennium Summit convened in New York for a special session. There, 8 goals known as Millennium Development Goals were approved. The first was to 'Eradicate Extreme Poverty and Hunger'. In line with this goal, many research organisations such as the World Bank's research group, the International Consortium of Agricultural Research Centers (CGIAR) and universities have begun to focus on defining and measuring poverty as well as assessing the impact of poverty reduction (Ravallion, 2012).

Developed countries' food policies have shifted from self-sufficiency to a different direction during this time period. This shift aims to establish standards for the safe and high-quality production and trade of food in a way that is not harmful to human health or the environment. However, the rise in international food prices in 2007-2008 significantly increased the number of hungry people around the world. Regardless of that, the food crisis occurred as a direct result of self-fulfilling expectations. The EU needs to develop innovative policy frameworks about food policy, especially since the food crises. Marsden (2015) **a new deal for food** states that the EU should consider 'a new deal for food'.

In 2015, more than 190 countries endorsed the UN Sustainable Development Goals (SDGs) at the United Nations General Assembly. To meet the SDGs, policy researchers studied various methods and improved new policy options to redesign food systems (Fan, 2020). With the adoption of Agenda 2030, an inclusive path has been charted to eradicate hunger and poverty, make agriculture sustainable, and eliminate inequality in food production and distribution.

Case 5.1. The Mansholt Plan.

The famous Mansholt Plan, prepared by Dr Sicco Mansholt, aimed to prevent a recurrence of the famine experienced by the Europeans at the end of World War II. Dr Mansholt, the first European Commissioner for Agriculture (1958-1972), was largely responsible for the Commission's 'Memorandum on the Reform of Agriculture', known as the Mansholt Plan and submitted in December 1968. The plan aims to encourage farming within the Community by gradually reducing the burden of subsidies on the economy, and by providing adequate support to farmers. During the 1970s, Mansholt became a defender of measures to protect the environment as a key element of agricultural policy. The plan sought to 'modernise' the EEC's early Common Agricultural Policy (CAP) by increasing productivity and self-sufficiency among European farmers, but farmers worried that it was threatening their livelihoods. The 'Mansholt plan' did not fare well. This initiative is considered to be one of the most controversial and unsuccessful attempts at European common policy-making to date (Garzon, 2006; Ingersent and Rayner, 1999; Marsden, 2003). Fifty years ago, nearly 100,000 European farmers and also farmer organisations protested against the Mansholt plan. Ever since, farmers have been prominent in their criticism of Europe's agricultural policy.

5.4 An overview of EU food policy

Food policy encompasses a wide range of issues, including food safety and food quality, production, consumption, distribution, transportation, culture and traditions, among others (Chatzopoulou, 2018). It also involves a variety of institutions and actors and adapts to specific decision-making processes and directives within the framework of the EU's multilevel governance. The concept of food security in the EU gained meaning in the process of establishing the Common Agricultural Policy after the war (see also Case 5.1). Food supply was one of the most pressing issues that arose in Europe during and immediately after World War II. The situation in agriculture after the War was so bad that there was not enough food. Food supply security and adequate food stability are among the sine qua non for the rebuilding of Europe. Since the founding of the

European Economic Community, the agricultural sector has played a central role in economic and trade policy. The first attempts to develop common agricultural behaviour patterns in Europe began in the 1950s. The concern about food insufficiency during the war, the fact that the agricultural sector employed most of the active population in the European continent, and the necessity of eliminating the serious differences between the national agricultural policies of the member countries, brought up the issue of establishing a common agricultural policy. Under this trajectory food policy is often cited as a kind of public good to secure EEC's food supply and contribute to global food production. CAP consolidated the productivist policy paradigm in the 1960s of the EEC. CAP, established in 1962 as the Community's first common policy with common goals, aimed to rebuild the European agriculture sector. The primary goal therefore was to support farmers' incomes, increase production, ensure food supply, and safeguard the European market and production. The Community, which expanded for the first time in 1973 with the addition of the United Kingdom, Denmark, and Ireland, implemented a series of policies aimed at balancing increased production with lower CAP budget expenditures. With the first expansion, the rate of product self-sufficiency increased. The financial need for a solution to the problems created by overproduction put a heavy burden on the budget. As a result, the end of the 1970s were spent looking for ways to change agricultural policies and developing solutions to problems caused by these policies. By the end of the 1980s, the conditions that existed when CAP was established had largely vanished, and it began to deviate from its basic goals. Furthermore, the decline in agricultural population and employment reduced both the agricultural sector's political importance and the power of the agriculture lobby within the Community. Since the Community's reforms to address the agricultural problem in the 1980s were insufficient, the most radical change was implemented in 1992, thus ending what some authors called 'thirty years of immobility' (Garzon, 2006).

Several reforms have been implemented to improve the CAP's environmental and social performance. As a result of these reforms, the EU abandoned policies that subsidised agricultural surplus exports to developing countries. Food security was explicitly identified as a key challenge in the Commission's 2010 CAP Communication, to which the reform must respond in order to address food security concerns at the EU and international levels. In any case, the CAP reforms, especially the decision to decouple subsidies from production, have significantly reduced the distorting impact on production and global markets compared to previous EU agricultural policies (Bureau and Swinnen, 2017). The CAP post-2020, as proposed by the European Commission (EC, 2021) to Member States on how they apply the CAP, addresses environmental and sustainability by offering a new Green Architecture and a delivery model that provides greater flexibility. While EU policies continue to have an impact on global prices and food security in developing countries, the current CAP has a much smaller impact on global markets than in the past. The current reform cycle may result in additional CAP adjustments for the post-2020 period.

food safety   In addition to CAP, EU food Policy includes the concept of food safety. In the 21st century, EU consumers are particularly concerned about food safety and quality. By the 1990s food safety was mainstream, aided by a series of crises and food scandals in Europe. The bovine spongiform encephalopathy (BSE) crisis was a catalyst for policy change, sparking new rules, norms, and beliefs regarding food safety and health standards. The EU food safety policies aim to protect consumer health by implementing traceability requirements throughout the EU food chain from farm to fork. While the 'Dioxin crisis' that emerged in Belgium in 1999 and the avian influenza disease that followed exposed institutional flaws, it also shook consumers' trust in the relevant institutions. The food scandals of the mid-1990s highlighted the need for institutional reform in the areas of food safety, public health, and consumer protection. EU legislation was revised to restore consumer confidence. The European Food Safety Authority, which plays a role in the protection of public health and food safety, was established with the 2002 regulation.

5.5 Conclusions and outlook

Food is pivotal topic in international policy debates. Food systems are the sum of global, national, and local actors and interactions along food value chains. The food crisis of the late 2000s reminded everyone that the impact of food prices on food security is complex, with often opposing consequences for developing-country food consumers and producers. The 2008 price crisis exposed flaws in the international food system, and the Covid-19 crisis highlighted the food system's fragility, supply chain, and supply-demand imbalance. While food prices were a major issue in 2008, the issue with COVID-19 is the disruption of global circulation (OECD, 2020). COVID-19 and the lockdown have put a tremendous strain on the global economy while also raising the risk of long-term food insecurity. The devastating effects of the COVID-19 pandemic triggered an EU-wide discourse about establishing a Common Food Policy. In this context, there are urgent calls for food systems to be reformed, or indeed more fundamentally transformed (De Schutter, 2017). There is even a public debate about whether CAP should be converted into an innovative common food policy for member states (EESC, 2016; Fresco and Poppe, 2016; IPES Food, 2017; Marsden, 2015). Food therefore has become an epicentre of global discussion and will continue to be of major interest in the post-COVID 19 period (Egwue *et al.*, 2020).

sustainable food policy

Despite these calls and developments, the transition to a truly integrated and substantive food policy is still a long way off (Candel and Biesbroek, 2016). A comprehensive food policy should be linked to food security, food safety, and food system structures. Lang *et al.* (2009: 46-52) proposed six goals for a sustainable food policy that will be critical to 21st century food policy. The first goal of ensuring production adequacy from an ecological standpoint is to change the existing food waste system to promote sustainable development. The second goal of preventing diet-related illness is to find (technical) solutions. The third goal is to include all sciences to meet the need for interdisciplinary and coordinated collection of evidence for developing food policy. The fourth goal addresses the impact of food on the environment; It produces strategies for living within environmental limits and reducing carbon footprints and greenhouse gas emissions. The fifth goal is to help consumers embrace the coexistence of ethics and morality with social justice by addressing inequalities within countries. The final goal is food democracy, which refers to a process of striving for improvements in food for everyone, including the rights and responsibilities that come with it. Briefly, globalisation is a key force that has radically changed the world's food and consumption habits. Although food has always been at the centre of all global collective efforts, it needs to be interpreted in a more comprehensive, broader, deeper, and more sustainable way in the 21st century.

References

Barrett, C.B. and Maxwell, D., 2005. Towards a global food aid compact. Food Policy 31: 2. Available at: https://ssrn.com/abstract=715002

Bureau, J. and Swinnen J., 2017. EU policies and global food security. Foodsecure working paper no. 58.

Candel, J.J.L. and Biesbroek, R., 2018. Policy integration in the EU governance of global food security. Food Security 10: 195-209. https://doi.org/10.1007/s12571-017-0752-5

Chambolle, M., 1988. Food policy and the consumer. Journal of Consumer Policy 11: 435-448. https://doi.org/10.1007/BF00411855

Chatzopoulou, S., 2018. The food policy of European Union. In: *The Oxford encyclopedia of European Union politics*. Oxford University Press, Oxford, UK. https://doi.org/10.1093/acrefore/9780190228637.013.595

Coff, C. and Kemp, P., 2014. Food ethics and policies. In: Kaplan, D.M. (ed.) Encylopedia od Food and Agricultural Ethics. Springer, Berlin, Germany, pp. 880-887.

De Schutter, O., 2017. The political economy of food systems reform. European Review of Agricultural Economics 44 (4): 705-731.

European Economic and Social Committee (EESC) 2017. Civil society's contribution to the development of a comprehensive food policy in the EU (own-initiative opinion). Available at: https://www.eesc.europa.eu/en/our-work/opinions-information-reports/opinions/civil-societys-contribution-development-comprehensive-food-policy-eu-own-initiative-opinion

European Commission (EC), 2021. The new common agricultural policy: 2023-27. Available at: https://ec.europa.eu/info/food-farming-fisheries/key-policies/common-agricultural-policy/new-cap-2023-27_en

Egwue, O.L, Agbugba, I.K. and Mukaila, R., 2020. Assessment of rural households food insecurity during Covid-19 pandemic in south-east Nigeria. International Journal of Research – Granthaalayah 8(12): 182-194.

Food and Agriculture Organization (FAO), n.d. Website. Available at: https://www.fao.org/about/en/.

Food and Agriculture Organization (FAO), 1996. Rome Declaration on World Food Security and World Food Summit Plan of Action. FAO, Rome, Italy. Available at: http://www.fao.org/3/w3613e/w3613e00.htm.

Food and Agriculture Organization (FAO), 2019b. FAO framework for the Urban Food Agenda. Leveraging sub-national and local government action to ensure sustainable food systems and improved nutrition. https://doi.org/10. 4060/ca3151en

Fan, S., 2020. Reflections of food policy evolution over the last three decades. Applied Economic Perspectives and Policy 42 (3): 380-394. https://doi.org/10.1002/aepp.13065

Fresco, L.O. and Poppe, K.J., 2016. Towards a common agricultural and food policy. Wageningen University & Research, Wageningen, the Netherlands.

Garzon, I., 2006. Reforming the Common Agricultural Policy. History of a paradigm change. Palgrave Macmillan, Basingstoke, UK.

Gupta, S.C. and Sethi, J.D., 1967. Food policy: objectives and instruments. Economic and Political Weekly 2(14): 683-693.

Hamilton, N.D., 2002. Putting a face on our food. Drake Journal of Agricultural Law 7(2): 408-454.

Harper, A., Alkon, A., Shattuck, A., Holt-Giménez, E. and Lambrick, F., 2009. Food policy councils: lessons learned. Development Report No. 21., Institute for Food and Development Policy. Available at: https://foodfirst.org/publication/food-policy-councils-lessons-learned/

Hoop, L.D., 2015. Food policy in the Netherlands. Wageningen University, Wageningen, the Netherlands. Available at: https://edepot.wur.nl/333360.

Ingersent, K.A. and Rayner, A.J., 1999. Agricultural Policy in Western Europe and the United States. Edward Elgar Publishing, Cheltenham, UK.

International Panel of Experts on Sustainable Food Systems (IPES Food), 2017. Updated concept note: October 2017 towards a common food policy for the European Union. A 3-year process of research, reflection and citizen engagement. Available at: http://www.ipes-food.org/images/Reports/CFP_ConceptNote.pdf

Lang, T. and Barling, D., 2012. Food security and food sustainability: reformulating the debate. Geographical Journal 178(4): 313-326.

Lang, T., Barling, D. and Caraher, M., 2009. Food policy: integrating health, environment and Society. Oxford University Press, Oxford, UK.

Lizzi, R. and Righettini, M.S., 2018. Food Policy in Italy. Reference Module in Food Science. Elsevier, Amsterdam, the Netherlands. https://doi.org/10.1016/B978-0-08-100596-5.21468-6

Lugo-Morin, D.R., 2022. Innovate or perish: food policy design in an indigenous context in a post-pandemic and climate adaptation era. Journal of Open Innovation 8: 34. https://doi.org/10.3390/joitmc8010034

Marsden, T., 2015. A common Food and Nutrition Policy for Europe, Available at https://transmango.wordpress.com/2015/10/19/commonfoodpolicy/

Maxwell, S. and Slater, R., 2003. Food policy old and new. Development Policy Review 21(5-6): 531-553. Basil Blackwell, Oxford, UK.

Pennington, T., 2003. When food kills: BSE, *E. coli* and disaster science. Oxford University Press, Oxford, UK.

Rothschild, E., 1976. Food politics. Foreign Affairs 54(2): 285-307. https://doi.org/10.2307/20039573

Ravallion, M. 2012. Benchmarking global poverty reduction. Policy Research Working Paper; No. 6205. World Bank, Washington, DC, USA. https://openknowledge.worldbank.org/handle/10986/12095

OECD, 2020. Food supply chains and COVID-19: impacts and policy lessons. Available at: https://read.oecd-ilibrary.org/view/?ref=134_134305-ybqvdf0kg9&title=Food-Supply-Chains-and-COVID-19-Impacts-and-policy-lessons.

Oosterveer, P., 2005. Global food governance. Wageningen University, Wageningen, the Netherlands.

Sen, A., 1981. Poverty and famines: an essay on entitlements and deprivation. Clarendon, Oxford, UK.

Shaw, D.J., 2007. World food security. Palgrave Macmillan, London, UK.

Timmer, C.P., 1983. Food policy analysis the World Bank. The Johns Hopkins University Press, Baltimore, MD, USA.

Timmer, C.P., 2001. Food policy. In: Semba, R.D. and Bloem, M.W. (eds) Nutrition and health in developing countries. Humana Press, Totowa, NJ, USA, pp. 781-792.

Zerbian, T. and de Luis Romero, E., 2021. The role of cities in good governance for food security: lessons from Madrid's urban food strategy. Territory, Politics, Governance. https://doi.org/10.1080/21622671.2021.1873174

6. Risk analysis for foods

This is what you need to know about risk analysis in EU food law

Alie de Boer

Food Claims Centre Venlo, Faculty of Science and Engineering, Maastricht University, Nassaustraat 36, 5911 BV Venlo, the Netherlands; a.deboer@maastrichtuniversity.nl

Abstract

In the EU, using scientific evidence as the basis for policies is highly encouraged, especially when it concerns regulating risks. One of the sectors in which policies and decisions are based on scientific evidence is food, to contribute to the main aim of European food law: achieving the highest level of consumer protection. Using scientific evidence for policy purposes is systematically organised by risk analysis, allowing for separating the scientific assessment of risks (risk assessment) from the political decision-making process (risk management). In this chapter, the historical development and application of risk analysis in European food law is described. We discuss how risk assessment, risk management and risk communication are conducted, and will show which novel methods and developments play important roles in these concepts. Finally, this chapter concludes with an outlook in which we show that whereas risk analysis has an important position in food law already, it is essential to ensure that new methods can be used, and more complex analyses can be conducted to support decision-making; and that for increasing their effective use, the functional separation of risk assessment and risk management warrants further investigation.

Key concepts

- Risk analysis is the structured approach to analysing risks and identifying measures that can be taken to deal with such risks.
- EU food legislation deals not only with risks but also with health effects or efficacy of food products. The use of risk analysis and risk assessment is therefore often called 'scientific analysis' or 'scientific assessment'.

Bart Wernaart and Bernd van der Meulen (eds)
Applied food science
DOI: 10.3920/978-90-8686-933-6_6, © Alie de Boer 2022

► Risk assessment is the first step in the process of risk analysis and deals with the scientific process of analysing whether something could present a risk. It is based on four key steps: hazard identification, hazard characterisation, exposure assessment and risk characterisation.

► New approach methods, i.e. methods that do not use animals for testing safety of compounds, play a key role in further advancing risk assessment.

► Risk management describes the process of combining the outcomes of risk assessment with other legitimate factors to weigh policy alternatives.

► Risk communication deals with the exchange of information between experts (e.g. between risk assessors and risk managers), as well as the exchange of information about risk assessment and/or management to other relevant stakeholders.

► EFSA is the scientific assessor for food and food safety considerations in the EU.

► Following the entry into force of the Transparency Regulation, risk assessment should become more transparent and sustainable.

Case 6.1. Titanium dioxide.

A well-known food additive, titanium dioxide (TiO_2), will probably be banned from foods across the EU in 2022. It has been used as a white pigment for colouring purposes in products such as soups and bakery products for a long time already, but now the European Commission is soon expected to propose a ban of this product in foods. This is attributable to the risk analysis cycle. Like any other food product or food ingredient, additives on the European market should be safe. TiO_2 has long been thought not to cause any safety issues, since it is rather insoluble (Jovanovi, 2015). In the EU, TiO_2 has therefore been authorised for use as a food improvement agent and was coded as E171. This authorisation already stems from before 20 January 2009 and, like other products authorised before that date, a re-evaluation of all available scientific evidence was required (under Regulation (EU) No 257/2010)[30]. In 2016, a first re-evaluation of all scientific evidence available on TiO_2 by the European Food Safety Authority, EFSA, did not raise concerns regarding exposure to the compound (Younes *et al.*, 2021). The risk assessment agency did, however, indicate that some uncertainty existed about what the potential health effects could be of a small fraction of TiO_2 that consists of nanoparticles. As risk manager, the European Commission did not have a reason

---

[30] Commission Regulation (EU) No 257/2010 of 25 March 2010 setting up a programme for the re-evaluation of approved food additives in accordance with Regulation (EC) No 1333/2008 of the European Parliament and of the Council on food additives. Official Journal of the European Union L 80: 19-27. *Consolidated version 27 March 2021.*

at that moment to conclude that there was a need to ban the product from the market. A next re-evaluation was planned for 2021. This risk assessment showed that new insights had been generated over the last few years. Not only did the data suggest that the substance may accumulate in the body even though the substance is rather insoluble, but findings on the effects of the nanoparticle fraction of TiO_2 were especially problematic. Studies into immunotoxicity, inflammation and neurotoxicity effects suggest that exposure to TiO_2 may negatively affect markers that are associated with these outcomes. Also, genotoxicity cannot be ruled out (Younes *et al.*, 2021). This means that the risk assessment findings call for action by the risk manager: the European Commission. On the same day that these results were published, Commissioner Stella Kyriakides tweeted about the next steps for proposing a ban on the use of E171 in foods in the EU.

6.1 Introduction

scientific evidence

The importance of scientific evidence as the basis of policy development is highly emphasised in the EU, especially concerning the regulation of risks (EC, 2015; Stibernitz, 2012). One of the sectors in which science is key in policy development is the field of food and food safety. Following various food scares in the 1990s (including the BSE crisis and the dioxin crisis), EU food law was completely reformed (Vos, 2000). The requirement to use independent scientific evidence became more outspoken in the newly developed framework regulation dealing with food and food safety within the EU: 'Regulation (EC) No 178/2002 laying down the general principles and requirements of food law, establishing the European Food Safety Authority and laying down procedures in matters of food safety' (abbreviated as General Food Law, GFL)[31]. This requirement should contribute to one of the two main aims of EU food law: achieving the highest level of consumer protection (Szajkowska, 2009; Vos, 2000).

risk analysis

The process of analysing scientific evidence related to food and food safety authorisations is often referred to as risk assessment (De Boer, 2019a). Risk assessment is the first step in a process known as 'risk analysis': the systematic approach upon which food policy and related decisions should be based, as defined in Article 6 of the GFL. Whereas scientific questions are addressed in risk assessment, the second component of risk analysis is focused on weighing policy alternatives to appropriately deal with the risk that needs to be dealt with in policy. The third constituent of risk analysis is risk communication:

---

[31] Regulation (EC) No 178/2002 of the European Parliament and of the Council of 28 January 2002 laying down the general principles and requirements of food law, establishing the European Food Safety Authority and laying down procedures in matters of food safety. Official Journal of the European Union L 31: 1-24. *Consolidated version 1 July 2018.*

the exchange of information and opinions between risk assessors, risk managers and other stakeholders including for example consumers, non-governmental organisations, and industry. This information exchange can take place throughout the process.

sectoral legislation

Even though the term 'risk' is an important term in describing these processes, EU food legislation does not only deal with risks. Sectoral legislation deals with more specific processes or products, such as the authorisation of health claims (Regulation (EC) No 1924/2006 on nutrition and health claims[32]), or the authorisation of new foods and food ingredients (Regulation (EU) 2015/2283 on novel foods[33]). In both procedures, scientific evidence plays a key role; whilst the authorisation procedure for novel foods requires scientific evidence for proving that a product is not unsafe. Also, for health claims scientific findings are required that establish the health effect or the efficacy of the product bearing the claim (De Boer, 2021; De Boer and Bast, 2018; Lenssen *et al.*, 2018). Official legislative documents describe these processes as risk analysis and its specific components, where the European Food Safety Authority EFSA, which conducts risk assessments in the EU, already describes its efforts more broadly as scientific assessments (EFSA, 2017b).

legislation

Whilst scientific evidence plays an important role in policy and legislation, there is a debate about the extent to which extent science and legislation are aligned. Nutritional sciences are developing at high speed and new insights regarding new, innovative products and technologies, as well as health effects of food and food ingredients, are rapidly evolving. And while nutritional sciences are thriving with ongoing critical appraisal of their concepts and ideas, food laws and subsequent technical standards and guidelines require definite statements concerning food safety and health. This results in the issue that legislation is often based on dated or even outdated scientific concepts and cannot make use of the most recent advancements (Bast and Hanekamp, 2017; De Boer *et al.*, 2020; Silano, 2009; Von Stackelberg and Williams, 2021; Zwietering, 2015). In subsequently developed sectoral legislation, scientific evidence is especially key: when health effects or safety of the products requesting market access are insufficiently established within a scientific dossier, the risk assessor will not provide a positive opinion to the risk manager, who subsequently needs to decide whether to allow such a claim or product on the European market (De Boer, 2021; De Boer *et al.*, 2014). The tension between nutritional sciences and

---

[32] Regulation (EC) No 1924/2006 of the European Parliament and of the Council of 20 December 2006 on nutrition and health claims made on foods. Official Journal of the European Union L 404: 9-25. *Consolidated version 13 December 2014.*
[33] Regulation (EU) No 2015/2883 of the European Parliament and of the Council of 25 November 2015 on novel foods. Official Journal of the European Union L 327: 1-22.

law becomes evident in the use of scientific findings under these specific food laws: scientific consensus must be reached between domain experts upon the established effects, and the causal relationships identified between ingredients and these effects. There is, however, a high degree of uncertainty about when new and innovative methods are sufficiently developed to be accepted under these regulatory procedures.

In this chapter, we will discuss the risk analysis concept in more detail and will provide insights into its application in EU regulatory procedures for foods. For that purpose, we will discuss the specific approaches to risk assessment and risk management in EU food legislation. In the final section of this chapter, we will reflect on the functional separation of risk assessment and risk management in the EU and the future implementation the risk analysis cycle for food authorisation decisions.

6.2 Risk analysis and their methods

independent scientific advice

Already in their 1997 Green Paper on European Food Law[34], the European Commission stressed the importance of independent scientific advice as the basis for all regulatory activities of the Community (Vos, 2000). Next to control and risk analysis, scientific evidence was described as the third essential instrument to ensure the development of effective consumer health policies (Vos, 2000). A few years later, the importance of scientific evidence was again highlighted in the GFL: Article 6 emphasises the need for scientific evidence to analyse the risk that a food poses, to ensure a high level of consumer protection. Whereas risk analysis and scientific evidence can be seen as two instruments to support the development of effective policies, the risk analysis approach also requires the usage of scientific evidence and the weighing-up of scientific information. Article 3 of the GFL defines the concepts of risk analysis and its underlying three components of risk assessment, risk management and risk communication. In risk analysis, the scientific assessment of a risk (risk assessment) is separated from the political (risk management) decision, in which other considerations are also considered next to the scientific evidence. Finally, risk communication needs to support the interaction and exchange of information and opinions between the assessors and the managers, as well as other relevant parties that need to be informed or who should be engaged in the process of risk analysis. These definitions closely follow the definitions that have been adopted by the Codex Alimentarius Committee (FAO/WHO, 2019).

---

[34] The general principles of food law in the European Union. Commission Green Paper (COM 97/176 final).

As previously described in the literature, risk analysis and risk assessment have become key principles of food law in general, as well as in the operations of international organisations working on standards including Codex Alimentarius, and the WTO Agreement on the Application of Sanitary and Phytosanitary Measures that are adopted by the World Trade Organisation (Henson and Caswell, 1999; Nauta *et al.*, 2018). Risk analysis has formed the structured approach by which to analyse a risk and to identify the measures that can be taken to deal with this risk (Henson and Caswell, 1999; Meyer, 2006). As highlighted in the recitals of the GFL (specifically recital 17), this systematic method should provide a solid basis for European policies upon which effective, proportionate, and targeted actions can be taken.

precautionary principle

This structured approach is expected to result in a more consistent approach to food safety regulations. Over various domains and in dealing with differing risks, it is essential that the full approach is based on scientific information (Henson and Caswell, 1999). In the rapid development of technologies and innovative products, however, insufficient information may sometimes be available to fully analyse the risk scientifically. In such cases of so-called 'scientific uncertainty', the GFL allows for risk managers to base a provisional management measure upon the precautionary principle (Article 7): only when a first risk assessment has identified the possibility of harmful effects, may a provisional measure be taken, whilst further scientific information should be gathered to conduct a comprehensive risk assessment. As a result, the risk manager is allowed to ensure a high level of consumer protection, even when the scientific evidence is incomplete. As described in Article 6 of the GFL on risk analysis, however, a risk manager should take the precautionary principle into account *next to* the results of risk assessment and not *instead of* such a risk assessment.

6.3 Applying risk analysis in food law

The term risk analysis became more widely known when it was used in dealing with environmental contaminants (Von Stackelberg and Williams, 2021). In the 1960s and 1970s, the increased use of chemical products in the environment, that were for example used as pesticides to support food production activities, gave rise to different controversies related to negative health and environmental effects that were associated with the use of these chemicals (Wu and Rodricks, 2020). In developing strategies to analyse what appropriate measures could be taken to deal with the increased probability of adverse health effects related to exposure to these environmental contaminants, it was acknowledged that it would not be possible to completely eliminate exposure to these contaminants and, subsequently, eliminate any health risk that originated from such exposure (Von Stackelberg and Williams, 2021). To develop appropriate policies, steps were taken to scientifically assess the risks related to exposure to these chemicals and identify what control measures would be appropriate to tackle these risks. In 1983, the National Research Council in the USA published a systematic procedure to analyse such risks, and to weigh up different policy options to address these identified risks (NRC, 1983; Von Stackelberg and Williams, 2021).

Red Book

This now well-known 'Red Book' defines risk assessment and risk management, describes a four-step procedure to risk assessment and shows how risk assessments can be organised and managed to support decision-making processes (NRC, 1983). The Committee emphasised that their approach to risk assessment was not limited to include merely quantitative risk assessment, relying on numerical results, but encompassed a broader approach in which quantitative and qualitative findings would be included. The four steps described in risk assessment therefore include hazard identification, dose-response assessment, exposure assessment and finally, risk characterisation.

hazard identification

Hazard identification focuses on identifying whether a specific agent or compound can increase the incidence of a health condition. In this first step, it is aimed at establishing a cause-and-effect relationship between this compound and the health condition. The second step of risk assessment described by the NRC is establishing a relationship between the dose of such a compound and the incidence of a health condition in a population that is exposed to the compound, known as the dose-response assessment. Together, hazard identification and dose-response assessment form the hazard assessment steps in risk assessment. Subsequently, exposure assessment allows for identifying who is exposed to a certain compound, for what time period, and in what dosage. This third step is defined as 'the process of measuring or estimating the intensity, frequency, and duration of human exposures to an agent currently present in the environment or of estimating hypothetical exposures that might arise from the release of new chemicals into the environment' (NRC, 1983). It should enable the identification of potential vulnerable groups, as well as potential ways to control the exposure. Risk characterisation finally is described as 'the process of estimating the incidence of a health effect under the various conditions of human exposure described in the exposure assessment' (NRC, 1983). In this final step in risk assessment, the estimated exposure is combined with dose-response information, to identify the magnitude of the potential public health problem and to offer risk estimates on mostly population levels. Any uncertainties and variabilities (e.g. related to inter-species differences) need to be considered in risk characterisation as well. Traditionally, the identified risk estimates are compared to health-based thresholds set for single chemicals (Von Stackelberg and Williams, 2021). This risk estimate subsequently serves as input for the risk manager, who decides what actions to take after weighing and selecting potential policy alternatives

dose-response assessment

exposure assessment

risk characterisation

(König, 2010). Decisions are made regarding what level of risk is acceptable, and how this level of risk can be controlled (FAO/WHO, 1997; Henson and Caswell, 1999).

food business operators

As defined in Article 3(3) of the GFL, food business operators are responsible for meeting and controlling the legal requirements of food law. When introducing new foods or new claims on the health effects of foods, the burden of proof also lies with the food business operator: they need to apply for authorisation of these new foods, food ingredients or food improvement agents, or new claims (De Boer, 2021; De Boer *et al.*, 2020). Figure 6.1 displays the general four steps in this process.

6.4 Risk assessment in EU food law

White Paper on Food Safety

European Food Authority

In EU food law, the tasks of risk assessment and risk management are functionally separated: Article 6(2) of the GFL describes that risk assessment must be based on 'available scientific evidence and undertaken in an independent, objective and transparent manner'. Risk assessment was already described in the White Paper on Food Safety[35] to encompass scientific advice and information analysis, that should be based on accurate and up-to-date scientific evidence. This same White Paper suggested that risk assessment and risk management should be functionally separated, for which the to-be-established European Food Authority (now: EFSA) would be essential in providing scientific advice as risk assessor. Following the entry into force of the GFL in 2002, EFSA has been officially tasked with risk assessment for food and food safety related questions in the EU. Within the relatively short period of time that EFSA exists, the Authority has become a well-recognised institution for scientific advice related to food and food safety issues (Ludden *et al.*, 2018). EFSA publishes all its scientific outputs, that include not only scientific opinions, but also various publications to support their work. This includes:

---

[35] European Commission (2000). White paper on Food Safety. COM(2019) 719.

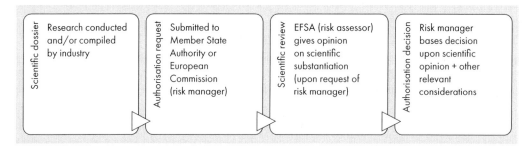

Figure 6.1. General process for authorisation requests for food and nutrition under EU Food law (adopted from De Boer, 2021).

(1) other scientific outputs of EFSA, either statements of the Scientific Committee or Scientific Panel (containing less details then opinions) or Guidance documents of the Scientific Committee or Scientific Panel (to explain the principles behind the approaches of EFSA); (2) other scientific outputs which include the results of a peer review process on scientific evidence related to pesticides; and (3) supporting publications that are either describing technical processes (so-called Technical Reports), External Scientific Reports related to risk assessment and its methodologies, or Event Reports (EFSA, 2010). All scientific outputs of EFSA can be found in the EFSA Journal, in which also various supporting publications are made available. On September 27th 2021, a total of 8533 publications were published in the EFSA Journal, all related to the following 15 topics: animal feed, animal health and welfare, biological hazards, chemical contaminants, corporate, cross-cutting science, data, emerging risks, food ingredients and packaging, GMO, methodology, nutrition, pesticides, plant health and scientific cooperation (available through http://www.efsa.europa.eu/en/publications).

scientific process to risk assessment

Following the four steps as described by the NRC in 1983 (NRC, 1983), the GFL defines the scientific process to risk assessment as follows in Article 3(11) of the GFL: hazard identification, hazard characterisation, exposure assessment and risk characterisation. A hazard (Article 3(14)) is purely the *potential* of an agent, such as a food or food ingredient, to cause an adverse health effect following consumption, whereas a risk is in Article 3(9) defined as the probability for such an adverse health effect to occur, as well as the severity of this effect. Hazard identification concerns the process to identify potential biological or chemical hazards that can be present in foods; the characterisation of hazards (also referred to as dose-response assessment) entails the specifications of the effects that these hazards could result in. By means of exposure assessment, the amount of the hazard present in foods and how much these foods are consumed by specific groups of people will be identified. In the fourth stage of risk assessment (risk characterisation) a conclusion on the risk is drawn, to determine whether there could be a safety concern for the general population or specific groups.

Traditionally, risk assessment has focused on assessing how likely it is that exposure to a certain compound will have an adverse health impact (König, 2010; Von Stackelberg and Williams, 2021). For foods, this has often been translated into identifying levels of exposure in which there is no risk identified, the so-called 'absence of risk' (Tijhuis *et al.*, 2012). This means that it is not the idea to find an absolute safety level in any circumstance, but to enable the identification of safe levels of exposure (Deluyker, 2017). This has greatly influenced the development and use of specific methods in risk assessment.

6.4.1 Hazard identification

Hazard identification includes the process of identifying what the effect of a compound can be (Tijhuis *et al.*, 2012). A hazard can be described as an agent that has the potential to cause an adverse health effect such as illness or injury when it is insufficiently controlled (FAO/WHO, 1997; Sperber, 2001). In general, such agents can be of biological nature (e.g. *Salmonella*), of chemical nature (acrylamide), or of physical nature (such as glass). When referring to the hazard identification process for pre-authorisation purposes specifically, hazards investigated are mainly considered the intrinsic potential of the food or food ingredients themselves to act as an agent that can potentially cause harm (International Programme on Chemical Safety, 2009; Kramer *et al.*, 2019). A detailed characterisation of the food (ingredient) and its potential impurities is supportive in this step, to ensure that all relevant aspects are considered in the risk assessment process (De Boer *et al.*, 2020). In this phase of risk assessment, it is not yet important how relevant these hazards are, as the main goal is to identify *all* potential hazards (Tollefsen *et al.*, 2014).

6.4.2 Hazard characterisation

The second step in risk assessment, the characterisation of the hazard, allows for providing insights into the quantitative nature of the relationship between a hazard and its effect (Dybing *et al.*, 2002). Hazard characterisation is therefore also referred to as dose-response characterisation or dose response assessment (Dybing *et al.*, 2002; Von Stackelberg and Williams, 2021). This step thereby provides insights into the nature of the adverse events and shows the relevance of the identified hazards, because of exposure levels. Traditionally, *in vivo* testing has been one of the key sources of information for hazard identification and hazard characterisation (Tijhuis *et al.*, 2012; Vinken *et al.*, 2020; Vrolijk *et al.*, 2020). Such animal tests support the identification of relevant apical endpoints: empirically verifiable outcomes of exposure to a substance, which include the development of specific anomalies, reproduction impairment, alterations in the histopathology of organs, or even death (Blaauboer *et al.*, 2016). At the same time however, the need to conduct such animal studies gives rise to ethical concerns for needing to use experimental animals for assessing toxicological and food safety risks. And next to ethical considerations (see also Chapter 3; Wernaart, 2022), also financial and scientific concerns exist when using traditional animal testing approaches in assessing food safety (Kramer *et al.*, 2019; Tollefsen *et al.*, 2014; Vrolijk *et al.*, 2020). It is increasingly acknowledged that, although animal models may support the identification of hazards and their characterisation, no model is perfect in predicting adverse effects (Knight *et al.*, 2021).

new approach methods

Increasingly, new approach methods (non-animal-based approaches) can be used to collect data for risk assessment purposes (Blaauboer *et al.*, 2016; De Boer *et al.*, 2020; Knight *et al.*, 2021; von Stackelberg and Williams, 2021). Three novel methods that can be particularly relevant in the first two steps of risk assessment, hazard identification and characterisation, include read-across approaches, Adverse Outcome Pathways (AOPs), as well as the use of high throughput methods. As described by Derek J. Knight *et al.* (2021), the read-across approach is based on the assumption that chemicals which have similar chemical compositions, and can therefore be seen as structurally similar, are also likely to have similar biological properties. The endpoint information for one substance, which is referred to as the target substance, is predicted by using data on the same endpoint from another substance or multiple other substances, called the source substances (De Boer *et al.*, 2020). Next to comparing the chemical structure of the compound, further scientific justifications to use the read-across approach are often required for safety assessments. These can, for example, be based on similarities in bioavailability and metabolism of both compounds (De Boer *et al.*, 2020; Knight *et al.*, 2021; Schultz *et al.*, 2015). The European research initiative EUToxRisk21 should further support the development of clear strategies to deal with uncertainties related to such similarity analyses and the use of read-across approach in safety assessments for chemicals, including foods (Rovida *et al.*, 2020). Another important development in new approach methods for hazard identification and characterisation are AOPs. AOPs allow for identifying the mechanism of action of a substance, by mechanistically describing how a sequence of events is initiated by exposure to a substance (Knight *et al.*, 2021; Vinken *et al.*, 2020). Every AOP starts with the molecular initiating event, where the exposed substance interacts with a biological target, and the subsequent intermediate key events that follow from this exposure ultimately lead to one or more adverse outcomes that are of interest for risk assessment (De Boer *et al.*, 2020; Von Stackelberg and Williams, 2021). AOPs thereby give insights into the consequences of exposure and how these are organised (Sakuratani *et al.*, 2018). Such mechanistic understanding may serve as the basis for establishing points of departure, but they can also serve as a source of information in 'integrated approaches to testing and assessment' strategies, known as IATA strategies (Sakuratani *et al.*, 2018; Tollefsen *et al.*, 2014). IATA is a science-based approach to analyse existing information and highlight requirements for which new information should be generated with other testing strategies (Knight *et al.*, 2021). In IATA, the overall evidence is weighed up for hazard characterisation purposes (De Boer *et al.*, 2020). By such a systematic combination of insights from different information sources, a more comprehensive picture can be built up on the safety of a compound (Hartung *et al.*, 2013). High-throughput screening techniques (HTS) are also increasingly used for hazard identification and characterisation purposes, as replacement for the traditionally used *in vivo*

Adverse Outcome Pathways (AOP)

high-throughput screening techniques (HTS)

animal studies to establish both hazardous substances as well as the relationship between concentration of such substances and their effect (Von Stackelberg and Williams, 2021). A large variety of *in vitro*, cell-based assays has been developed under the US EPA's ToxCast and Tox21 programmes, that for example include reporter assays for the activation of hormone receptors or assays to analyse kinase activity (De Boer *et al.*, 2020; Von Stackelberg and Williams, 2021). In combination with computational, *in silico* approaches to analyse the data, rapid insights have been gained into specific effects that are elicited by specific chemical compounds, including those used in foods (Karmaus *et al.*, 2016). These effects are subsequently suggested to be used as molecular initiating events and key events in AOPs (De Boer *et al.*, 2020), even though their mapping may be challenging (Von Stackelberg and Williams, 2021).

6.4.3 Exposure assessment

In analysing whether the hazard is relevant to the population that consumes a certain product, it is essential to establish their exposure to this compound. This is the third step in risk assessment, that will allow for identifying the amount of the hazardous compound present in the food and how much of this product is consumed by a specific population. In the case of food efficacy and safety, exposure will be assessed based on the intended use of a product, as described by the applicant in the dossier (Deluyker, 2017). Specifically, for health claims, pesticides, GMOs, feed additives, food improvement agents (additives, enzymes and flavourings) and smoke flavourings, information on exposure and intended use is essential (De Boer, 2021; Deluyker, 2017). At the same time, also for emerging food safety risks, such exposure insights are important to, for example, estimate the risk of migrating compounds from food packaging (De Fátima Poças and Hogg, 2007).

data

Even though ideally, data is available on what the levels of consumption are of a certain product, or the level of exposure to a certain compound, mostly this data is incomplete or not yet available in the case of new products (Deluyker, 2017; NRC, 1983). In such cases, measurement data from small groups can be extrapolated to the population they represent, but also modelling can be used to estimate levels of exposure. Today, exposure assessments can include insights that have been generated through biomonitoring (Choi *et al.*, 2015). Developments in -omics technologies (including but not limited to proteomics – studying a large number of proteins in cells, tissues and organisms (biological systems, McArdle and Menikou, 2021) at once – and metabolomics – studying metabolites produced by such biological systems, Ordovas *et al.*, 2018) can further support improving exposure assessments (EFSA, 2017c). Also, food supply or consumption data is important, originating from databases such as (inter)national food consumption databases including the EFSA Comprehensive European Food Consumption Database[36] (De Boer *et al.*, 2020; Deluyker, 2017; EFSA, 2011; Kukk and Torres, 2020).

exposure

Since the actual risk of a compound depends on exposure, when exposure is considered low, it may even be questioned whether it is relevant to perform additional tests to identify toxicity. A recent development in risk assessment is therefore the use of the 'threshold of toxicological concern', the TTC concept (Kroes *et al.*, 2005). This decision tree supports identifying whether there is a level of exposure to a given substance below which no significant risk is expected to occur. In case the identified or predicted exposure is below this established threshold level, no further testing would be needed. Threshold levels have been established for different chemicals already, including food flavourings (Knight *et al.*, 2021).

threshold of toxicological concern

6.4.4 Risk characterisation

In the fourth and final stage of risk assessment (risk characterisation), a conclusion on the risk is drawn, to conclude whether there could be a safety concern for the general population or for specific groups. For that purpose, all findings from the first three steps of risk assessment (hazard identification, characterisation and exposure assessment) are combined into what the NRC called 'a final expression of risk' (NRC, 1983; Renwick *et al.*, 2003). As described by researchers involved in the FOSIE project, risk characterisation is affected by the purpose of the risk assessment: in case of dealing with hazards, a threshold exposure level below which no risks are expected will allow for translating the findings from the previous risk assessment steps into guidance values such as acceptable daily intakes, ADIs, that can be compared with the estimated intake (Renwick *et al.*, 2003). However, when dealing with no-threshold effects in case of substances that are expected to immediately affect health, reducing the exposure to as low as reasonably achievable (ALARA) or as low as reasonably practicable (ALARP) may be advised (Renwick, 2004). Finally, in this last stage of risk assessment, it is essential to consider and describe uncertainty of the findings of the assessment. Such uncertainties highly depend on the nature and context of the assessment undertaken, but efforts have been undertaken to allow for systematically analysing such uncertainties in scientific assessments by EFSA (EFSA, 2018).

---

[36] Comprehensive Food Consumption Database, available via https://www.efsa.europa.eu/en/data-report/food-consumption-data.

6.5 Risk management in EU food law

As described above, the roles of risk assessor and risk manager are explicitly separated in EU food law. Based on the submitted scientific evidence with its uncertainty and variability, EFSA's assessment proposes a range of policy measures with differing levels of scientific support (EFSA, 2017a; Zwietering, 2015). In political risk management decisions, these findings are combined with legitimate other concerns to decide upon a policy measure, that includes *inter alia* public health considerations (Meyer, 2006). As Article 3 of the GFL lays down, weighing the policy alternatives (conducted in consultation with interested parties) should allow for the selection of appropriate prevention and/or control options (Van Kleef *et al.*, 2006). For this broad range of tasks (from prevention to control measures), a wide range of decisions are available: from commissioning research projects or recommendations, to informing the general public, and of course to legally binding measures such as administrative decisions (Meyer, 2006).

risk management role

The risk management role is fulfilled by the different bodies that develop European policies and by those bodies that take decisions upon prevention and control of risks within the food system: the European Commission, the European Parliament, and individual Member States. The European Commission is the sole body that can take the initiative for making EU law (known as the full legislative initiative), as defined in Article 17(2) of the Treaty on European Union (TEU)[37]. Since the European Parliament and the Council need to adopt the laws that are drawn up by the Commission in the ordinary legislative procedure (Article 294 of the Treaty on the Functioning of the European Union (TFEU)[38]), these institutions can also be considered risk managers. With European Union law being part of the legal systems of the Member States that these Member States enforce within their own jurisdictions, the Member States (or their competent authorities) also function as risk manager in EU food law.

risk management bodies

The roles of the risk management bodies involved in decision-making processes in the European Union thus result from the fact that their roles and responsibilities are laid down in the European Treaties. Although the description and especially the separation of tasks within the risk analysis procedure becomes very clear from the GFL, this regulation does not provide many details on the European bodies involved in risk management. Only Article 22(8) and Article 40(1) of the GFL specify the organisations involved in risk analysis: Article 22(8) requires the different organisations (EFSA, the Commission and the Member States) to cooperate to ensure that risk assessment, risk management and risk

---

[37] Treaty on European Union. Consolidated version 1 September 2016.
[38] Treaty on the Functioning of the European Union. Consolidated version 1 September 2016.

communication functions are well connected; from Article 40(1) it becomes clear that EFSA is able to provide communications related to its mission, whilst it is defined as the Commission's role to communicate upon its risk management decisions. Article 17(2) touches upon the responsibility of the Member States, to enforce food law throughout the food chain. As specified in OpenEFSA[39], the portal that is the successor of EFSA's Register of Questions, most questions addressed to EFSA originate from the European Commission and relate to scientific opinions on specific regulated products.

6.6 Risk communication

The third element of risk analysis is risk communication: the exchange of information and opinions between experts, as well as the exchange of information about risk assessment and/or management to other relevant stakeholders. These other stakeholders are not only official organisations such as food business operators, non-governmental organisations, media and academics, also individuals themselves should have the possibility to interact within the risk analysis process, following the definition of the FAO/WHO on risk communication (FAO/WHO, 1998). When it comes to communicating to and with consumers, it has been suggested that transparency and independence of assessors and managers, as well as transparency in risk communication and public participation can support public trust in food safety (Cope *et al.*, 2010; Lofstedt, 2007). Even though the actual effects of these communication efforts have not all been established yet (De Boer, 2019b; Lofstedt, 2007), the next step towards increasing transparency in this risk analysis process even further has been taken by the entry into force of the Transparency Regulation[40]. Next to specific adjustments to eight sectoral regulations and directives, this Regulation amends the GFL by, among other things, further clarifying risk communication activities and roles (in Article 8a, 8b and 8c). Other adjustments following from the entry into force of the Transparency Regulation mainly relate to risk assessment practices for applicants (as reviewed in De Boer (2019a) and Chatzopoulou *et al.* (2020) among others). The adoption and entry into force of the Transparency Regulation has prompted EFSA to publish several reports on technical information and best practice advice for applicants and other stakeholders, which efforts have been described as a step towards shaping a future EU-wide food safety risk communication plan (EFSA, 2021). These

---

[39] OpenEFSA, available via https://open.efsa.europa.eu/.
[40] Regulation (EU) 2019/1381 of the European Parliament and of the Council of 20 June 2019 on the transparency and sustainability of the EU risk assessment in the food chain and amending Regulations (EC) No 178/2002, (EC) No 1829/2003, (EC) No 1831/2003, (EC) No 2065/2003, (EC) No 1935/2004, (EC) No 1331/2008, (EC) No 1107/2009, (EU) 2015/2283 and Directive 2001/18/EC.

different steps show how the different institutions involved aim for active engagement with a range of stakeholders, to ensure that risk communication is more than merely a one-way route of information from risk assessors and risk managers to other interested parties.

6.7 Looking ahead: the future of risk analysis

As exemplified in the case of titanium dioxide, new scientific insights about foods require specific actions from policymakers. Whether these products are new to the market or already available to EU consumers, the risk analysis process has proved to be an effective instrument with which to carefully analyse risks and develop policies for dealing with these risks. The previous sections show that risk analysis is at the core of EU food law. Risk assessment allows for a critical assessment of scientific evidence by independent experts, that can support decisions made by risk managers who need to weigh up different policy options and considerations in dealing with risks and health effects of foods. The Transparency Regulation should support that risk assessment and risk communication efforts can become more transparent for all involved stakeholders (without infringing upon commercial interests), and it should ensure that the process is made more sustainable and therefore future proof.[10] This should further strengthen consumer trust. It can therefore be expected that risk analysis will remain an important tool in EU food law.

The adoption of the Transparency Regulation, related clarifications on risk assessment requirements and risk communication efforts of, for example, the European Food Safety Authority, as well as previous guidance for analysing and communicating uncertainty are important steps towards further improving the use of scientific evidence and the risk analysis cycle. Still, questions about its effective implementation, its effect on stakeholders and subsequent effects on the actual goals of EU food law still exist (Chatzopoulou *et al.*, 2020; De Boer, 2019a; Ní Chearnaigh, 2021; Von Stackelberg and Williams, 2021). As outlined in the previous sections, it is therefore important that next to increasing transparency and sustainability of risk assessment and communication efforts, other steps are also taken to further strengthen risk analysis. We will highlight three essential elements that call for further deliberation, addressing both policy and the generation and use of scientific data.

four-step risk assessment paradigm

Firstly, the methods used in risk assessment today to establish that foods have a certain effect or that they are not unsafe, still follow the four-step risk assessment paradigm that was already being used in the 1980s (Von Stackelberg and Williams, 2021). At the same time however, the complexity of risk assessment has increased. The increased availability of technology, the development of novel methods and increasing amounts of data have become available to study hazards, exposure and gaining insights into risks (Brock *et al.*, 2003; Von Stackelberg and Williams, 2021). Some methods have been touched upon shortly in this chapter, but there is a large range of technologies and methods available that can provide insights into food health and safety effects. It is therefore essential to critically analyse the validity of these new methods and their putative role in scientific and regulatory assessments. This may also lead to a careful reflection on the current aims and goals of risk assessment studies, and potential adjustments to study results that need to be reported in such dossiers (De Boer *et al.*, 2020; Knight *et al.*, 2021).

Next to the input data, the aim of risk assessment (and subsequently, risk management) has also become more complex (Wu and Rodricks, 2020). **complex input data** Whereas traditionally, risk assessment aimed for a risk characterisation of single chemicals, nowadays the safety of complex food mixtures needs to be analysed. At the same time, increasing attention is given to the concept of benefit-risk assessment approaches: a process to estimate benefits and risks that originate from (a lack of) exposure to a certain food (component), integrated in comparable measures (Tijhuis *et al.*, 2012). Benefit-risk assessments already have a long history in the assessment of safety and efficacy of medicinal products, chemicals that are intentionally consumed for a certain health benefit expected from consumption. As consumption may result in beneficial effects, but could also cause side effects, benefit-risk assessments provide a means to weigh up these potential outcomes. Whereas in nutrition no side effects are allowed after consumption, also in this field this approach has gained interest relatively recently (Tijhuis *et al.*, 2012; Verhagen *et al.*, 2012). For different nutrients, optimal consumption ranges are established that support maintaining healthy bodily functions, whilst consuming too little or too much of them would induce a risk for adverse health effects. Benefit-risk assessments could also address whole foods, e.g. weighing risks and benefits of consuming fish that contains environmental pollutants as well as healthy ingredients, omega-3 fatty acids and vitamin D (Tuomisto *et al.*, 2020). So far however, no clear method has been established to systematically use benefit-risk assessments in regulatory procedures for foods. Difficulties that still need to be tackled arise from, among others, discussions on how to weigh up and express the outcomes of the assessment, as well necessary interaction between risk assessors and risk managers to ensure that the outcomes are used well in policy considerations (Tijhuis *et al.*, 2012).

This immediately touches upon the final consideration to be addressed in this **functional separation** chapter: the functional separation between risk assessment and risk management. **between risk** In EU food law, this means that EFSA should work independently of any **assessment and risk** political influence of either the risk manager or other parties (Deluyker, 2017). **management** This strict separation should ensure that the scientific advice is completely independent, and is purely an objective, transparent assessment of scientific evidence related to the question that needs to be answered. The functional separation of the assessment from the management decision has however been debated in literature: the practical reality of the risk analysis procedure requires a certain amount of interaction between risk assessors and risk managers to ensure that questions and outcomes are well understood, and the separation is therefore not always clear-cut (Ely *et al.*, 2009; Houghton *et al.*, 2008; Pagano, 2017). The influence of the risk manager on the work of EFSA as assessor is highly relevant, already shown by the risk manager providing the mandate to EFSA and thereby the framing or terms of reference within which the Authority needs to operate (Dreyer *et al.*, 2009; Houghton *et al.*, 2008; Vos and Wendler, 2009). The effect of risk assessment on risk management decisions can also not be underestimated: although scientific opinions issued by EFSA are not binding for the European Commission, motivations have to be provided when the EFSA opinion is not complied with (Silano, 2009; Szajkowska, 2009). The importance attributed to knowledge underlying EU food policy enlarges the dependence on EFSA's scientific advice (Szajkowska, 2009). The separation of tasks within the GFL, however, does give rise to various effects that help unravelling assessment and management, such as ensuring that individuals involved in risk assessment are not engaged in risk management (Houghton *et al.*, 2008). A transparent dialogue to ensure understanding of, on the one hand, the risk manager's questions asked to the assessor and on the other hand the assessor's advice provided to the manager, becomes even more essential for useful risk analysis (Gabbi, 2007). To what extent risk assessment and risk management are still intertwined in spite of this legal functional separation laid down in the GFL, requires further review.

6.8 Conclusions

Risk analysis is a key process in EU food law. Over the last twenty years, since the enactment of the GFL, it has become even more important in different regulatory procedures dealing with food. As displayed in this chapter, risk analysis is not merely used when dealing with risks; the process also plays an important role when regulating health effects or efficacy of food products. Separating the science-based risk assessment process from the political risk management procedure and ensuring an interactive exchange of information and opinions by stakeholders in risk communication, should support effective policy-making contributing to a high level of consumer protection.

This chapter highlights that different aspects of the risk analysis process can benefit from careful review and potentially from further improvements. Most importantly, even though risk assessment (EFSA) and risk management (mostly the Commission) are separated on paper in the GFL, this functional separation may be less strict in practice. Even though risk communication plays an important role in risk analysis, this separation may also cause difficulties in the understanding of questions asked and scientific opinions provided. Additionally, risk assessment procedures would specifically benefit from the swift uptake of well-developed non-animal based or new approach methods. As soon as scientific consensus is reached, these methods should find their way to guidance documents issued by risk assessors. However, the Transparency Regulation is already expected to support the idea that risk assessment (and risk communication) become more transparent and future-proof.

References

Bast, A. and Hanekamp, J.C., 2017. 'The policy of truth' – anchoring toxicology in regulation. In: Bast, A. and Hanekamp, J.C. (eds) Toxicology: what everyone should know. Elsevier Academic Press, London, UK, pp. 71-78.

Blaauboer, B.J., Boobis, A.R., Bradford, B., Cockburn, A., Constable, A., Daneshian, M., Edwards, G., Garthoff, J.A., Jeffery, B., Krul, C. and Schuermans, J., 2016. Considering new methodologies in strategies for safety assessment of foods and food ingredients. Food and Chemical Toxicology 91: 19-35. https://doi.org/10.1016/J. FCT.2016.02.019

Brock, W.J., Rodricks, J. V., Rulis, A., Dellarco, V.L., Gray, G.M. and Lane, R.W., 2003. Food safety: risk assessment methodology and decision-making criteria. International Journal of Toxicology 22: 435-451. https://doi.org/10.1177/109158180302200605

Chatzopoulou, S., Eriksson, N.L. and Eriksson, D., 2020. Improving risk assessment in the European Food Safety Authority: lessons from the European Medicines Agency. Frontiers in Plant Science 11: 349. https://doi.org/10.3389/fpls.2020.00349

Choi, J., Mørck, T.A., Polcher, A., Knudsen, L.E. and Joas, A., 2015. Review of the state of the art of human biomonitoring for chemical substances and its application to human exposure assessment for food safety. EFSA Support Publ EN-724. https:// doi.org/10.2903/sp.efsa.2015.EN-724

Cope, S., Frewer, L.J., Houghton, J., Rowe, G., Fischer, A.R.H. and De Jonge, J., 2010. Consumer perceptions of best practice in food risk communication and management: Implications for risk analysis policy. Food Policy 35: 349-357. https:// doi.org/10.1016/J.FOODPOL.2010.04.002

De Boer, A., 2019a. Scientific assessments in European food law: making it future-proof. Regulatory Toxicology and Pharmacology 108: 104437. https://doi.org/10.1016/j. yrtph.2019.104437

De Boer, A., 2019b. Transparency and consumer trust in scientific assessments under European food law. In: Urazbaeva, A. (ed.) The functional field of food law: reconciling the market and human rights. Wageningen Academic Publishers, Wageningen, the Netherlands, pp. 65-76. https://doi.org/10.3920/978-90-8686-885-8_17

De Boer, A., 2021. Fifteen years of regulating nutrition and health claims in Europe: the past, the present and the future. Nutrients 13: 1725. https://doi.org/10.3390/nu13051725

De Boer, A. and Bast, A., 2018. Demanding safe foods – Safety testing under the novel food regulation (2015/2283). Trends in Food Science and Technology 72: 125-133. https://doi.org/10.1016/j.tifs.2017.12.013

De Boer, A., Krul, L., Fehr, M., Geurts, L., Kramer, N., Tabernero Urbieta, M., Van der Harst, J., Van de Water, B., Venema, K., Schütte, K. and Hepburn, P.A., 2020. Animal-free strategies in food safety & nutrition: what are we waiting for? Part I: food safety. Trends in Food Science and Technology 106: 469-484. https://doi.org/10.1016/j.tifs.2020.10.034

De Boer, A., Vos, E. and Bast, A., 2014. Implementation of the nutrition and health claim regulation – The case of antioxidants. Regulatory Toxicology and Pharmacology 68: 475-487. https://doi.org/10.1016/j.yrtph.2014.01.014

De Fátima Poças, M. and Hogg, T., 2007. Exposure assessment of chemicals from packaging materials in foods: a review. Trends in Food Science and Technology 18: 219-230. https://doi.org/10.1016/J.TIFS.2006.12.008

Deluyker, H., 2017. Is scientific assessment a scientific discipline? EFSA Journal 15: e15111.

Dreyer, M., Renn, O., Ely, A., Stirling, A., Vos, E. and Wendler, F., 2009. Summary: key features of the general framework. In: Dreyer, M. and Renn, O. (eds) Food safety governance – integrating science, precaution and public involvement. Springer-Verlag, Berlin, Germany, pp. 159-166.

Dybing, E., Doe, J., Groten, J., Kleiner, J., O'Brien, J., Renwick, A.G., Schlatter, J., Steinberg, P., Tritscher, A., Walker, R. and Younes, M., 2002. Hazard characterisation of chemicals in food and diet: dose response, mechanisms and extrapolation issues. Food and Chemical Toxicology 40: 237-282. https://doi.org/10.1016/S0278-6915(01)00115-6

Ely, A., Stirling, A., Dreyer, M, Renn, O, Vos, E., Wendler, F., 2009. The need for change. In: Dreyer, M. and Renn, O. (eds) Food safety governance – integrating science, precaution and public involvement. Springer-Verlag, Berlin, Germany, pp. 11-28.

European Commission (EC), 2015. Strengthening evidence based policy making through scientific advice reviewing existing practice and setting up a European science advice mechanism. Brussels, Belgium.

European Food Safety Authority (EFSA), 2010. Definitions of EFSA Scientific Outputs and Supporting Publications. Available at: https://www.efsa.europa.eu/en/efsajournal/scdocdefinitions.

European Food Safety Authority (EFSA), 2011. Use of the EFSA Comprehensive European Food Consumption Database in exposure assessment. EFSA Journal 9: 2097. https://doi.org/10.2903/j.efsa.2011.2097

European Food Safety Authority (EFSA), 2017a. Guidance on the use of the weight of evidence approach in scientific assessments. EFSA Journal 15: 4971.

European Food Safety Authority (EFSA), 2017b. Scientific motivations and criteria to consider updating EFSA scientific assessments. EFSA Journal 15: 4737.

European Food Safety Authority (EFSA), 2017c. Scientific Opinion of the PPR Panel on the follow-up of the findings of the External Scientific Report 'Literature review of epidemiological studies linking exposure to pesticides and health effects.' EFSA Journal 15: 5007. https://doi.org/https:// doi.org/10.2903/j.efsa.2017.5007

European Food Safety Authority (EFSA), 2018. Guidance on uncertainty analysis in scientific assessments. EFSA Journal 16: 5123.

European Food Safety Authority (EFSA), 2021. EFSA reports set to inspire future risk communications in Europe. Available at: https://www.efsa.europa.eu/en/news/efsa-reports-set-inspire-future-risk-communications-europe.

Food and Agriculture Organization of the United Nations / World Health Organization (FAO/WHO), 1997. Risk management and food safety – Report of a Joint FAO/WHO Consultation. FAO, Rome, Italy.

Food and Agriculture Organization of the United Nations / World Health Organization (FAO/WHO), 1998. The application of risk communication to food standards and safety matters: report of a joint FAO/WHO expert consultation, Rome, 2-6 February 1998. FAO, Rome, Italy.

Food and Agriculture Organization of the United Nations / World Health Organization (FAO/WHO), 2019. Codex Alimentarius Commission – Procedural Manual twenty-seventh edition. FAO, Rome, Italy.

Gabbi, S., 2007. The interaction between risk assessors and risk managers. European Food and Feed Law Review 2: 126-135.

Hartung, T., Luechtefeld, T., Maertens, A. and Kleensang, A., 2013. Food for thought ... integrated testing strategies for safety assessments. ALTEX 30: 3-18. https://doi.org/10.14573/ALTEX.2013.1.003

Henson, S. and Caswell, J., 1999. Food safety regulation: an overview of contemporary issues. Food Policy 24: 589-603.

Houghton, J.R., Rowe, G., Frewer, L.J., Van Kleef, E., Chryssochoidis, G., Kehagia, O., Korzen-Bohr, S., Lassen, J., Pfenning, U. and Strada, A., 2008. The quality of food risk management in Europe: Perspectives and priorities. Food Policy 33: 13-26.

International Programme on Chemical Safety, 2009. Principles and methods for the risk assessment of chemicals in foods – Chapter 2: Risk assessment and its role in risk analysis. Environmental Health Criteria 240. FAO/WHO Publication. Available at: https://apps.who.int/iris/bitstream/handle/10665/44065/WHO_EHC_240_eng.pdf

Jovanovi, B., 2015. Critical review of public health regulations of titanium dioxide, a human food additive. Integrated Environmental Assessment and Management 11: 10-20. https://doi.org/10.1002/ieam.1571

Karmaus, A.L., Filer, D.L., Martin, M.T. and Houck, K.A., 2016. Evaluation of food-relevant chemicals in the ToxCast high-throughput screening program. Food and Chemical Toxicology 92: 188-196. https://doi.org/10.1016/J.FCT.2016.04.012

Knight, D.J., Deluyker, H., Chaudhry, Q., Vidal, J.M. and De Boer, A., 2021. A call for action on the development and implementation of new methodologies for safety assessment of chemical-based products in the EU – A short communication. Regulatory Toxicology and Pharmacology 119: 104837. https://doi.org/10.1016/j.yrtph.2020.104837

König, A., 2010. Compatibility of the SAFE FOODS risk analysis framework with the legal and institutional settings of the EU and the WTO. Food Control 21: 1638-1652. https://doi.org/10.1016/J.FOODCONT.2009.11.018

Kramer, N.I., Hoffmans, Y., Wu, S., Thiel, A., Thatcher, N., Allen, T.E.H., Levorato, S., Traussnig, H., Schulte, S., Boobis, A., Rietjens, I.M.C.M. and Vinken, M., 2019. Characterizing the coverage of critical effects relevant in the safety evaluation of food additives by AOPs. Archives of Toxicology 93: 2115-2125. https://doi.org/10.1007/S00204-019-02501-X

Kroes, R., Kleiner, J. and Renwick, A., 2005. The threshold of toxicological concern concept in risk assessment. Toxicological Sciences 86: 226-230. https://doi.org/10.1093/toxsci/kfi169

Kukk, M. and Torres, D., 2020. Risk assessment related to food additives and food processing-derived chemical contaminants exposure for the Portuguese population. EFSA Journal 19: e181110. https://doi.org/10.2903/j.efsa.2020.e181110

Lenssen, K.G.M., Bast, A. and De Boer, A., 2018. Clarifying the health claim assessment procedure of EFSA will benefit functional food innovation. Journal of Functional Foods 47: 386-396. https://doi.org/10.1016/j.jff.2018.05.047

Lofstedt, R.E., 2007. How can we make food risk communication better: where are we and where are we going? Journal of Risk Research 9: 869-890. https://doi.org/10.1080/13669870601065585

Ludden, V., Godfrey, E., Kobilsky, A., Hahn, F. and Jansen, L., 2018. The 3rd independent external evaluation of EFSA 2011-2016. final report. Available at: https://www.efsa.europa.eu/sites/default/files/3rd-Evaluation-of-EFSA_Final-Report100818.pdf

McArdle, A.J. and Menikou, S., 2021. What is proteomics? Archives of Disease in Childhood – Education & Practice 106: 178. https://doi.org/10.1136/archdischild-2019-317434

Meyer, A.H., 2006. Risk analysis in accordance with Article 6, Regulation (EC) No. 178/2002. European Food and Feed Law Review 3: 146-153.

National Research Council (NRC), 1983. Risk assessment in the federal government: managing the process. National Academies Press, Washington, DC, USA. https://doi.org/10.17226/366

Nauta, M.J., Andersen, R., Pilegaard, K., Pires, S.M., Ravn-Haren, G., Tetens, I. and Poulsen, M., 2018. Meeting the challenges in the development of risk-benefit assessment of foods. Trends in Food Science and Technology 76: 90-100. https://doi.org/10.1016/J.TIFS.2018.04.004

Ní Chearnaigh, B., 2021. Piecemeal Transparency: An Appraisal of Regulation (EU) No. 2019/1381 on the Transparency and Sustainability of the EU Risk Assessment in the Food Chain. European Journal of Risk Regulation 12: 699-710. https://doi.org/10.1017/ERR.2020.110

Ordovas, J.M., Ferguson, L.R., Tai, E.S. and Mathers, J.C., 2018. Personalised nutrition and health. BMJ 361: bmj.k2173. https://doi.org/10.1136/bmj.k2173

Pagano, M., 2017. The Italian Xylella case: the role of EFSA in the EU decision-making on risk. European Journal of Risk Regulation 8: 599-605.

Renwick, A.G., 2004. Risk characterisation of chemicals in food. Toxicology Letters 149: 163-176. https://doi.org/10.1016/J.TOXLET.2003.12.063

Renwick, A.G., Barlow, S.M., Hertz-Picciotto, I., Boobis, A.R., Dybing, E., Edler, L., Eisenbrand, G., Greig, J.B., Kleiner, J., Lambe, J., Müller, D.J.G., Smith, M.R., Tritscher, A., Tuijtelaars, S., Van Den Brandt, P.A., Walker, R. and Kroes, R., 2003. Risk characterisation of chemicals in food and diet. Food and Chemical Toxicology 41: 1211-1271. https://doi.org/10.1016/S0278-6915(03)00064-4

Rovida, C., Barton-Maclaren, T., Benfenati, E., Caloni, F., Chandrasekera, P.C., Chesné, C., Cronin, M.T.D., Knecht, J. De, Dietrich, D.R., Escher, S.E., Fitzpatrick, S., Flannery, B., Herzler, M., Bennekou, S.H., Hubesch, B., Kamp, H., Kisitu, J., Kleinstreuer, N., Kovarich, S., Leist, M., Maertens, A., Nugent, K., Pallocca, G., Pastor, M., Patlewicz, G., Pavan, M., Presgrave, O., Smirnova, L., Schwarz, M., Yamada, T. and Hartung, T., 2020. Internalization of read-across as a validated new approach method (NAM) for regulatory toxicology. ALTEX 37: 579-606. https://doi.org/10.14573/ALTEX.1912181

Sakuratani, Y., Horie, M. and Leinala, E., 2018. Integrated approaches to testing and assessment: OECD Activities on the development and use of adverse outcome pathways and case studies. Basic & Clinical Pharmacology & Toxicology 123: 20-28. https://doi.org/10.1111/BCPT.12955

Schultz, T.W., Amcoff, P., Berggren, E., Gautier, F., Klaric, M., Knight, D.J., Mahony, C., Schwarz, M., White, A. and Cronin, M.T.D., 2015. A strategy for structuring and reporting a read-across prediction of toxicity. Regulatory Toxicology and Pharmacology 72: 586-601. https://doi.org/10.1016/J.YRTPH.2015.05.016

Silano, V., 2009. Science, risk assessment and decision-making to ensure food and feed safety in the European Union. European Food and Feed Law Review 4: 400-405.

Sperber, W.H., 2001. Hazard identification: from a quantitative to a qualitative approach. Food Control 12: 223-228. https://doi.org/10.1016/S0956-7135(00)00044-X

Stibernitz, B., 2012. A brief comment on science-based risk regulation within the European Union. European Journal of Risk Regulation 3: 86-91.

Szajkowska, A., 2009. From mutual recognition to mutual scientific opinion? Constitutional framework for risk analysis in EU food safety law. Food Policy 34: 529-538. https://doi.org/10.1016/j.foodpol.2009.09.004

Tijhuis, M.J., De Jong, N., Pohjola, M. V., Gunnlaugsdóttir, H., Hendriksen, M., Hoekstra, J., Holm, F., Kalogeras, N., Leino, O., van Leeuwen, F.X.R., Luteijn, J.M., Magnússon, S.H., Odekerken, G., Rompelberg, C., Tuomisto, J.T., Ueland,

White, B.C. and Verhagen, H., 2012. State of the art in benefit-risk analysis: Food and nutrition. Food and Chemical Toxicology 50: 5-25. https://doi.org/10.1016/J. FCT.2011.06.010

Tollefsen, K.E., Scholz, S., Cronin, M.T., Edwards, S.W., de Knecht, J., Crofton, K., Garcia-Reyero, N., Hartung, T., Worth, A. and Patlewicz, G., 2014. Applying adverse outcome pathways (AOPs) to support Integrated approaches to testing and assessment (IATA). Regulatory Toxicology and Pharmacology 70: 629-640. https://doi.org/10.1016/J.YRTPH.2014.09.009

Tuomisto, J.T., Asikainen, A., Meriläinen, P. and Haapasaari, P., 2020. Health effects of nutrients and environmental pollutants in Baltic herring and salmon: a quantitative benefit-risk assessment. BMC Public Health 20: 64. https://doi.org/10.1186/S12889-019-8094-1

Van Kleef, E., Frewer, L.J., Chryssochoidis, G.M., Houghton, J.R., Korzen-Bohr, S., Krystallis, T., Lassen, J., Pfenning, U. and Rowe, G., 2006. Perceptions of food risk management among key stakeholders: results from a cross-European study. Appetite 47: 46-63.

Verhagen, H., Tijhuis, M.J., Gunnlaugsdóttir, H., Kalogeras, N., Leino, O., Luteijn, J.M., Magnússon, S.H., Odekerken, G., Pohjola, M. V., Tuomisto, J.T., Ueland, White, B.C. and Holm, F., 2012. State of the art in benefit-risk analysis: introduction. Food and Chemical Toxicology 50: 2-4. https://doi.org/10.1016/J.FCT.2011.06.007

Vinken, M., Kramer, N., Allen, T.E.H., Hoffmans, Y., Thatcher, N., Levorato, S., Traussnig, H., Schulte, S., Boobis, A., Thiel, A. and Rietjens, I.M.C.M., 2020. The use of adverse outcome pathways in the safety evaluation of food additives. Archives of Toxicology 94: 959-966. https://doi.org/10.1007/S00204-020-02670-0

Von Stackelberg, K. and Williams, P., 2021. evolving science and practice of risk assessment. Risk Analysis 41: 571-583. https://doi.org/10.1111/RISA.13647

Vos, E., 2000. EU food safety regulation in the aftermath of the BSE crisis. Journal of Consumer Policy 23: 227-255. https://doi.org/10.1023/A:1007123502914

Vos, E. and Wendler, F., 2009. Legal and institutional aspects of the general framework. In: Dreyer, M. and Renn, O. (eds) Food safety governance – integrating science, precaution and public involvement. Springer-Verlag, Berlin, Germany, pp. 83-110.

Vrolijk, M., Deluyker, H., Bast, A. and De Boer, A., 2020. Analysis and reflection on the role of the 90-day oral toxicity study in European chemical risk assessment. Regulatory Toxicology and Pharmacology 117: 104786. https://doi.org/10.1016/j. yrtph.2020.104786

Wernaart, B.F.W., 2022. Food ethics. In: Wernaart, B.F.W. and Van der Meulen, B.M.J. (eds) Applied Food Science. Wageningen Academic Publishers, Wageningen, the Netherlands, pp. 45-64.

Wu, F. and Rodricks, J. V., 2020. Forty years of food safety risk assessment: a history and analysis. Risk Analysis 40: 2218-2230. https://doi.org/10.1111/RISA.13624

Younes, M., Aquilina, G., Castle, L., Engel, K.H., Fowler, P., Frutos Fernandez, M.J., Fürst, P., Gundert-Remy, U., Gürtler, R., Husøy, T., Manco, M., Mennes, W., Moldeus, P., Passamonti, S., Shah, R., Waalkens-Berendsen, I., Wölfle, D., Corsini,

E., Cubadda, F., De Groot, D., FitzGerald, R., Gunnare, S., Gutleb, A.C., Mast, J., Mortensen, A., Oomen, A., Piersma, A., Plichta, V., Ulbrich, B., Van Loveren, H., Benford, D., Bignami, M., Bolognesi, C., Crebelli, R., Dusinska, M., Marcon, F., Nielsen, E., Schlatter, J., Vleminckx, C., Barmaz, S., Carfí, M., Civitella, C., Giarola, A., Rincon, A.M., Serafimova, R., Smeraldi, C., Tarazona, J., Tard, A. and Wright, M., 2021. Safety assessment of titanium dioxide (E171) as a food additive. EFSA Journal 19: 6585. https://doi.org/10.2903/J.EFSA.2021.6585

Zwietering, M.H., 2015. Risk assessment and risk management for safe foods: Assessment needs inclusion of variability and uncertainty, management needs discrete decisions. International Journal of Food Microbiology 213: 118-123. https://doi.org/10.1016/j.ijfoodmicro.2015.03.032

7. Genetics

This is what you need to know about genetics and its application in breeding our food crops and animals

Niels P. Louwaars

Law Group, Wageningen University, Hollandseweg 1, Wageningen, the Netherlands; and Plantum, Gouda, the Netherlands; niels.louwaars@wur.nl

Abstract

Genetics is the study of heredity and the functioning of hereditary material, the genes. Genetics is relevant to food in two distinct ways: (1) Firstly, the genetic makeup of the organisms producing food (plants, animals and microorganisms) determines to a significant extent the composition of the food in nutritional and anti-nutritional terms. It furthermore relates to other quality components, such as processing qualities, shelf life, and sensitivity to the growth of toxic microorganisms in or on the consumed product. (2) Secondly, on the side of the consumer there may be human-genetic preconditions that determine sensitivities to certain foods. This section focuses on the former and draws examples mainly from food plants. This introduction to genetics is also relevant because debates are arising again about certain genetic techniques that can be used in breeding plants and animals. We therefore introduce the concept of breeding, including its technological developments over time. Breeding is the use and creation of genetic diversity to select plants and animals that respond better to the needs of farmers and consumers.

Key concepts

- ► Genetics is the science of heredity and the genes that regulate it.
- ► Chromosomes are thread-like structures in the nucleus of a cell, consisting of DNA and proteins.
- ► DNA: 'deoxyribonucleic acid'; the very large molecules consisting of a string of four nucleotides, positioned in a double helix structure. DNA is the molecule that harbours genes.

Bart Wernaart and Bernd van der Meulen (eds)
Applied food science
DOI: 10.3920/978-90-8686-933-6_7, © Niels P. Louwaars 2022

- ▶ Genes are the starting point of all metabolic pathways in the cell; how plants and animals grow and the composition of the food they produce depends on the interplay between the genes and the environment. Genes are a sequence of nucleotides in DNA (or RNA) that encodes a chemical process in a living cell.
- ▶ Allele: a sequence variation of the same gene at the same location on a chromosome, e.g. coding for either blue or brown eye colour.
- ▶ Mutation and segregation are the major sources of genetic diversity, which together with selection shape evolution.
- ▶ All the food we eat is the result of millennia of rearranging the genetics of crops and animals, distancing them from their origins in nature.
- ▶ Breeding is the applied science that uses genetics to shape crops and animals to the needs of humankind: farmers, value chain operators and consumers; all our food originates from bred plants and animals, with the exception of those caught in the wild.
- ▶ Breeding methods are increasingly scrutinised by society, including food specialists.

Case 7.1. The discovery of DNA.

'Professor, the images are getting sharper', Raymond Gosling said. Rosalind Franklin looked at the pictures. Photo 51 was intriguing. 'Can't we publish this, professor? I truly think this is a key to the structure of DNA.' 'I guess, but we need to understand how this structure is biologically functional before we go public.' X-ray crystallography was a great technology and Franklin's lab at King's College London was a forerunner. Individual molecules and their structure could be made visible, that is, when large enough. And DNA is one such large molecule. Her colleague, the physicist Maurice Wilkins, and not exactly her best friend, presented the photo soon after at a conference and sent the picture to James Watson, 100 km north in Cambridge a few months later in 1952.

The young American showed it to his colleague Francis Crick. 'Something wrong with your camera, James? Is this a flock of geese in a winter sky, using the wrong diaphragm?'. 'No, no, it's DNA, I got it from Wilkins and I think it is one of the keys to our research.' Together they gave it a closer look – and started to use it to build a model of the DNA molecule – a double helix. 'Wow: this explains a lot – the two strands can connect with weak molecular bonds between the individual bases. In the helix configuration guanine fitted neatly with cytosine and adenine with thymine. The two strands were thus mirrors of each other, which could explain both the solid configuration, and also that it could replicate when the strands separated. How would that work? What makes the strand separate?' 'Don't know, but let's get this published as quickly as possible before anyone else does.' Their publication in 1953 on the structure of deoxyribonucleic acid (DNA), which failed to credit Franklin, was possibly not the best example of scientific ethics, but it became the basis of the science of molecular genetics and for their Nobel Prize together with Wilkins nine years later (Franklin had passed away by then). The structure of DNA was thus explained.

The history of genetics is built on many more stories involving both transpiration (Franklin) and inspiration (Crick), from Mendel's gardens in Brno to Charpentier and Doudna's laboratories at Berkeley. All these scientists stand on the shoulders of many others, but there's only space for so many names in the history book of science. Each section of this chapter helps provide an understanding of genetics and its relevance to food. Therefore, here is a brief summary.

7.1 History of genetics as a field of science

heredity

That there is something like heredity has been known for millennia. Children take after their parents, and some features, such as eye colour, are even predictable to a certain level without any knowledge about how this may work. Plants and animals have been selected since the dawn of agriculture, attempting to bring about certain traits in the offspring. Darwin (1859) based a significant part of his theory of evolution on the observation of domestic dog breeds. He could explain that breads could be kept 'pure' when mating was restricted to animals of the same breed; he was, however, quite puzzled by the enormous diversity of dogs when it was assumed that all dogs stem from domesticated wolves. He further sharpened his ideas when he came across the diverse finches on different islands of the Galapagos.

Jean-Baptiste Lamarck had postulated in 1809 that traits could be acquired during the lifetime of an organism, and transferred to the next generation, a theory that remained popular under Stalin's regime. It fitted in very well with his idea to create the perfect socialist society. It led to a 'war' between his chief geneticist Trofim Lysenko, and the already established ideas about heredity by the director of the Lenin All-Union Academy of Agricultural Sciences at Leningrad, Nikolaj I. Vavilov. The latter lost and succumbed in the Gulag. The balance between 'nature' (genetics) and 'nurture' (environment) is, however, still the subject of many debates.

segregation

The search for understanding heredity was on in the 19th century. It was Gregor Mendel who identified in 1866, in his experiments crossing peas with other crops, that a segregation ration of 1:3 was very common, or 1:3:3:9 when two independent dominant characteristics are studied. This led to the conclusion that whatever causes heredity, it behaves in duplicate form, and a character (allele) is either the same in both (homozygous) or different (heterozygous).

The observed ratios are found when one allele is dominant over the other; otherwise a 1:2:1 ratio is expected (for example in white, red and pink flowering individuals). Mendel's publication from 1866 remained unnoticed until three scientists independently (or not) rediscovered it in 1900. It created a revolution **population genetics** in biological sciences, notably in plants and after its extension to population genetics by Hardy and Weinberg in 1908 also in animal breeding. Plant and animal breeding became (applied) sciences using prediction through calculation rather than trial and error only.

In the meantime, Hugo de Vries, one of the discoverers of Mendel's publication in 1900, had already identified in 1889 that heredity has a material origin, **genes** which he called 'pangenes'. This was later shortened to 'genes' by Wilhelm Johannsen in 1909, who also made the important distinction between genotype (the genetic identity) and phenotype (the visible expression thereof). This means that the environment in which an organism lives influences the expression of the genotype and that the traits that Mendel looked at, which were not affected by the environment, were either lucky shots or an example of serendipity. The genetic basis of many traits can thus not be identified by simply looking at them. On that basis, Ronald Fisher developed the science of quantitative genetics through his analysis of variance (ANOVA) methods of statistical analysis in the 1920s. He thus bridged the divide between Mendel's and Darwin's theories.

chromosomes Nobody knew what this material origin was. Chromosomes, threads in cells that became visible after dying dividing cells under a microscope had already been observed in 1879 by Walther Flemming, but the pieces still had to be put together. Following the material origin of heredity, however, De Vries also hypothesised that genetic diversity arises through heritable changes that occur naturally, which **mutation** he called 'mutation'. We now know that the large changes that de Vries observed were likely the result of large chromosomal abnormalities, such as chromosome duplications, but the concept and the term 'mutation' is still very important. It is currently mainly used to indicate small changes in the DNA. Nobel prize winner Hermann J. Muller identified chromosome changes in fruit flies when these were subjected to irradiation as recently as 1927. This knowledge was quickly put to use in plant breeding. The technique of mutagenesis was born. Large-scale mutagenesis trials in barley resulted in improved brewing quality and disease resistance. Probably all the beer that we drink nowadays is brewed from barleys that have these mutants in their pedigree. However, the trouble with such randomly induced mutations is that often tens of thousands of individuals need to be screened to identify such a 'lucky shot', which is only possible for easily selectable traits such as a flower colour. Almost all chrysanthemums are mutants; a breeder creates a new flower type or better vase life through crossing and selection, and then mutates the new plant to create all colour sorts.

These hypotheses proved very useful in practice, but they kept us in the dark about the material structures responsible for heredity, and how these actually create the observed expression of traits.

James Watson and Francis Crick published the structure of DNA as a double helix in 1953, based on X-ray crystallography work done by Rosalind Franklin. The Cambridge team won the Nobel Prize for their chemical and structural clarification of the material basis of heredity. The double helix explains how DNA is able to replicate: when the double strands separate, individual bases connect into a new strand completing the double helix again as a true copy. It also laid the foundation for the next question: that is, how such a molecule could be the basis of all life functions in an organism. DNA itself had already been chemically identified in 1929 by Phoebus Levene. Based on his work on RNA in 1909, he identified the deoxyribose sugar and suggested that DNA consists of a string of four nucleotide units linked together through the phosphate groups. Nikolai Koltsov had already proposed in 1927 that there should be a 'giant hereditary molecule' made up of 'two mirror strands'. These two suggestions were thus confirmed in 1953 by Watson and Crick.

molecular genetics The science of molecular genetics developed soon after, and became an applied science when Marc van Montagu in Ghent identified the bacterial Ti-plasmid and the way it can by nature transport DNA fragments and include them in plant DNA. And so, genetic modification was born. Transgenic crops are grown in many countries, but in Europe on a limited scale only. Research on transgenic animals (Herman the bull producing offspring that could produce human lactoferrin) is limited due to ethical concerns (see also Chapter 3; Wernaart, 2022), but a wide range of animals have been created for research purposes, e.g. mice for cancer research, and also GM salmon, malaria mosquitos and others. GM technology is widely used with little public concern in medicine, including vaccine development. Molecular genetics also led to the development of various gene-editing systems like TALEN (Zhang *et al.*, 2014) and ODM (Dalbadie-McFarland *et al.*, 1982), and the easiest and therefore most widely used technologies based on CRISPR (clustered regularly interspaced short palindromic repeats) identified in bacteria by the group led by Van der Oost in Wageningen (Brouns *et al.*, 2008), and put to practical use by Emmanuelle Charpentier and Jennifer Doudna in Berkeley for which they were recognised with the Nobel Prize in Chemistry in 2020.

There is much more to say about the discoveries in the field of genetics and the inventions based on them. For example, genetic linkages and chromosomal crossovers may confuse expected outcomes of a cross based on Mendel's laws. All kinds of 'mistakes' occur in cell division (mitosis) and even more so in the development of haploid sex cells (the meiosis). There is also DNA outside the cell nucleus, i.e. in cell organelles such as chloroplasts and mitochondria, which are commonly inherited only through the maternal line, similar to the genes in the X-chromosome in mammals. There is also much more to say about ploidy levels, where Mendel's peas (and humans) have pairs of chromosomes (diploid), potato has four and strawberry has eight sets. In vegetatively propagated species also aneuploidy occurs, when sets are incomplete and still the plants look healthy and thrive both in nature and in our greenhouses. And, some species don't have DNA at all, but live on the basis of RNA (ribonucleic acid).

7.2 How do genes regulate biological processes?

Deoxyribonucleic acid (DNA) is a long molecule that consists of four nucleotides: adenine (A), cytosine (C), guanine (G), and thymine (T). These form a string of quite indefinite combinations of letters that replicate when cells divide. So, in principle, every cell in an organism has the same genetic code. Particular combinations of letters can form a functional gene, that includes 'start and stop' codes. Such genes can be 'read' by a similarly constructed molecule, RNA (in which uracil replaces thymine), in a process known as transcription. Such 'messenger RNA (mRNA)' can carry the information out of the cell nucleus towards other organelles in the cell, called ribosomes, where the mRNA code is 'read' to form proteins, in a process known as translation. Each triplet of RNA bases binds with one of 20 possible amino acids, and when connected in an RNA-defined sequence forms a very specific protein, each creating a special 3-dimensional structure. These are responsible for biological functions, as starters of metabolic pathways, for example by acting as enzymes, facilitating chemical reactions in the cell. It is these pathways that determine the biological functions and chemical composition of a cell, and thus – among many other things – the nutritional characteristics of a plant or animal product, and indeed also the occurrence of toxins and allergens. These pathways also determine the biological activity of microbes used in food processing, as well as the functions of pathogens and the response of cells to invading pathogens.

messenger RNA (mRNA)

However, it is important to realise that even though every cell has the same DNA, cells in an organism have very different functions and require very different parts of the DNA to be activated – and others not. Similar to gene × environment interactions, which made the simple Mendelian genetics much more complex, neither knowledge of the chemical structure nor the exact sequence of the roughly three billion base pairs of the human genome, help us 'understand the language in which God created life' as Bill Clinton expressed it when celebrating the finalisation of the Human Genome Project in 2000. It was indeed a milestone in the history of genetics, but there is much more to learn.

So, both pressures outside (neighbouring cells or physical stresses) but also inside the cell may switch transcription off or on and thus affect gene expression. A cell's DNA harbours thousands of genes that need to be active at any given moment. A liver cell needs to express different genes from a brain cell. In an organism, cells affect each other so much that cell functions are commonly fixed depending on the functions they have to perform in a certain organ. Much medical research is currently being done to 'unlearn' cells, i.e. to make them function as totipotent stem cells that can still differentiate in new directions and replace damaged tissue.

The cell has several methods to control the expression of genes. Transcription factors are regulatory proteins that 'cover' DNA or make it available for transcription into mRNA, so they determine which parts of DNA are activated at any point in time. Knowledge of the ways such transcription factors work allows us to greatly influence the cell processes without changing the genetics. There are also structural features where DNA is more permanently inhibited from transcription, e.g. by binding (methylation) to the DNA molecule. Such bonds may even continue to exist after cell division and even into a next generation. Such 'epigenetic' (Weinhold, 2006) effects change expression while leaving the DNA sequence intact. This, for example, is used to explain why the prevalence of obesity is high in people whose parents have suffered from undernutrition.

7.3 How can genetic changes occur?

The material nature of DNA also explains the changes that are required to illustrate evolution (Darwin) and mutations (de Vries). Changes may occur either through natural (physical or biological) means or human intervention.

natural genetic changes

Mutation: small or larger changes in the DNA, often through external influences like (cosmic) radiation, and through biological means, i.e. 'mistakes' in the duplication of DNA when cells split (mitosis) and particularly in the meiosis, when sex cells are created with half the amount of DNA. Mutations can also be the effect of transposons (or jumping genes), a DNA sequence that can move in a genome, identified by Barbara McClintock (1950). Mutations occur every day in plants and animals. Most mismatches and breaks are resolved by the natural repair mechanisms of the cell. They are only inherited when they occur in gametes (sex cells) or their precursors. Natural changes can also occur when cells are attacked by microorganisms; bacterial genes can enter plant cells and stay there. Sweet potato is the best-known example of a crop harbouring bacterial genes that apparently have evolutionary advantages.

Recombination: when plants or animals cross sexually, the genes of the parents mix, creating diversity among individuals in a next population. In regions where wild relatives of the species occur, occasional interspecific crosses may also occur, bringing new diversity into the species (introgression). This diversity through recombination (and introgression) and mutation, combined with Darwin's concept of survival of the fittest individuals in such diverse populations, feeds evolution.

human intervention   Human intervention to obtain diversity to select from in plant and animal breeding builds on these same principles:

Targeted cross breeding, which has been done since the domestication of animals, was increasingly pursued in (ornamental and fruit) plants after Rudolf Jacob Camerarius published in 1694 that plants also have male and female organs. The breeding of major food crops starting in the 19th century; interest in crop improvement flourished and greatly intensified in the 20th century when genetics became a science after Mendel and others. Targeted crossing aims at combining 'positive' traits in the progeny.

Random mutagenesis is the wilful stimulation of mutations throughout the DNA using irradiation (known in the 1920s, but popular from the 1950s onwards) and treatments with chemical mutagens, notably colchicine and ethyl methane sulfonate (EMS) (Talebi *et al.*, 2012). These may cause breaks or other changes in the DNA that may, when they occur in a functional gene, appear as visual or physiological changes. For example, 'variegated' ornamental plants with dark and light green parts of leaves, are the effect of mutations in certain cell layers.

Transgenesis is the transfer of functional genes between species, based on molecular genetics experiments in the 1970. And cisgenesis is the transfer of genes within the species, initiated in the early 21st century to improve difficult-to-breed crops, such as polyploids (potato) and tree species (apples), where the generation time makes repeated (back)crossing impractical. Initially, such genes were randomly inserted into the host DNA; more recently these genes can be placed in particular locations. Transgenesis evolved from the observation that bacteria like *Bacillus thuringiensis* (Bt) naturally exchange bits of their DNA with their hosts. Such random transgenesis can also be done by connecting DNA to particles that are 'shot' into the cell. Such techniques make it possible to channel specific genes from basically any origin into crop plants.

Targeted mutagenesis is the creation of changes (mutations) in a particular location in the DNA, either a deletion or a specific change of one or a few base pairs. This has been done since the early 21st century, but is greatly facilitated by the developments of CRISPR technologies developed from 2014 onwards.

Genome editing is a term used for both targeted mutagenesis and targeted gene transfer. For example, CRISPR Cas9 was developed from the observation that bacteria had a cluster of similar genetic sequences (repeats) that appeared to be a library to help the bacteria identify invading viruses, which then could be cut and destroyed. This natural phenomenon is now used to identify and cut DNA in a specific location. It can be used simply to cut the DNA allowing the natural repair mechanisms to do their job, which may lead to minor mistakes, or mutations, for example making the whole gene dysfunctional (silencing, Zhang *et al.*, 2020). This is particularly useful for the study of gene functions. Alternatively, CRISPR can 'cut and replace', creating a specific predefined mutation in the gene. When 'cut and replace' involves a whole functional gene, then we can speak of targeted trans (or cis-) genesis.

next step – selection The plants and animals that provide the food we eat are quite different from those found in nature. A wild tomato is a tiny little berry in a family (Nightshade) known to harbour quite toxic species. Almost all cabbages, from Brussels sprouts to broccoli and kohlrabi, originate from one wild species selected for different growth habits and edible plant organs. This illustrates that selection, which follows the creation (or collection) of genetic diversity, is a major force shaping our food. Also, selection has undergone quite some changes with the inclusion of molecular biology in plant and animal breeding. Originally, mass selection was the only tool farmer-breeders had in order to select in their crops and breeds: removing poor-looking individuals (negative mass selection) or allowing only the best to multiply (positive mass selection). Line selection and family selection looked at the progeny of a certain cross to significantly speed up selection and to establish the breeding value of certain individuals (notably male animals). Genetics came in as a science after Mendel and even more so after Ronald Fisher brought mathematical statistics (analysis of variance) into breeding.

The applied science of plant breeding subsequently 'absorbed' new sciences. Molecular biology greatly improved not only the creation of diversity, but also the selection methods that breeders use. Marker-assisted selection uses genetic markers to predict important traits in segregating populations (after a cross). A tiny leaf can tell the breeder whether the individual plant carries the desired trait(s) – all other plants can be removed and will not take part in further multiplication. This significantly speeds up the selection processes. When sequencing (actual 'reading') of DNA became cheaper and quicker, genomic selection was introduced (Lin *et al.*, 2014), identifying the 'perfect' set of parents for a cross in order to reach the desired outcome, and selecting at the DNA level. This development requires the analysis of big data, and artificial intelligence is becoming part of the breeder's toolbox.

Obviously, breeders need a lot of knowledge about the species they breed; their physiology, pathology and organic chemistry are important for the users (processors, consumers) of the crops and animals they breed. This makes breeding an applied science building on quite different basic sciences.

However, we are not just interested in the genetics, but in the expression of the genetics into useful characteristics of plants and animals, useful for the farmers (productivity, resilience to diseases and environmental factors), the processors (brewing and baking qualities) and consumers (taste, looks, nutrition, reduced decay during storage). Production systems, obviously, also influence the final outcome – genotype-by-environment interactions remain important. When particular soils do not contain sufficient chemical compounds, the levels in plants (e.g. the anti-carcinogenic glucosinolates in broccoli requiring sulphur) are also bound to be reduced irrespective of the genes. When grain is harvested and stored in moist conditions, fungi are likely to render it unsafe for consumption.

7.4 Plant breeding and food quality and safety

Plants can be seen as complex biochemical factories producing large numbers of compounds. We need several such compounds as food in order to live. We need carbohydrates (cereals), proteins (legumes), oils (oilseed rape), fibres (potato) and various micro-nutrients and vitamins (vegetables).

toxins

Mankind has selected plant species that produce nutritious food, and bypassed most of the obviously poisonous ones. However, several food plants still produce such toxic compounds, such as cassava and almonds (cyanides), potatoes (glycoalkaloids), beans and especially castor beans (lectins), soybeans and quinoa (saponins) (Centre for Food Safety, 2007). We have learned to deal with these in our food processing; e.g. boiling dry beans to disable the action of lectins. All these compounds are the result of the genetic makeup of the plant and the environment in which it grows.

Our food plants have changed considerably due to farmer selection and scientific breeding. Almost none of our food crops would survive well in nature. Such changes continue to occur; mutations that may affect the metabolism of the cell occur every day. In view of this, it is quite remarkable that we don't experience such mutations leading to either the known toxins that protected the forefathers of our crops like the nightshade family (tomato, peppers, potato and others), or to totally new compounds that may be toxic to us. Mutations occur in the breeding stage, in the multiplication of seeds and in the food production fields, but we apparently don't need to test each lot of wheat, potato or lettuce after harvesting. Plant breeders may screen for the known ('solanine' in potato) toxins when they have crossed their potato with wild ancestors that are known to produce high levels of this toxin (Sinden *et al.*, 1984), and even effectively blocked the pathway of erucic acid production in oilseed rape making the oil edible (Stefansson and Haugen, 1964). However, breeders have never screened their varieties for all possible (unknown) toxins without any negative effects on food safety. We rightfully trust our food that is derived from known crops, even though we know that their DNA is scrambled during mating, mutated due to exposure to UV sunlight both in nature and in breeding. This scrambling could affect metabolic processes in the living cell during plant growth, storage and food processing in the factory or the kitchen. Louwaars (2019) claims that crop plants producing toxins do not have a competitive advantage in the same way that plants in nature have, because we care for the crop and keep animals out. Since mutations leading to such toxic compounds would come at a cost for the plant, they would not be selected by the breeder. The taxonomic knowledge of a professional breeder furthermore prevents 'accidents' with toxins that are known in the species or its relatives.

nutrition Plant breeding started primarily with the needs of farmers: how to support increasing yields and make crops more resilient to pests and diseases, and environmental stresses like drought. Processing qualities became important for some crops, and in horticulture the looks (flowers) and taste (vegetables) also became relevant. For example, cauliflower size was genetically reduced following the reduction of household size, and the browning of lettuce is avoided now that most lettuce in the supermarket is sold ready-cut. Nutritional content was not a primary breeding objective until recently. Some studies indicate that the level of micronutrients has been reduced, partly as an effect of increased yields (Cursio, 2018). Others, however, conclude that there is no difference when old and new varieties are cultivated side by side under current agronomic conditions, but that there can be significant changes among varieties in general[41], which could be an excellent basis for breeders. It is likely though that fruit-vegetables like tomato and cucumber contain more water and fewer vitamins per kilo, when bred for yield capacity only. And crops like Brussels sprouts in which the bitter taste is bred out have reduced concentrations of the compounds that cause bitterness but potentially have an anti-carcinogenic effect.

---

[41] See for instance https://ag.umass.edu/news-events/highlights/growing-nutrient-dense-vegetables.

On the other hand, breeders working for poor consumers did start to focus on important minerals and vitamins (biofortification, see Bouis, 2002), and those aiming to create health products raised levels of antioxidants (Hanson *et al.*, 2004) and healthy glucosinolates. Breeding for oil composition in crops like soybean and oilseed rape also contributes to healthier products (Abbadi and Leckband, 2011).

7.5 Concerns about genetic technologies

Plant breeding has operated away from the public eye for a very long time. Breeders produce varieties and seeds for farmers, and further down the chain very little is known about the genetic origin of food. In the 1970s breeding became known as the solution to the world food situation following the Nobel Prize awarded to Norman Borlaug, a wheat breeder in Mexico, and initiator of what became known as the Green Revolution.

transgenics

Concerns did arise though with the development of transgenics. Natural processes like crossing and selection and even mutagenesis have not caused any food-related concerns, but transferring functional genes from one species to an unrelated one stirred up debates in the 1980s. We wouldn't know how the inserted gene would act in its new genetic environment, notably because we couldn't know where and how many copies of the gene were inserted during the transformation process. The 'precautionary principle' became a leading concept, but with very different interpretations (Hanson *et al.*, 2004; Verkerk *et al.,* 2009). Close to 35 years of experience have not allayed this concern, and genetically modified organisms (GMOs) are legally required to be extensively tested for safety for humans and the environment in most countries before being admitted to cultivation and to the food chain.

genome editing

A similar debate is currently ongoing with regard to genome editing. Should products of targeted mutagenesis be regarded as regulated GMOs and as a result tested and labelled? There are various discussions around this topic. Just as with conventional GMOs, legal specialists, plant scientists, environmentalists, food safety specialists, ethicists and social scientists all debate aspects of the technologies and their socio-economic impact.

A legal debate with different outcomes in different countries: The European Court of Justice ruled in favour of including gene-edited plants as regulated GMO in 2018[42]; In Argentina, Australia and Japan targeted mutagenesis is not or hardly regulated. A debate among plant scientists seems to have resulted

---

[42] See for instance this prejudicial procedure: Europan Court of Justice, 2018, 18 January. https://curia.europa.eu/jcms/upload/docs/application/pdf/2018-01/cp180004en.pdf.

in quite a broad consensus to question the need to extensively test such food plants, since they claim that the mutations could evolve naturally and such natural mutants would not be regulated by the GM laws. The European Food Safety Authority EFSA came to the conclusion in 2012 that risks of cisgenesis are comparable to those after conventional plant breeding (EFSA Panel on Genetically Modified Organisms (GMO) (EFSA, 2012)). Among food safety specialists there is a debate about process-based (where technology has been used) and product-based approaches to risk management (Mampuys 2019); and on the basis of the fact that mutants occur every day and that they might initiate different metabolic processes, there are those who think that, in fact, all products of plant breeding should undergo more extensive risk management. Finally, different ethicists claim that gene editing damages the integrity of the cell and the genome, that species barriers should not be crossed, or that the technology should not lead to greater influence from multinational corporations (Louwaars and Jochemsen, 2021). On the latter argument, opponents say that GM-regulation has actually significantly contributed to the emergence of such multinationals in plant breeding[43]. As explained in the chapter on legal and regulatory affairs in this book, authorisation procedures are costly, lengthy and their outcome is uncertain. Finally, there are also ethicists who say that in a world with food insecurity and climate change, it is unethical NOT to use such technological tools.

7.6 Looking ahead

Some people thought we knew all there was to know about life when the structure of DNA was unravelled in 1953 and the Human Genome Project was finalised in 2000, publishing the 'code of life'. However, as with any scientific field, the more we know, the better we know what we don't know. But also, as with many other areas of science: the more we know, the better we can use that knowledge to solve problems and to make life better. The same is true for genetics. We know how to identify functional genes; we know how to edit them. But there is still a lot to learn about transcription and translation, and the genetic aspects of cell differentiation, and interactions with the environment in both common (natural) biology and man-made uses of our knowledge. We also increasingly know how to identify the chemical and physiological effects of genetics and its interaction with the environment. Technology, such as phenotyping using real-time image analysis (Hartmann *et al.*, 2011) and the application of sensors, connected to big data analysis and artificial intelligence, will continue to teach us how life processes work, and I am sure that they will continue to reveal new unknowns. Many of these developments originate

---

[43] For instance http://www.genewatch.org/sub-568236.

from medical research, but such knowledge can help us to produce food more sustainably in the face of climate change, population growth and changing consumer needs. This obviously includes making sure that food optimally contributes – together with processing science, behavioural sciences and many more – to human, animal, and environmental health. The science of genetics will continue to be an important basis for food availability, quality and diversity in order to sustain a good life.

References

Abbadi, A. and Leckband, G., 2011. Rapeseed breeding for oil content, quality, and sustainability. European Journal of Lipid Science and Technology 113: 1198-1206. https://doi.org/10.1002/ejlt.201100063

Bouis, H.E., 2002. Plant breeding: a new tool for fighting micronutrient malnutrition. Journal of Nutrition 132:491S-494S. https://doi.org/10.1093/jn/132.3.491S

Brouns, S.J.J., Jore, M.M., Lundgren, M., Westra, E.R., Slijkhuis, R.J.H., Snijders, A.P.L., Dickman, M.J., Makarova, K.S., Koonin, E.V. and Van der Oost, J., 2008. Small CRISPR RNAs guide antiviral defense in prokaryotes. Science 15: 960-964.

Centre for Food Safety, 2007. Natural toxins in food plants. Risk Assessment Studies Report No. 27. Food and Environmental Hygiene Department, Hong Kong. http://www.cfs.gov.hk/english/programme/programme_rafs/files/ras27_natural_toxin_in_food_plant.pdf.

Cursio, S., 2018, 25 November. What's behind the invisible decline in nutrient density? Medium. Available at: https://medium.com/@stacey_59725/whats-behind-the-invisible-decline-in-nutrient-density-b2227306992f.

Dalbadie-McFarland, G., Cohen, L.W., Riggs, A.D., Morin, C., Itakura, K. and Richards, J.H., 1982. Oligonucleotide-directed mutagenesis as a general and powerful method for studies of protein function. Proceedings of the National Academy of Sciences of the USA 79: 6409-6413. https://doi.org/10.1073/pnas.79.21.6409

Darwin, C., 1859 On the origin of species by means of natural selection, or the preservation of favoured races in the struggle for life. Murray, London, UK.

European Food Safety Authority (EFSA), 2012. EFSA Panel on Genetically Modified Organisms (GMO). Scientific opinion addressing the safety assessment of plants developed through cisgenesis and intragenesis. EFSA Journal 10: 2561. https://doi.org/10.2903/j.efsa.2012.2561

Hanson, P.M., Yang, R-Y., Wu, J., Chen, J-T., Ledesma, D., Tsou, S.C.S. and Lee, T-C., 2004. Variation in antioxidant activity and antioxidants in tomato. Journal of the American Society for Horticultural Science 129: 704-711. https://doi.org/10.21273/JASHS.129.5.0704

Hartmann, A., Czauderna, T., Hoffmann, R. Stein, N and Schreiber, F. 2011. HTPheno: An image analysis pipeline for high-throughput plant phenotyping. BMC Bioinformatics 12: 148. https://doi.org/10.1186/1471-2105-12-148

Louwaars, N., 2019. Food safety and plant breeding; why are there no problems in practice? Chapter 5. In: Urazbaeva, A., Szajkowska, A., Wernaart, B., Tilkin Franssens, N. and Spirovska Vaskoska, R. (eds) The functional field of food law. European Institute of Food Law Series Vo 11. Wageningen Academic Publishers, Wageningen, the Netherlands, pp 89-101.

Louwaars, N.P. and Jochemsen, H., 2021. An ethical and societal analysis for biotechnological methods in plant breeding. Agronomy 11: 1183. https://doi. org/10.3390/agronomy11061183

Lin, Z., Hayes, B.J. and Daetwyler, H.D., 2014 Genomic selection in crops, trees and forages: a review. Crop and Pasture Science 65: 1177-1191. https://doi.org/10.1071/ CP13363

Mampuys, R., 2019. No rose without thorns; Implications of a product-based regulatory system for GM crops in the European Union. Netherlands Commission on Genetic Modification, COGEM Policy Report CGM/191010-01, 72 p. https://cogem.net/en/ publication/no-rose-without-thorns-2/.

McClintock, B., 1950. The origin and behavior of mutable loci in maize. Proceedings of the National Academy of Sciences of the USA 36: 344-355. https://doi.org/10.1073/ pnas.36.6.344

Sinden, S.L., Sanford, L.L., Webb, R.E.,1984. Genetic and environmental control of potato glycoalkaloids. American Potato Journal 61, pages 141-156. https://doi. org/10.1007/BF02854035

Stefansson, B.R. and Haugen, F.W., 1964. Selection of rape plants (*Brassica napus*) with seed oil practically free from erucic acid. Canadian Journal of Plant Science 44: 359-364. https://doi.org/10.4141/cjps64-069

Talebi, A., Talebi, A. and Shahrokhifar, B., 2012. Ethyl methane sulphonate (EMS) induced mutagenesis in malaysian rice (cv. mr219) for lethal dose determination. American Journal of Plant Sciences 3: 1661-1665. https://doi.org/10.4236/ ajps.2012.312202

Verkerk, R., Schreiner, M., Krumbein, A., Ciska, E., Holst, B., Rowland, I., De Schrijver, R., Hansen, M., Gerhäuser, C., Mithen, R. and Dekker M., 2009. Glucosinolates in *Brassica* vegetables: the influence of the food supply chain on intake, bioavailability and human health. Molecular Nutrition and Food Research 53: S219-S265. https:// doi.org/10.1002/mnfr.200800065

Weinhold, B., 2006. Epigenetics: the science of change. Environmental Health Perspectives 114: A160-A167. https://doi.org/10.1289/ehp.114-a160

Wernaart, B.F.W., 2022. Food ethics. In: Wernaart, B.F.W. and Van der Meulen, B.M.J. (eds) Applied Food Science. Wageningen Academic Publishers, Wageningen, the Netherlands, pp. 45-64.

Zhang, D.Q., Zhang, Z.Y.,Unver, T. and Zhang, B.H., 2020. CRISPR/Cas: a powerful tool for gene function study and crop improvement. Journal of Advanced Research 29: 207-221. https://doi.org/10.1016/j.jare.2020.10.003

Zhang, X., Ferreira, I.R.S. and Schnorrer, F., 2014. A simple TALEN-based protocol for efficient genome-editing in *Drosophila*. Methods 69: 32-37. https://doi.org/10.1016/j.ymeth.2014.03.020

Further reading

While Wikipedia is indeed an excellent source of information on all the technical subjects mentioned, an introduction to the information via the inventors of concepts and technologies is advised.

https://thebiologynotes.com/microbial-genetics

There are excellent YouTube films on genetics from National Geographic and others.

8. Systems biology

This is what you need to know about adaptive strategies of food microorganisms from a systems biology perspective

Bas Teusink

Systems Biology Lab/AIMMS, Vrije Universiteit Amsterdam, O2 building, Location code 2E51, De Boelelaan 1085, 1081 HV Amsterdam, the Netherlands; b.teusink@vu.nl

Abstract

Microbes, in particular the ones that thrive in foods, are just like humans: they feast and waste, collaborate, cheat, seduce, and go to war. These behaviours are concerned with their growth and metabolism, and affect the properties of fermented foods, their structure, taste, acidity and shelf life. Obviously, the microbes are unaware of this; they are simply trying to win in the battle of the fittest. In this chapter I will discuss such behaviours for lactic acid bacteria and Baker's yeast, two important classes of food microbes. I will explain under what conditions they can be considered adaptive strategies, i.e. how these strategies contribute to fitness. For this I use ideas and concepts from systems biology and game theory. Systems biology combines systems theory with molecular biology to understand how phenotypes (behaviour) arise from the interplay of the molecular components of cells. It can be used to compute the costs and benefits of different metabolic strategies for the growth rate of an organism, an important determinant of fitness. Fitness, however, is always relative to competitors, especially your own mutated progeny, and to evaluate winning strategies requires game theory. It turns out that the fundamental questions – why yeast makes ethanol and lactic acid bacteria make lactic acid, and why both make pleasant flavours – are deceivingly complex and require interdisciplinary systems biology research that combines physiology, evolutionary biology, molecular biology and ecology. I hope to give the reader a hint of what this research entails, and what the answers might be.

Bart Wernaart and Bernd van der Meulen (eds)
Applied food science
DOI: 10.3920/978-90-8686-933-6_8, © Bas Teusink 2022

Key concepts

▶ Adaptive strategies are specific phenotypes that may or may not confer a fitness benefit to the organism using them.
▶ Fermentation is the conversion of sugars into products, such as ethanol, lactate, acetate or formate, typical 'fermentation products'.
▶ Respiration (in the context of this chapter) is the process of complete combustion of sugars with oxygen to produce carbon dioxide and water.
▶ Top-down systems biology is an inductive, data-driven approach that seeks patterns in data.
▶ Bottom-up systems biology is a deductive approach that makes predictions of emergent behaviour.
▶ Biomarkers are specific compounds that expose a certain state.
▶ (Evolutionary) Game Theory explains winning strategies based on the reciprocal behaviour of competitors.

8.1 Introduction

About 5-40% of all foods that humans consume are fermented by microorganisms. Lactic acid bacteria are the most common microbes encountered in fermented foods, but other bacteria, such as bacilli, yeast and moulds are also used to ferment foods (Tamang and Kailasapathy, 2010). Since we also find often the same (combination of) species in spontaneous fermentations, such as yeasts and lactic acid bacteria (sour dough, kimchi, kefir, and many more), these microorganisms must have adapted really well to nutrient-rich environments such as milk and plant material (vegetables) – and possibly to each other as well. The adaptive strategies followed by food microbes, however, are not always intuitive and still food for research in industry and academia.

adaptive strategies But what do we mean by adaptive strategies? Strategies in microbiology are phenotypes, or behaviours. The use of the word 'strategy' rather than phenotype implies an objective, which is indeed what I wish to imply. It means that there are different options for cells, and they have been selected to use the one that confers the highest fitness benefit under a certain condition – this is what adaptive implies. In this chapter I will discuss a number of such strategies from the perspective of systems biology. Systems biology is a relatively new field that tries to understand how molecular interactions between cell components give rise to cell behaviour. Just like lower-level interactions between ants or neurons can give rise to much more complex or even intelligent *emergent* behaviour, molecules such as DNA, RNA, proteins and metabolites interact and somehow create cellular phenotypes. It is truly amazing if you think about it: how can a cell that contains thousands of different molecules manage all the processes to create an exact copy of all these components, within roughly 1-24 hours, or often even less in foods!

top-down systems biology

Systems biology comes in two different flavours: *top-down* systems biology is data-driven and uses data science (biostatistics, machine learning, AI) to integrate the massive amounts of data generated by different high-throughput functional genomics technologies that measure the molecular composition of cells, such as DNA and RNA (through sequencing), proteins and enzymes (via proteomics) or the small molecules (called metabolomics). Integrating such data to generate new understanding is (still) a challenge. Top-down systems biology is inductive: 'here is the evidence, what is the hypothesis?' (Kell and Oliver 2004). Typical results of such analysis are networks of (potential) interactions between proteins, metabolites or genes, or so-called biomarkers for health or disease. Biomarkers are specific compounds that expose a certain state, think of glucose in urine as a biomarker for diabetes, or complex combinations thereof in the form of a pattern, such as the MammaPrint that predicts breast tumour type and disease and treatment outcome (Sun *et al.*, 2021). But similarly biomarkers can be used to monitor food fermentations (Wang *et al.*, 2021). With more and more data and the advent of powerful AI approaches, this field will become more and more important. However, the question is whether the underlying networks or neural nets will be understandable or explainable. They are often largely a black box.

bottom-up systems biology

Bottom-up systems biology opens up the lid of the box and uses existing knowledge of components and their interactions to study the emergence of systems behaviour, and is therefore in essence deductive. It uses mechanistic models and computation, scenario testing, and systems theory. Many of these computational techniques are also used in engineering, physics but also ecology, as emergent properties are also studied in these systems.

Good systems biology practice is to combine both approaches in an iterative fashion, where data lead to hypotheses about mechanisms, and models lead to discriminative experiments that then generate (preferably quantitative) data, and so forth. Given increased technology push and high-tech specialisation, however, we see too often that the powers of the two approaches are not combined to their full potential; it is a challenge to maintain a large group of multidisciplinary scientists for long enough to make this happen.

food microbiology

A systems biology approach in food microbiology combines traditional quantitative microbial physiology – the study of growth rate, yield, conversion efficiencies, thermodynamics, kinetics, metabolism – with computational modelling, functional genomics and data science. In the examples I will give, it also includes evolutionary thinking, to understand adaptive strategies of food microbes from a systems biology perspective. *Saccharomyces cerevisiae*, or Baker's yeast (Case 8.1), and lactic acid bacteria (Case 8.2) will be the heroes of my story. They will have a lot in common, it turns out, and so they will regularly meet in the examples.

Case 8.1. The case of Baker's yeast: respiration versus fermentation.

S. cerevisiae is obviously used for dough leavening and alcoholic beverages but is also an important workhorse for the industrial production of biobased chemicals. In fundamental science, it is the model organism for cell biology of complex (eukaryotic) cells: most of what we know about basic molecular mechanisms in human cells was first figured out in yeast. Yeast is really good at turning glucose (or other sugars) into ethanol and carbon dioxide. We need a little bit of basic biochemistry and metabolism here to appreciate this feature, which is surprisingly less trivial than it may seem.

All Bachelor students in the life sciences have had to study glycolysis (Greek for breaking down (lysis) glucose). This metabolic pathway breaks down glucose in a series of ten chemical conversions all the way to pyruvate. Intermediates of the pathway branch out to essential components of the cell, such as amino acids for proteins, riboses for DNA and RNA, or reducing equivalents (electrons) to fight oxidative stress or make lipids. The largest chemical flux through glycolysis (the rate of the pathway), however, is to generate adenosine triosephosphate (better known as ATP), the energy currency of the cell. Whenever a cell needs to do chemical work that costs energy, it uses ATP to drive the reaction. ATP is made from the energy generated through the breakdown of nutrients – *in casu* glucose. Converting glucose to pyruvate provides enough energy to generate only a rather miserable 2 ATP per glucose. If yeast ferments, which means it turns the pyruvate into ethanol (and CO_2), no extra ATP can be formed and therefore only 2 ATP per glucose is produced. Since we use ethanol as a fuel for making fondue or to drive Brazilian cars, it is obvious that the yeast cells that produce it do not harvest all energy available to them in the form of glucose. Indeed, when there is oxygen around, pyruvate can be combusted completely to CO_2, which yields a lot more energy, enough for another 10-15 ATP[44]. This process is called respiration.

Now, if you are sitting in grape must under anaerobic conditions, there is no oxygen and so there is no choice but to produce ethanol and to do it fast to supply the ATP needed for the cells. This is why we selected this species. However, the puzzling

---

[44] Yeast is not very good at respiration and is a bit sloppy when it comes to energetic efficiency: we humans can generate 36 ATP per glucose from the combustion of glucose to CO_2.

thing is that *S. cerevisiae* also produces ethanol when there is plenty of oxygen – if there is an excess of glucose. This phenomenon is called the Crabtree effect (Dijken *et al.*, 1993). In cancer biology, a strikingly similar behaviour of tumour cells that ferment lactate under aerobic conditions is called the Warburg effect (Diaz-Ruiz *et al.*, 2011). If there is excess glucose, and all other essential nutrients, yeast cells can grow fast, and so they need lots of energy, ATP. Why not make use of the much more efficient respiratory strategy? Other yeasts, so-called Crabtree-negative yeasts, can grow equally fast or even faster by respiration than fermentation.

Case 8.2. *Lactococcus cremoris*, **chemical warfare or jealousy.**

One hypothesis is that ethanol kills other microbes, and as we will also see later, chemical warfare is often used when nutrients are plentiful. The problem with this hypothesis is, however, that it seems vulnerable to cheaters: if growth is really faster with respiration than with fermentation, then a mutant that could not help fermenting and killing others, but would rather respire, would grow faster and eventually take over the population (Goel *et al.*, 2012). When *Saccharomyces cerevisiae*, Baker's yeast, is cultivated in the absence of competitors, it also does not stop making ethanol, not even after prolonged cultivations. This suggests that killing others is not the reason for making ethanol.

Another hypothesis is also based on ecological reasoning: if there is excess glucose and you are limited by something else, you may wish to consume as much of the glucose as possible to prevent others from having it. So, if the objective is to consume the glucose as fast as possible, there may be too much ATP if it were fully respired. Indeed, glycolysis of yeast is regulated by ATP demand and will shut down under ATP excess (Van Heerden *et al.*, 2015), according to the same principle (negative feedback) as the heater shutting down if the temperature in the room is high enough. The reason I doubt this as a universal explanation, is that I also study lactic acid bacteria, and this explanation seems highly unlikely.

So, allow me to briefly explain the second case, the situation for *Lactococcus cremoris* – a cheese starter lactic acid bacterium formerly known as *Lactococcus lactis* – before returning to yeast again. Like yeast and cancer cells, *L cremoris* has two options for pyruvate. In this case, however, since *L. cremoris* grows in the absence of oxygen, the options are different fermentation products that give different ATP yields. The most efficient fermentation route is called mixed acid fermentation (acetate, formate and ethanol). The energy that is available from the conversion of glucose to these products is enough for 3 ATP per glucose. The alternative pathway is to run the glucose into lactate – like our muscles do. This conversion results in only 2 ATP per glucose. Like yeast, *L. cremoris* also uses the less efficient strategy (making lactate in this case) under glucose excess, rather than the mixed acids that would yield more ATP per glucose. However, for *L. cremoris* we know that they do not make lactate under glucose excess to avoid ATP excess: it appears that glycolysis is simply working at its maximal rate. When an ATP-consuming reaction is introduced (an ATP drain), glycolysis in *L. cremoris* does not go faster, but rather growth rate is reduced. Thus, *L. cremoris* does not seem to shift to the less efficient lactate fermentation to dump excess ATP.

8.2 An economic perspective

So, although we cannot completely exclude ATP excess as an explanation for yeast, there is another theory that most researchers in the field are willing to accept. For this we need to view the cell as an economy, or perhaps better, as a special cell factory that is making copies of itself. So far, we have only discussed the benefits of respiration, and its higher yield of ATP. But to harvest energy requires machinery, enzymes, that catalyse the chemical reactions. The question of which strategy is best, respiration or fermentation, then becomes: what are the costs and benefits of these strategies, and which one gives me the highest return on investment?

For yeast to be able to respire means it has to invest in lots of machinery, in particular mitochondria. In general, there appears to be a trade-off between metabolic benefit (yield of ATP) and the costs of implementing the enzymes to reap that benefit. It turns out that enzymes (proteins) are very costly: they are made by complex protein synthesis machines called ribosomes, and each peptide bond between amino acids in a protein costs about 5 ATP, and an average protein contains 300 of them. So, what is better: efficient but expensive mitochondria, or lousy but very cheap fermentation machinery (only two enzymes for yeast)?

computer models This is a quantitative question, and here systems biology comes in. We can make computer models to try to compute the costs and benefits. Over the last 20 years systems biologists have collected all the knowledge on metabolic reactions, and the enzymes that catalyse them, as well as the genes that encode these enzymes, and turned this information into computable knowledge bases that we call genome-scale metabolic models (O'Brien *et al.*, 2015). They consist of 500 reactions for a simple bacterium, to over 7,000 for humans (Thiele *et al.*, 2013). These models are being used to predict its metabolic potential, based on the genome of a microbe (or cell type). With such models, we can ask how many copies a cell can make if it were to respire or ferment. They compute the benefits, therefore, in terms of biomass (cell) yield. These models have been hugely successful in microbiology, and are also frequently used in food sciences, as we have reviewed (Somerville *et al.*, 2022).

To compute the costs of strategies, however, we need to augment these models with the costs associated with the synthesis of the enzymes. Such models exist now for a handful of species, pioneered in *Escherichia coli* (O'Brien *et al.*, 2013), but now also available for L. cremoris (Chen *et al.*, 2021) and *Saccharomyces cerevisiae* (Elsemman *et al.*, 2022). Thus, for a reaction in the model to carry flux requires the synthesis of the corresponding enzyme, and this requires ribosomes and ATP and amino acids. With these models we can compute the costs and benefits. It turns out that, in these models, the amount of ATP per glucose may be lower for fermentation than for respiration, but the amount of ATP per protein is higher.

This is key for understanding fermentation! We can now ask the model what strategy is best for fitness. For this we use growth rate as the best option, as the average growth rate over conditions is the definition for fitness for microbes (Bruggeman *et al.*, 2020). The model (and theory (De Groot *et al.*, 2019, 2020)) has taught us: it depends on the constraints! What we basically solve is an optimisation problem: how can you grow as fast as possible? The problem only has an optimum if there are constraints that are hit, otherwise the growth rate would be infinite. It turns out that, at low glucose levels, growth is only limited by the rate at which cells can take up glucose, and the only thing that matters is the efficiency of glucose conversion. Growth is slow, there are not many ribosomes and enzymes needed and there is plenty of space in the cell for mitochondria. So, the costs in terms of protein usage is not relevant, and the model picks respiration (Elsemman *et al.*, 2022). At high levels of glucose, however, it is not the metabolic efficiency that matters (there is plenty of glucose), but the associated protein costs determine the optimal strategy. Cells are full of enzymes and ribosomes making cell components, and sacrificing some metabolic efficiency for protein 'space' allows faster growth.

I have posed a deceptively simple question: why does *S. cerevisiae* ferment in the presence of oxygen? The answer is not completely clear, but it seems that we can explain ethanol formation as the pleasant result of a constrained optimisation problem of the cellular economy that evolution has solved. Systems biology provides the mechanistic underpinning and quantification of that economy and can make predictions of optimal strategies. Let us now shift to other aspects of food microbiology that provide equally puzzling observations in need of an explanation. For this we focus on lactic acid bacteria, but yeast will return as well.

8.3 Lactic acid bacteria: lactate instead of ethanol

Lactic acid bacteria (LAB) are so-called Gram-positive bacteria that can be separated into two groups based on their shape and genomes: *lactobacilli* (rod-shaped) and *streptococci* (spheres, often in chains). The taxonomy and corresponding nomenclature were revisited in 2020 and adjusted according to the latest scientific insights (Zheng *et al.*, 2020). Most LAB are Generally Recognized as Safe (GRAS), but there are also some nasty pathogens, such as *Streptococcus pyogenes* and *Streptococcus pneumoniae*. Like yeast, LAB are specialists in exploiting nutrient-rich environments, such as milk, (damaged) plants, meat, fish, but can also be found in the microbiomes associated with many mammals, including humans. Some LAB appear to have positive effects on (intestinal) health and are used as probiotic species in different food products.

conversion

The characteristic behaviour of lactic acid bacteria is the conversion of sugars into lactic acid through homolactic fermentation. Some LAB can, like *L. cremoris* we saw earlier, also perform mixed acid fermentation. Acidification of foods prevents the growth of spoilage organisms, results in a longer shelf life of food products, and is the basis of the use of LAB by humans. Just as for ethanol formation, many researchers believe that the acidification and killing of others is the main reason to make lactate, but the combination of ethanol, formate and acetate appears equally if not more toxic. However, if lactate production were to be used for warfare at the cost of growth rate compared to mixed acids, it would again be overtaken by cheaters.

Our genome-scale model of *L. cremoris*, with protein synthesis included, predicts that lactate gives higher ATP per protein than mixed acid fermentation (Chen *et al.*, 2021). This would explain homolactate formation under high sugar conditions if protein costs are indeed constraining growth rate under these conditions. But the picture is not so clear if we combine these predictions with experimental observations, because we previously saw that energy seems limiting, and we also have indications that non-functional proteins that only bear a cost can be overexpressed in *L. cremoris* without an impact on growth rate (Douwenga, Bachmann, personal communication). Rather, there seems to be an inherent limitation on the glycolytic flux, and that maximal flux is somehow coupled to lactate formation. It hints to some thermodynamic limit that we have not yet fully identified. And so, the deceivingly simple question (why do lactic acid bacteria produce lactate?) is not at all simple!

8.4 Other compounds and properties: flavours

Apart from lactate or mixed acids, some compounds made by LAB and other microbes are responsible for the typical flavours that we have learnt to appreciate, such as diacetyl (buttery flavour) and acetaldehyde (yoghurt flavour), but also more odorous molecules that characterise certain cheeses, for example isovaleric acid, 3-methylbutanal and many others (Smit *et al.*, 2005). It is important to realise that many different strains of the same species exist that can differ rather substantially in terms of acidification rate, flavour formation, texturing properties, optimal temperature and metabolism of lipids, proteins and vitamins. The combination of nutrients and strains provides an almost infinite number of different tastes and textures that make fermented foods and drinks so special and unique, and difficult to model (Somerville *et al.*, 2022). When it comes to these variations, things get really complicated and top-down data-driven approaches are needed – and used – to link genotypic variation of strains to phenotypic diversity (Fuhren *et al.*, 2020). Here the models that we have discussed before, the genome-scale metabolic models, do contribute to understanding, but as a data analysis tool: through its mapping of genes to proteins to metabolic reaction, it provides a metabolic context for data integration. For example, in a study of *Lactiplantibacillus plantarum* it was found that when grown under vigorously aerated conditions, growth was delayed that could be explained by a loss of carbon dioxide from the growth medium, and some CO_2 is needed for biosynthesis. This explanation was based on integration of gene expression data with a metabolic model (Stevens *et al.*, 2008).

8.5 Amino acid metabolism

Other examples of how these models may guide applications even when it comes to flavours, can be found in understanding not sugar metabolism, but amino acid metabolism. Lactic acid bacteria have lost the ability to synthesise many compounds that they need to be able to grow, in particular vitamins and amino acids (Teusink and Molenaar, 2017). Why spend resources on making these amino acids if you can simply eat them? In that sense, LAB require essential amino acids in their diet, just like we humans do. However, they often eat more protein than they need – another resemblance with men – and the excess amino acids are metabolised into all sorts of breakdown molecules that are perceived by us as odorous and tasty.

We have used a model of metabolism and protein synthesis to see which of these amino acids really contribute to growth. It is known ATP can be directly or indirectly generated from the metabolism of some of these, e.g. arginine: its metabolisation yields 1 ATP per arginine. We asked the model if it would like to take up more arginine than that measured experimentally. As with yeast, the answer was: it depends on the constraints! At high glucose levels, *L. cremoris* does not bother about arginine, it prefers glucose and turns it into lactate: the ATP per protein investment is lousy compared to lactate. But at lower glucose levels and thus growth rate, other sources of ATP are more than welcome. This goes for other amino acids as well.

Since the excess uptake of amino acids contributes to flavour, it appears that this LAB and possible others will only produce these smelly compounds when conditions are unfavourable. This gives another, possibly complementary to energetic, reason why they evolved to produce flavours. Perhaps these properties have been evolved to please us: we are largely responsible for their selection and propagation (Pollan, 2001) so this is quite possible. Alternatively, some of these metabolic conversions are like pheromones in the animal world and attract insects that may lead to dispersion to a better place. Typical fruity flavours also produced by many yeasts appear particularly suitable (Dzialo *et al.*, 2017). Additionally, some of the products of LAB may kill other bacteria – not us! – and others have been shown to resemble plant hormones and affect plant growth (Goffin *et al.*, 2010), possibly leading to leakage of nutrients from plants to LAB.

8.6 The dilemma of macromolecules

But where do the amino acids come from in the media of many foods? From the breakdown of macromolecules. This results in other fascinating adaptation strategies and moral dilemmas. In milk, proteins are abundant and there are only limited free amino acids available. The protein molecules are big (macromolecules), too big to take up and digest inside the cell. Therefore, these molecules first need to be cut into smaller fragments – peptides or even amino acids. This happens outside the cell, and this creates the dilemma: it costs quite a lot of energy to make the protease enzymes that do the cutting, and the resulting peptides are on the outside, ready for other cells to take up as well! So, a neighbour does not need to make the protease if you do so, and such a cheater can fully invest in growth and outcompete the 'poor' protease producers (Bachmann *et al.*, 2011, 2016). Therefore, protease expression is what we call an unstable evolutionary trait and if you are not careful, microbes can lose it and the fermentation will become much poorer. It is an intriguing and counterintuitive example where evolution can result in a less fit phenotype!

8.7 Evolutionary games

game theory

This type of behaviour can be understood from a theory called (evolutionary) game theory. Game theory was initially developed in economics to understand the outcome of different, conflicting strategies, and was later adapted for biology (Maynard-Smith and Price, 1973). In such 'games' there are often two strategies, one of cooperators and one of cheaters. The most famous example is the Prisoner's dilemma: two suspects have to come to court for a crime of which they are accused. They have two options: stay silent or accuse the other one. They know the odds: if they both stay silent, they face imprisonment for one year; if they both accuse each other, they both go to jail for five years. However, if one accuses the other, and the other stays silent, the accuser goes free and the accused one faces 20 years! What should they do? The answer is not too difficult: they should always accuse the other: no matter what the other one does, accusing will always be the best answer. So, they both go to jail for five years, even though they could have done only one year.

Translated to fitness, punishment is a decrease in fitness, or pay-off an increase in fitness.[45] Protease expressing LABs go extinct, and everybody loses. Because of this, it has been quite a challenge to explain how evolution can evolve cooperation, which it does, and there are still protease-expressing LAB around. It all depends on the investment costs and the pay-offs. And here the models that we have described come into the picture again – although admittedly they have not yet been used in a game theoretical setting, but they should have been! In the case of the protease, if the number of cells in a culture is low, then all cells are far apart from each other, and the benefit of cheating is low: the peptides have to diffuse a longer distance and the concentrations in the milk will be lower. Then the rules of the game change; in fact, the game is no longer a Prisoner's dilemma, and keeping the protease is stable (Bachmann *et al.*, 2011).

For yeast a similar study was done for the breakdown of sucrose (Gore *et al.*, 2009). This is a bit puzzling as sucrose is 'just' a disaccharide consisting of a fructose and a glucose molecule, and so it is rather small and most microbes can take it up, privatise it and break it down internally. *S. cerevisiae*, however, expresses an enzyme called invertase (or sucrase) and sticks it in its cell wall, and so breakdown occurs outside the cell. It was estimated that only 1% of the released sugars are taken up by the cell itself; the rest is common good (Gore *et al.*, 2009). It appears enough to maintain this strategy. Interestingly, the regulation of the synthesis of the enzyme shows signs of checks and balances:

---

[45] Richard Dawkins described evolutionary game theory very eloquently in his famous book, The Selfish Gene (Dawkins, 1976). However, Dawkins does not use examples from microbiology, where many such games are being played.

only if the benefits are high enough, and the costs not, will yeast make the invertase enzyme. This is done by sensing both the sucrose and the resulting glucose levels through intricate molecular networks that can be studied with systems biology tools.

antibiotic-like compounds

One other interesting example of a game that is played by food microbes has to do with another type of chemical warfare: the production of antibiotic-like compounds called bacteriocins. These are meant to kill other bacteria. Indeed, one of these bacteriocins, nisin, is widely used as an ingredient to prevent spoilage of our foods (Hansen, 1994). However, a producer of a bacteriocin has to be resistant to the compound it produces, otherwise it dies. So, if it encounters a resistant bacterium that does not bear the cost of producing it, the bacteriocin producer will lose. Obviously, it will win against a sensitive bacterium, which will be killed. A resistant bacterium, however, withstands the cost of resistance, and if it meets a sensitive strain, the sensitive strain wins! This is, therefore, an example of a rock-scissors-paper game played by bacteria (for an experimental example of this game in action, see Kerr *et al.*, 2002).

8.8 Conclusions

I explained how food microbes are important organisms for human health and nutrition, not only because they make many foods tasty and healthy, but also because their evolutionary history and adaptive strategies to be fit in nutrient-rich environments teach us fundamental lessons about evolution and physiology of these fascinating microbes. I gave examples of how systems biology can provide quantitative underpinning of the costs and benefits of the different strategies. It is clear that survival and prosperity in rich environments requires cheating, seduction and chemical warfare – strategies that we recognise all too well!

References

Bachmann, H., Bruggeman, F.J., Molenaar, D., Branco dos Santos, F. and Teusink, B., 2016. Public goods and metabolic strategies. Current Opinion in Microbiology 31: 109-115. https://doi.org/https://doi.org/10.1016/j.mib.2016.03.007

Bachmann, H., Molenaar, D., Kleerebezem, M. and van Hylckama Vlieg, J.E.T., 2011. High local substrate availability stabilizes a cooperative trait. The ISME Journal 5: 929-932. https://doi.org/10.1038/ismej.2010.179

Bruggeman, F.J., Planque, R., Molenaar, D. and Teusink, B., 2020. Searching for principles of microbial physiology. FEMS Microbiology Reviews 44: 821-844. https://doi.org/10.1093/femsre/fuaa034

Chen, Y., van Pelt-KleinJan, E., van Olst, B., Douwenga, S., Boeren, S., Bachmann, H., Molenaar, D., Nielsen, J. and Teusink, B., 2021. Proteome constraints reveal targets for improving microbial fitness in nutrient-rich environments. Molecular Systems Biology 17: e10093. https://doi.org/10.15252/msb.202010093

Dawkins, R., 1976. The selfish gene. Oxford University Press, Oxford, UK.

De Groot, D.H., Hulshof, J., Teusink, B., Bruggeman, F.J. and Planque, R., 2020. Elementary Growth Modes provide a molecular description of cellular self-fabrication. PLoS Computational Biology 16: e1007559. https://doi.org/10.1371/journal.pcbi.1007559

De Groot, D.H., Van Boxtel, C., Planque, R., Bruggeman, F.J. and Teusink, B., 2019. The number of active metabolic pathways is bounded by the number of cellular constraints at maximal metabolic rates. PLoS Computational Biology 15: e1006858. https://doi.org/10.1371/journal.pcbi.1006858

Diaz-Ruiz, R., Rigoulet, M. and Devin, A., 2011. The Warburg and Crabtree effects: On the origin of cancer cell energy metabolism and of yeast glucose repression. Biochimica et Biophysica Acta 1807: 568-576. https://doi.org/10.1016/j.bbabio.2010.08.010

Dzialo, M.C., Park, R., Steensels, J., Lievens, B. and Verstrepen, K.J., 2017. Physiology, ecology and industrial applications of aroma formation in yeast. FEMS Microbiology Reviews 41: S95-S128. https://doi.org/10.1093/femsre/fux031

Elsemman, I.E., Rodriguez Prado, A., Grigaitis, P., Garcia Albornoz, M., Harman, V., Holman, S.W., van Heerden, J., Bruggeman, F.J., Bisschops, M.M.M., Sonnenschein, N., Hubbard, S., Beynon, R., Daran-Lapujade, P., Nielsen, J. and Teusink, B., 2022. Whole-cell modeling in yeast predicts compartment-specific proteome constraints that drive metabolic strategies. Nature Communications 13: 801. https://doi.org/10.1038/s41467-022-28467-6

Fuhren, J., Rosch, C., Ten Napel, M., Schols, H.A. and Kleerebezem, M., 2020. Synbiotic matchmaking in *Lactobacillus plantarum*: substrate screening and gene-trait matching to characterize strain-specific carbohydrate utilization. Applied and Environmental Microbiology 86: e01081-20. https://doi.org/10.1128/AEM.01081-20

Goel, A., Wortel, M.T., Molenaar, D. and Teusink, B., 2012. Metabolic shifts: a fitness perspective for microbial cell factories. Biotechnology Letters 34: 2147-2160. https://doi.org/10.1007/s10529-012-1038-9

Goffin, P., van de Bunt, B., Giovane, M., Leveau, J.H., Hoppener-Ogawa, S., Teusink, B. and Hugenholtz, J., 2010. Understanding the physiology of *Lactobacillus plantarum* at zero growth. Molecular Systems Biology 6: 413. https://doi.org/10.1038/msb.2010.67

Gore, J., Youk, H. and van Oudenaarden, A., 2009. Snowdrift game dynamics and facultative cheating in yeast. Nature 459: 253-256. https://doi.org/10.1038/nature07921

Hansen, J.N., 1994. Nisin as a model food preservative. Critical Reviews in Food Science and Nutrition 34: 69-93. https://doi.org/10.1080/10408399409527650

Kell, D.B. and Oliver, S.G., 2004. Here is the evidence, now what is the hypothesis? The complementary roles of inductive and hypothesis-driven science in the post-genomic era. Bioessays 26: 99-105. https://doi.org/10.1002/bies.10385

Kerr, B., Riley, M.A., Feldman, M.W. and Bohannan, B.J., 2002. Local dispersal promotes biodiversity in a real-life game of rock-paper-scissors. Nature 418: 171-174. https://doi.org/10.1038/nature00823

O'Brien, E.J., Lerman, J.A., Chang, R.L., Hyduke, D.R. and Palsson, B.O., 2013. Genome-scale models of metabolism and gene expression extend and refine growth phenotype prediction. Molecular Systems Biology 9: 693. https://doi.org/10.1038/msb.2013.52

O'Brien, E.J., Monk, J.M. and Palsson, B.O., 2015. Using genome-scale models to predict biological capabilities. Cell 161: 971-987. https://doi.org/10.1016/j.cell.2015.05.019

Pollan, M., 2001. The botany of desire. Random House, New York, NY, USA.

Smit, G., Smit, B.A. and Engels, W.J., 2005. Flavour formation by lactic acid bacteria and biochemical flavour profiling of cheese products. FEMS Microbiology Reviews 29: 591-610. https://doi.org/10.1016/j.femsre.2005.04.002

Maynard-Smith, J. and Price, G.R., 1973. The logic of animal conflict. Nature 246: 15-18. https://doi.org/10.1038/246015a0

Somerville, V., Grigaitis, P., Battjes, J., Moro, F. and Teusink, B., 2022. Use and limitations of genome-scale metabolic models in food microbiology. Current Opinion in Food Science 43: 225-231. https://doi.org/https://doi.org/10.1016/j.cofs.2021.12.010

Stevens, M.J., Wiersma, A., de Vos, W.M., Kuipers, O.P., Smid, E.J., Molenaar, D. and Kleerebezem, M., 2008. Improvement of *Lactobacillus plantarum* aerobic growth as directed by comprehensive transcriptome analysis. Applied and Environmental Microbiology 74: 4776-4778. https://doi.org/10.1128/AEM.00136-08

Sun, L., Wu, A., Bean, G.R., Hagemann, I.S. and Lin, C.Y., 2021. Molecular testing in breast cancer: current status and future directions. Journal of Molecular Diagnostics 23: 1422-1432. https://doi.org/10.1016/j.jmoldx.2021.07.026

Tamang, J.P. and Kailasapathy, K. (eds), 2010. Fermented foods and beverages of the world. CRC Press, Boca Raton, FL, USA.

Teusink, B. and Molenaar, D., 2017. Systems biology of lactic acid bacteria: for food and thought. Current Opinion in Systems Biology 6: 7-13.

Thiele, I., Swainston, N., Fleming, R.M., Hoppe, A., Sahoo, S., Aurich, M.K., Haraldsdottir, H., Mo, M.L., Rolfsson, O., Stobbe, M.D., Thorleifsson, S.G., Agren, R., Bolling, C., Bordel, S., Chavali, A.K., Dobson, P., Dunn, W.B., Endler, L., Hala, D., Hucka, M., Hull, D., Jameson, D., Jamshidi, N., Jonsson, J.J., Juty, N., Keating, S., Nookaew, I., Le Novere, N., Malys, N., Mazein, A., Papin, J.A., Price, N.D., Selkov, E., Sr., Sigurdsson, M.I., Simeonidis, E., Sonnenschein, N., Smallbone, K., Sorokin, A., Van Beek, J.H., Weichart, D., Goryanin, I., Nielsen, J., Westerhoff, H.V., Kell, D.B., Mendes, P. and Palsson, B.O., 2013. A community-driven global reconstruction of human metabolism. Nature Biotechnology 31: 419-425. https://doi.org/10.1038/nbt.2488

Van Dijken, J.P., Weusthuis, R.A. and Pronk, J.T., 1993. Kinetics of growth and sugar consumption in yeasts. Antonie Van Leeuwenhoek 63: 343-352. https://doi.org/10.1007/BF00871229

Van Heerden, J.H., Bruggeman, F.J. and Teusink, B., 2015. Multi-tasking of biosynthetic and energetic functions of glycolysis explained by supply and demand logic. Bioessays 37: 34-45. https://doi.org/10.1002/bies.201400108

Wang, Y., Wu, Y., Shen, Y., Li, C., Zhao, Y., Qi, B., Li, L. and Chen, Y., 2021. Metabolic footprint analysis of volatile organic compounds by gas chromatography-ion mobility spectrometry to discriminate mandarin fish (*Siniperca chuatsi*) at different fermentation stages. Frontiers in Bioengineering and Biotechnology 9: 805364. https://doi.org/10.3389/fbioe.2021.805364

Zheng, J., Wittouck, S., Salvetti, E., Franz, C., Harris, H.M.B., Mattarelli, P., O'Toole, P.W., Pot, B., Vandamme, P., Walter, J., Watanabe, K., Wuyts, S., Felis, G.E., Ganzle, M.G. and Lebeer, S., 2020. A taxonomic note on the genus *Lactobacillus*: description of 23 novel genera, emended description of the genus *Lactobacillus* Beijerinck 1901, and union of *Lactobacillaceae* and *Leuconostocaceae*. International Journal of Systematic and Evolutionary Microbiology 70: 2782-2858. https://doi.org/10.1099/ijsem.0.004107

9. Food processing

This is what you need to know about the beginnings, advantages, threats, and opportunities of food processing

Ole G. Mouritsen, Karsten Olsen and Vibeke Orlien[*]

Department of Food Science, Faculty of Science, University of Copenhagen, Rolighedsvej 26, 1958 Frederiksberg C, Denmark; vor@food.ku.dk

Abstract

Food processing is about transforming raw food materials into food for consumption. Processing is carried out for multiple purposes, such as improving for palatability in terms of flavour and texture, improving nutritional value and bioavailability, eliminating possible toxic factors, as well as providing for conservation and storage ability. Food processing takes place both in home kitchens, restaurants, and industrial settings. It is necessary for providing food enough for the world's increasing population. However, food processing is also one of the important stressors on the global health of our planet and therefore needs to be carefully conducted to lead to a more sustainable global food system. In this chapter we illustrate the use of various fundamental food processing methods needed to produce the components of a meal according to a hypothetical menu. The menu is designed to place the focus on techniques which develop those flavours and textures that will facilitate the advancement of a green cuisine.

Key concepts

- ▶ Food processing is essential for nutrition, palatability, and flavour (taste, texture) of foodstuff.
- ▶ Food processing involves a wide range of different technologies, such as mechanical, thermodynamic (heat and pressure), and microbiological.
- ▶ Flavour is essential for food preference and eating behaviour.
- ▶ Food processing is required for providing nutritious, healthy, and sustainable food for a growing global population.

Bart Wernaart and Bernd van der Meulen (eds)
Applied food science
DOI: 10.3920/978-90-8686-933-6_9, © Ole G. Mouritsen *et al.* 2022

- ▸ An understanding of the relationship between properties of raw food ingredients and the impact of food processing is needed to promote a green transition in the global food systems.
- ▸ Food processing in combination with culinary sciences provides the foundation for food design and food innovation.

9.1 Introduction

Whereas some other primates do use tools to capture and process living or dead biological material for food (Dunn and Sanchez, 2021), *Homo sapiens,* who is thought to date back about 200,000 years (Wrangham, 2009), is unique in eating a diet that is rich in cooked and non-thermally processed food (Organ *et al.,* 2011). This is a universal trait for humans across all cultures and continents. Man is a cook, cooking has made us human, and food processing is one of the most important human technologies.

evolutionary biology Evolutionary biologists consider the period 1.8-1.9 million years ago as a time of transition from ape species (*Australopithecines*) to the hominins (*Homo erectus*) with a smaller brain than modern humans. It has been argued that an essential driving force for this evolution was that, during this transition period, the ancestors (the habilines) of *Homo erectus* became hunter-gatherers and changed from herbivores into omnivores. It has been surmised that the hominins learned to use fire and heat to prepare food 1.9 million years ago, which provided for an energy- and protein-rich diet (Comody *et al.,* 2011; Wrangham, 2009; Wrangham and Conklin-Brittain, 2003). This in turn fuelled the development of a larger brain, facilitated by access to large quantities of super-unsaturated fatty acids from marine sources that are instrumental for building the neural circuitry of a large and complex brain (Crawford, 2010; Cunnane and Crawford, 2014; Mouritsen, 2016). Our universal craving for sweet and umami is closely linked to the co-evolution of taste and olfactory receptors and food preferences, which have given us those evolutionary advantages that calories and proteins provide (Mouritsen, 2016; Dunn and Sanchez, 2021).

Along with the heat treatment of both plant-based foods and meat followed not only calories and accessible nutrients but also a softer texture (mouthfeel) (Mouritsen and Styrbæk, 2017) due to gelatinisation of carbohydrates, denaturation of proteins, and dissolution of fibres. The softer texture required less mastication, leading to the evolution of the peculiar face, oral cavity, and head of our species (Liebermann, 2011).

food processing
Today almost all our food is processed one way or another, either by producers, in our kitchens, or in industrial settings, and a very wide range of processing techniques has been designed involving the use of thermodynamic (temperature, pressure, water activity), physical (mechanical, emulsifying, gelation, etc.), physico-chemical (acids, salts), chemical (additives, preservatives, flavours), as well as microbial/enzymatic (fermentation, ageing) procedures. The aims of the processing take account of nutrition and health aspects, consumers' preference, as well as marketing and economic factors. Furthermore, processing is often intimately integrated with food safety, hygiene, packaging, storage, etc.

industrial food production
Effective industrial food production and food processing has made it possible for the world's population to increase dramatically over recent centuries. However, the success has not come without severe consequences. The 2019 EAT-Lancet Commission on Healthy Diets from Sustainable Food Systems has analysed the conditions for providing for a healthy, nutritious, and sustainable diet for a global population that is set to reach almost ten billion people in 2050 (Willett *et al.*, 2019). The report concludes that food production is a main cause for the anthropogenic changes in the Earth's ecosystems, including climate changes, and that there is an urgent need for global changes in the food production systems. Agriculture is responsible for using 40% of the land, 30% of greenhouse emissions, and 70% of freshwater use. Foresting for agriculture, and wide-spread monoculture has led to a great loss in biodiversity, damage to whole ecosystems, as well as the emission of excess nutrients and greenhouse gases. Global cycles of carbon, phosphorus, and nitrogen have been disturbed. At the same time, food waste is skyrocketing in the entire food chain from production and processing to consumption.

Paradoxically, the current global production of food provides for enough calories to feed everyone, but at the same time more than 10% of the 7.9 billion people on the planet are starving and 2 billion people suffer from diet-related diseases (Searchinger, 2019; Willett *et al.*, 2019). One reason is that the distribution of food is very uneven across the globe, and second, modern diets are often dominated by a high consumption of energy-dense foods, refined foods, animal foods, oils and fats, as well as too much salt (Sproesser *et al.*, 2019). Recent meta-analyses indicate a correlation between ultra-processed foods and the risk of cardiovascular disease as well as all-cause mortality (Rico-Campà *et al.*, 2019; Srour *et al.*, 2019). The epidemic growth of obesity is likely to be influenced by poor food quality, flavour additives leading our primordial senses astray, and ultra-processed foodstuffs (Schatzker, 2015).

But as much as current food systems and food processing have claimed their toll on human health and the health of our planet, those same systems and new processing technologies will have to come to the rescue if we are to implement the necessary global transition towards a more sustainable situation. The EAT-Lancet Commission proposes a diet with more plant-based food, including 500 g vegetables and fruit every day and little or no red meat (Willett *et al.*, 2019). The specific recommendations involve a diet consisting mainly of vegetables, fruit, whole grain, legumes, nuts and unsaturated fats, only moderate or small amounts of fish, poultry, and dairy, and no or very little red meat, processed meat, added sugars, refined cereals and starchy vegetables. We are talking about **green transition** a green transition.

This transition will require a monumental global effort with respect to both developing a sustainable food system and producing food that, in addition to being healthy and nutritious, is also delicious, i.e. food that people are actually going to eat and enjoy. Science by means of food processing and technology will have to show how to accomplish this transition with a focus on reaping the benefits of the many unexplored opportunities of old and new ways of processing and preparing delicious food.

In the present topical chapter, we will indicate some plausible routes to accomplish the mission outlined above by illustrating the use of various food processing methods needed to produce the components of a hypothetical meal according to a menu. The menu is designed to focus on techniques which develop those flavours and textures that will facilitate the advancement of a green cuisine. Although we attempt to cover a certain range of processing techniques, they are not described in technical detail and our list is far from being complete. We therefore refer the reader to authoritative and comprehensive reviews and specialised articles on food processing (Burke *et al.*, 2021; Clark *et al.*, 2014; Fellows, 2011; McGee, 2004; Varzakas and Tzia, 2016).

9.2 The menu

Whereas the following menu is a construct to illustrate a range of different food processing techniques, it could well be served as dinner at home or in a restaurant in the same or a related version. Furthermore, the processing techniques are fairly general and can be executed both in a small-scale kitchen as well as in an industrial setting with the necessary caveats.

9.2.1 Condiment

Condiments are small side dishes or prepared ingredients that can serve a purpose on their own as an appetiser, amuse bouche, and palate cleanser or as a topping and accompaniment to a main dish. We have chosen here an example from the Japanese cuisine to describe processing of vegetables, so-called *tsukemono*, for enhancement of taste and texture for any of these end uses. Preparation of *tsukemono* can involve drying, marinating/pickling, fermentation, and umamification (Mouritsen and Styrbæk, 2021a) and the processes are usually carried out at room temperature or lower.

osmotic effects

Fundamentally, the preparation can be compared to ordinary pickling and may also serve as a means of preservation. The water activity is manipulated by dehydration, by osmotic effects via salt and pH, and possibly with the intervention of microorganisms and/or enzymes. Taste is modified by salts, sugar, and acids and in particular by umami from marinades based on dashi. Dashi is an extract of appropriate seaweeds (contributing free glutamate) combined with fish products like katsuobushi or dried fungi, such as shiitake, contributing free nucleotides like inosinate and guanylate, thereby eliciting umami synergy (Mouritsen and Styrbæk, 2014). We refer to this last step as

umamification

umamification, providing umami to vegetables which by nature have little free glutamate or nucleotides to provide on their own (Mouritsen and Styrbæk, 2020).

dehydration

Due to the dehydration step(s), *tsukemono* generally turn out extremely crunchy (Mouritsen and Styrbæk, 2021a), in particular when using air-drying. Firming of the texture can be increased by using divalent salts (Ca^{++} or Mg^{++}) in the marinade, leading to enhanced cross-binding of pectin in the vegetables.

Figure 9.1A shows a selection of *tsukemono* consisting of two types of daikon, two types of cucumber, aubergine, and ginger root. The crunchy perception can be investigated quantitatively by texture analysis as shown in Figure 9.1B.

crunch

The auditory response, the crunch, is reflected here in a series of successive events of partial fracture of a *tsukemono*-style radish before it finally ruptures.

An interpretation of the phenomena can be provided by microscopy imaging (Clausen *et al.*, 2021) of the cell network as shown in Figure 9.1C. The images reveal the cellular structure of a radish in the fresh state, after drying

rehydration

(dehydration), and after rehydration (in a marinade). During dehydration the cells shrink, and the stiff and regular network of cell walls becomes crumpled

dehydration

and more flexible. During subsequent rehydration (in a marinade), the cells take up some water again, but the network remains irregular and flexible. This implies that the perceived texture in the mouth is one where the vegetable is pliable and deformable, but when the teeth cut through the vegetable, it

auditory sensation

fractures with a crunchy mouthfeel and a noise, i.e. an auditory sensation.

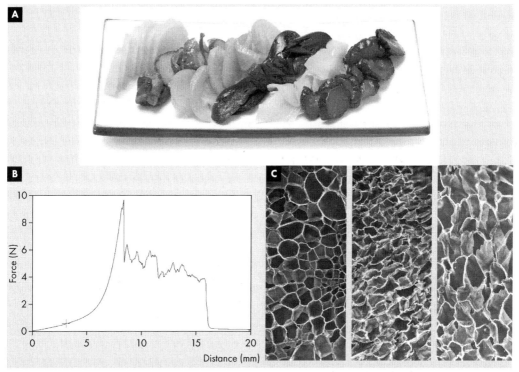

Figure 9.1. Pickled vegetables (*Tsukemono*-style). (A) Selection of vegetable *tsukemono* (photo courtesy of Jonas Drotner Mouritsen). (B) Texture analysis (force vs deformation) of *tsukemono*-style radish (dried and rehydrated/marinated) (courtesy of Dr Mathias Porsmose Clausen). (C) Light microscopy images of radish: fresh (left), dried (middle), rehydrated/marinated (right). Each image corresponds to 0.6×1.6 mm. Plant cell walls were stained with CalcoFluor white (courtesy of Dr Mathias Porsmose Clausen).

9.2.2 Bread

Bread is culturally one of the most staple and versatile food products in the world, served for many purposes as a first meal in the morning, as the main part of lunch (e.g. sandwich) or dinner (as a burger), an appetiser before dinner or accompanying the dinner course (especially used as an absorbent for soup). Bread consists of three vital ingredients: flour, water, and raising agent like baker's yeast or sourdough. The typical bread-making process involves three steps: (1) mixing, (2) fermentation, and (3) baking. Step 1: The critical step where all the ingredients are mixed and incorporated in the dough in order to develop a gluten network. This is normally achieved by kneading and to facilitate the hydration of flour. Step 2: The yeast will ferment sugar components and produce CO_2 (carbon dioxide gas), which is trapped and retained by the gluten matrix, making the dough rise. Step 3: Baking will further enhance the gas expansion resulting in the bread's final volume, taste and appearance.

gluten matrix

Different breads will of course have different/additional ingredients, like salt, sugar, and oil/butter, and follow different bread-making processes. Hence, there is no ideal or standard bread-making process. It depends on the ingredients and the type of bread you seek to achieve. All the steps in the bread making will influence the sensory and storage quality of the final bread.

dough development The crucial part in the mixing step is the increase in consistency of the dough, the dough development. During this stage of mixing, the flour, water and yeast (often also salt and sugar) is converted from a thick viscous mass into a smooth viscoelastic mass. This is important because then the dough will have the ability to be extended without breaking apart. The crucial practical parameter during the mixing process is the mixing time, which is dependent on and equal to the total energy input. In general, it is the time required for the dough to reach the peak consistency, and thereby the maximum resistance. It is very important to mix the dough at the right time to reach the maximum loaf volume; if the dough is under-mixed or over-mixed, the loaf volume will be affected negatively. Freshly mixed dough contains a high number of very small air cells in the dough. It is the purpose of the yeast to increase the number of gas cells in the dough. The dough expands when the continuous CO_2 from the yeast fermentation is released during leavening. The gas cells in the dough increase and consequently the volume of the bread is enlarged. Therefore, under-mixing and too little time leads to a weak gluten structure, the dough is unable to retain gas bubbles, and the bread turns out flat. On the other hand, over-mixing leads to too strong a dough, which is unable to expand properly during fermentation and baking. Another property of the fermentation is the production of flavour compounds. The flavour intensity and the special bread flavour notes continue to change (some will disappear and others are formed) with increasing fermentation time. The purpose of the final baking is to convert the dough into a stable product ready for consumption, thus to form a stable structure and develop flavour components. First, the surface of the dough is heated, and the heat is conducted into the product, whereupon the temperature in the centre of the bread rises slowly. Upon further heating, the yeast becomes deactivated at 55 °C, the starch begins to gelatinise around 60 °C and the formation of the bread structure begins when the gluten proteins are denatured around 70 °C. If you have tried baking bread at home, you will have experienced that the bread volume continues to increase (to a certain point) during baking. Because of the ideal gas law, the already formed gas cells will expand due to the increasing temperature, and the expanding gas bubbles will push the dough aside, thus expand in volume. When the centre of the bread is 98-99 °C, the crumb formation is finished. However, the temperature of the bread surface continues to rise as water evaporates. When the surface reaches temperatures well above 100 °C, the browning of the bread crust begins.

wheat flour

For the techno-functional role, wheat flour is the most commonly used flour for bread due to its gluten content. In fact, gluten is essential for dough formation, which means wheat gluten is unique for its bread-making ability. The protein content is very important for the baking properties and needs to be over 12% in normal wheat flour for good bread making. Gluten does not exist in free form but is a complex of two proteins: gliadin and glutenin. These components are responsible for the ability to form a viscoelastic network in a bread dough. When wheat flour is mixed with water, they are hydrated and kneaded, and they form a viscoelastic matrix that is a three-dimensional, continuous network. It is exactly this gluten network that gives the necessary cohesiveness to form a dough. In addition, the relatively strong network prevents the diffusion of the gas bubbles through the dough during leavening, hence ensuring the rising before and during baking. Starch is another important flour component that influences bread quality. When starch is heated during baking, the granules swell and lose their crystalline structure, leaking free amylose out of the granules into the water in the dough. This free amylose forms a gel that is incorporated in the gluten network and, thereby, increases the viscosity of the dough. However, it is also the starch that is responsible for the staling of bread due the retrogradation.

starch

heating of the dough

For the sensory role, the heating of the dough is the important step, since both caramelisation and the Maillard reaction takes place, and both contribute to the development of crust colour and aroma compounds. The aroma compounds produced by the Maillard reaction are diverse and are influenced by the type of sugar and amino acids of the proteins, hence the type of flour. For example, the heterocyclic compound 2-acetyl-1-pyrroline is one of the most important odorants in the crust, contributing to the characteristic popcorn, roast, cracker-like aroma notes of the crust. In addition, the aldehydes 2-(E)-nonenal and 3-methylbutanal are the second most perceived odorants in wheat bread crust, and contribute with odour notes, such as malty, roasted, fatty and paper. Common to these compounds is the low odour thresholds, meaning that we can detect them even at very low concentration – think about the wonderful smell of freshly baked bread.

Flour produced from sustainable plant materials, e.g. locally grown cereal or legume varieties, in combination with other bread-making processing, e.g. sour dough, may facilitate the development of bread with different shapes, styles, and flavours. For example, the Japanese bread Shokupan has a sweet and subtle flavour, accompanied by an aroma of butter and milk. Moreover, the colour is pale and white, but it is its mouthfeel – upon chewing the very soft crumb – which makes Shokupan a unique eating experience. Figure 9.2 shows a similar fluffy bread. Such types of bread may complement the vegetable soup or dish with additional flavours, thus supporting the consumption of green foods.

Figure 9.2. Bread which is extremely soft and pillowy, with a subtle sweetness and a delicate white colour. Air bubbles are small and homogenously distributed, the crumb is moist and tender (photo courtesy of Aran Raventos).

9.2.3 Soup

Soups can be served at any given time during a meal, depending on the actual menu and the type of soup. Soups are basically an aqueous liquid, possible with some dissolved fats, which are enriched in flavours, both taste and aroma. It can have different consistencies from very thin to highly viscous, thickened by thickening and/or gelation substances. Moreover, the soup can contain particulate material and even small or large pieces of solid foodstuffs. Some soups are served warm and others cold.

stock

The basis of a soup is a stock that is an aqueous cooking extract of meat, bones, vegetables, fish, shellfish, or fungi being boiled in water often for extended times. After sieved to remove the solid pieces the stock primarily consists of water and contains dissolved carbohydrates, proteins, and fats, in addition to taste and aroma substances. In its purest form, it has little taste. If seasonings, for example, salt, pepper, and other spices and herbs have been added to it, it is usually referred to as a broth or by the French term, bouillon (from the French *bouilli*, meaning 'cooked'). Sometimes the stock is clarified using an egg white, as this causes the water-soluble proteins in the liquid to coagulate. Consommé is a broth that has been concentrated by boiling off some of the liquid, referred to as reduction.

The general idea behind a stock is to capture as many different and flavourful, soluble substances in a suitable and palatable concentration. Some stocks contain an oily fraction that can capture hydrophobic aroma compounds. Because the ingredients to form the stock contain both soluble as well as more or less volatile compounds, preparing and reducing a stock is often a delicate balance of extracting as many compounds as possible while avoiding evaporating too many of the volatile aroma compounds (Snitkjær *et al.*, 2010). Soups of different kinds also serve as the starting point for sauces, some dressings, and the flavouring of sausages and patés.

sautéing

A light stock is usually made from white meat and vegetables. For a dark stock, bones, meat, and vegetables are sautéed before the liquid is added. The sautéing releases flavourful Maillard compounds. In a broth, small pieces of the raw ingredients are usually left in, and egg or some starches, such as rice or barley, may be added to thicken it. At the other end of the scale is the Japanese stock *dashi*, to be described below, which is simply an aqueous extract (without any fat) and not boiled at all.

umami

The Japanese soup stock, *dashi*, deserves a special mention because it is, so to speak, the motherlode of umami and umami synergy, being extremely clear in its taste and with very little aroma. Using the general concept of umami is the key to umamifying green dishes (Mouritsen and Styrbæk, 2020). There are several ways of making *dashi* using a variety of ingredients, but they are all based on extracting glutamate from a brown seaweed species (*konbu*, *Saccharina japonica*) to obtain basal umami (Mouritsen *et al.*, 2019). Extraction temperatures up to 60 °C are preferable to minimise the extraction of bitter compounds. After removing the seaweed, a second extraction is made of either the fish product *katsuobushi* or, for a vegan *dashi*, dried shiitake. These two ingredients contain respectively large quantities of water-soluble free nucleotides inosinate and guanylate. Together with free glutamate these free nucleotides enter into a powerful umami synergy.

The easiest and most traditional way to draw umami taste out of meat is to boil both the meat and the bones together with vegetables to make a soup stock. Simmering for hours leads to a very robust stock. The umami content increases over time as the soup cooks, because a greater proportion of the proteins in the meat are broken down. As there are many more taste substances in this type of stock than in *dashi*, the taste is not as clean and is more complex. Hence, Western or Chinese soup is made using a rather brutal cooking process that breaks down unprocessed raw ingredients, namely meat and vegetables, resulting in a broad spectrum of taste substances, not just umami. Glutamic acid still predominates, but in addition there is a broad spectrum of other amino acids, with both sweet and bitter tastes. Consequently, the taste is much more complex than that of *dashi*. Japanese *dashi* is exactly the opposite, being a gentle extraction at lower temperatures from just two ingredients that have only a few taste substances.

fresh meat   Fresh meat is a major source of umami, as it contains both glutamate and the nucleotides that create synergy. Poultry meat, particularly from chicken, duck, and turkey, has a high concentration of glutamate, and the same is true for game birds and veal. Pork and chicken are rich in inosinate, while beef has a reasonable amount. As lamb is a poor source of umami, it benefits from being prepared in duck stock, which has a great deal of it.

fish stock   A fish stock can be made using only fish and shellfish, as they contain both the glutamate and nucleotides needed to impart umami. The combination of shellfish with vegetables that have a good quantity of glutamate yields a particularly robust stock with intense umami. A good example is green pea soup with scallops, where the green peas contribute free glutamate, and the scallops contain large quantities of the free nucleotide adenylate (cf. Figure 9.3). If this soup is prepared on the basis of a *dashi*, the umami synergy is even more powerful.

vegetarian   Purely vegetarian soups can derive glutamate from a variety of sources, such as cooked potatoes, green peas, soybeans, maize, green asparagus, cauliflower, mushrooms, and, of course, tomatoes. Fungi, especially dried shiitake mushrooms, have good quantities of the nucleotides that can interact synergistically, but there are generally very few vegetables that can play this role.

Figure 9.3. Soup of green peas with scallops and pieces of kelp (photo courtesy of Jonas Drotner Mouritsen).

salt

Soups taste good only if they have a salt content of about 0.75%, but if the soup has a great deal of umami, this can be reduced by a half.

thickening of
the soup

Soups are thickened in the same way as sauces, with the addition of starch, roux, gelatine, gelling agents, eggs, milk, cream, cheese, and mashed potato (Mouritsen and Styrbæk, 2017). For the soups to have an interesting texture and coat the mouth in a pleasing way, it is necessary for the thickened soup to be able to mix well with the saliva. This is provided for easily when starch or gelatine, for example, are used as thickeners. But soups thickened with some of the more complex polysaccharides, such as xanthan gum, do not mix readily with the saliva. The result is that the saliva does not receive the taste substances sufficiently quickly, and in addition, the soup may feel a little sticky.

mouthfeel

To further elaborate on the mouthfeel of a stock to produce a soup, as exemplified in Figure 9.3 above, basically anything solid or jelly can be added: meat, fish, shellfish, vegetables, grains, peas, mushrooms, noodles, seaweed, miso paste, or tofu. It is often the contrast between the textures of the liquid element and the solid ingredients that is interesting and a defining characteristic of soup. The liquid coating of the oral cavity also helps intensify the taste of the solid particles. As some of these solid pieces may need to be prepared in a different way or for longer or shorter periods of time, the soup base/stock and the solid elements are often prepared separately. Provençal bouillabaisse is an excellent illustration of how to achieve textural variations using a range of ingredients that need to be prepared independently and added to the soup at different times.

9.2.4 Vegetable dish

starchy vegetables

Vegetables are usually classified according to being starchy or not. Starchy vegetables such as tubers (e.g. potatoes) are usually cooked in water (heat treated) leading to a gelatinisation of the starches, thereby rendering the mouthfeel soft, either creamy or mealy/dry. Leafy green vegetables (like cabbages) and many root vegetables (like carrots and kohlrabi) can in some cases be consumed fresh and raw, are often cooked, but are also processed as marinated, fermented, or pickled.

leafy green
vegetables

brassica species

Our example of processing vegetables has been chosen to illustrate how one can control the texture and the flavour of a brassica species (broccolini, *Brassica oleracea* × *alboglabra*) that is often considered less delectable due to a certain bitterness. Moreover, this vegetable shares the fate of most other vegetables in having very little umami taste and being seldom sweet (with some notable exceptions like carrots). We will highlight a low-temperature treatment of the raw vegetable using a certain mould culture, *koji*, based on the microorganism

Aspergillus oryzae. This treatment is inspired by Japanese cuisine where it is known as *koji-zuke* and it leads to a crispy, sweet, and umami-tasting product with a substantially reduced bitterness. The processing described can be applied to a very wide range of vegetables and is particularly powerful for bitter vegetables.

enzymes

Koji is well known as it is key in the traditional production of soy sauce, miso, and sake. *Aspergillus oryzae* contains a host of different enzymes that turn carbohydrates into sugars and proteins into small peptides and free amino acids. Certain small peptides can elicit *kokumi* (Nishimura and Kuroda, 2019) and free amino acids like glutamic acid (glutamate) can elicit umami. Both *kokumi* and umami act to enhance sweetness and reduce bitterness.

It is possible to avoid working with the intact *koji* culture using a commercial preparation known as *shio-koji* that is a salty medium (fluid or paste-like) only containing the enzymes of the microorganism. Treating fresh vegetables like broccolini with *koji* or *shio-koji* at 5 °C for 1-3 days leads to a product that has a deep umami taste, an increased sweetness, and a concomitant reduced bitterness (Mouritsen and Styrbæk, 2021a). As a consequence, this cold-treated vegetable, cf. Figure 9.4, becomes much more delectable and can serve as vegetable dish in itself or as a condiment to, e.g. fish dishes. An appropriate dressing could be sweet, white miso or a miso-mayo.

9.2.5 Seafood dish

cephalopods

As the example of a seafood dish for our menu we have chosen a challenging one: cephalopods (Mouritsen and Styrbæk, 2021b). Processing cephalopods, in particular octopuses, can be tricky due to the large and highly cross-linked collagen content of their muscle tissue. In the context of sustainability, future

Figure 9.4. Cold preparation of broccolini using enzymatic treatment via shio-koji (photo courtesy of Jonas Drotner Mouritsen).

food, and a better use of marine food resources cephalopods have come into focus because, in contrast to most global fish populations, cephalopod populations have been on the rise for the last sixty years (Mouritsen and Schmidt, 2020).

squid Processing of squid for optimal texture and taste qualities may take advantage of recent insight into squid muscular structure and how it depends on heat treatment (Schmidt *et al.*, 2020, 2021a), specifically sous-vide cooking but also cooking in oil (confit). As for other types of meat, optimal preparation involves a delicate balance between protein denaturation, collagen-gelatine transformation, and water loss, which have their different dependences on temperature and preparation time. However, due to the large content of collagen, typically three times as much as in fish and in meat from land animals, the situation is rather complex. Consumer preference studies must eventually be performed to decide on the best processing protocol (Schmidt *et al.*, 2021b).

Examples of micrographs of squid mantle before and after sous-vide treatment at 55 °C are shown in Figure 9.5A. In the raw tissue (to the left) the collagen is organised in a dense network of long linear bundles that are entangled in the lateral direction in a roughly perpendicular manner. After sous-vide treatment (to the right), the signal intensity decreases, indicating that the heat treatment has changed the molecular structure of the collagen as the squid becomes more tender. A few intact collagen fibres are seen, surrounded by what appears to be **denatured collagen** partly or fully denatured collagen. Corresponding texture analyses under the same conditions as well as for several other cooking times are also shown in Figure 9.5A.

The results of consumer tests revealed that, even if the longest cooking time resulted in a more tender texture profile, this preparation was not the most liked. Rather, the evaluation demonstrated that a relatively short cooking time (but not raw) was preferred, as it yielded a better taste and a more stimulating **mastication** texture during mastication, whereas longer cooking times yielded an overcooked surface, a bitter taste, and a dull mouthfeel throughout mastication (Schmidt *et al.*, 2021b).

As an example of a simple squid dish, Figure 9.5B shows squid mantle strips that have been cooked in sunflower seed oil at 85 °C for 35-40 min. The resulting curly structure is caused by the peculiar stratification of the muscle fibres in the squid mantle. The curly mantle confit is silky soft and can be served as is with lemon juice, chopped parsley, salt, and pepper. This curly squid that is extremely tender can also be served cold, like pasta, in a cold salad, or in most ways in which pasta is used.

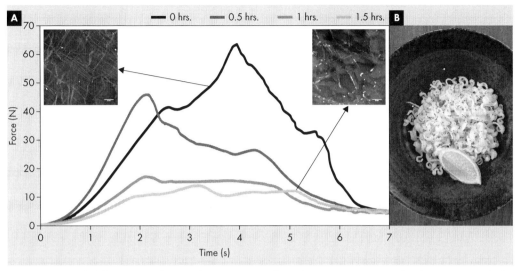

Figure 9.5. Squid. (A) Texture analysis and microstructure of squid. Force versus time at loading rate 2 mm/ s for a squid mantle sample before and after sous-vide treatment (55 °C for 0.5, 1.0, and 1.5 h). Insets show second-harmonic microscopy images of the collagen structure when raw (left, orange curve) and after sous-vide treatment (right, blue curve). The scale bars correspond to 50 μm (Faxholm *et al.*, 2018). (B) 'Silky squid' (curly mantle confit) of squid mantle strips (dish created by Peter Lionet Faxholm; photo courtesy of Jonas Drotner Mouritsen).

9.2.6 Meat dish

myofibrillar proteins Muscles from mammal and poultry have a very unique three-dimensional structure; the muscle fibres, formed by the myofibrillar proteins, allow the muscle to contract and expand. The myofibrillar proteins constitute two thirds of the total proteins in a muscle and consist mostly of myosin and actin proteins. Hence, myosin and actin are the most important proteins that contribute to meat texture. A muscle cell is surrounded by connective tissue at different organisation levels, mainly consisting of different types of collagen and elastin. The total collagen content can vary from 1 to 15% of the muscle. Collagen has a rather rigid triple-helix structure and is the reason behind the toughness of meat. When animal muscle becomes meat for eating, these proteins will together influence the physical, technological and sensory features of meat thereby defining the eating quality.

proteins Meat is a good source of valuable proteins, since 100 g of meat contains approximately 20 g of high biological-value proteins. However, meat is rarely eaten raw (except for a few culinary preparations for beef, like steak tartar or carpaccio) because it is rather tough (and for safety reasons), and the taste of raw meat is not appreciated. Hence, meat is cooked in many different ways and to different extents before consumption or is processed in various ways and with different ingredients and additives into ready-to-eat or convenience meat products. The cooking of meat produces the three most important characteristics of eating quality: texture, flavour and juiciness. Several cooking methods for meat exist based on different foundations: water methods (e.g. boiling/simmering, *sous vide* and steaming), air methods (e.g. grilling and roasting), oil methods (e.g. pan or deep-frying) and combinations like braising and stewing.

cooking

Plain cooking, i.e. boiling meat in water, changes the texture and tenderness of meat in accordance with the changes in the two structural components, muscle fibre and connective tissue. As the temperature of the water increases, a gradual change in meat toughness follows (Christensen *et al.*, 2000). An increase in meat toughness occurs upon reaching the temperature 40-50 °C because of a thermal shrinkage of the connective tissue. Then follows a decrease in toughness between 50-60 °C likely caused by denaturation and solubilisation of the collagen. When the meat temperature exceeds 60 °C, meat toughness increases again due to denaturation and aggregation of myofibrillar proteins. The textural quality can be controlled by temperature. However, cooking in boiling water is an uneven process, making the desired effects (collagen denaturation is desirable and myofibrillar protein denaturation and aggregation is undesirable) for the whole piece of meat difficult to control. To overcome this challenge, the low-temperature long-time cooking or sous-vide cooking methods are useful. This involves cooking vacuum-packed meat at low temperatures (55-60 °C) in thermostated water baths for several hours or even days. This method makes it easy to control the temperature- and time-dependent processes that alter texture, flavour, and colour, thereby achieving the desired sensory properties (Mortensen *et al.*, 2012). Unfortunately, as seen in Figure 9.6, the two most important sensory properties, tenderness and juiciness, could not be optimised simultaneously by sous-vide cooking of beef eye of round (Mortensen *et al.*, 2012). Cooking at 56 °C for 3 hours gave maximum juiciness, while it took 12 hours to obtain maximum tenderness. A compromise between temperature and time is needed to balance these desired eating qualities.

high pressure processing

A new way of processing meat is high pressure processing (HPP). For example, HPP can be an alternative method to the sous-vide treatment, but with much shorter processing times (within the order of magnitude of minutes for HPP rather than hours for sous-vide), to achieve tender and juicy meat products. In addition, HPP can be used for manufacturing processed meat types, e.g. low-salt sausages, due to the ability of HPP to improve the functional properties of myofibrillar proteins, thereby affecting textural properties, water-holding capacity and sensory perception (Bolumar *et al.*, 2021).

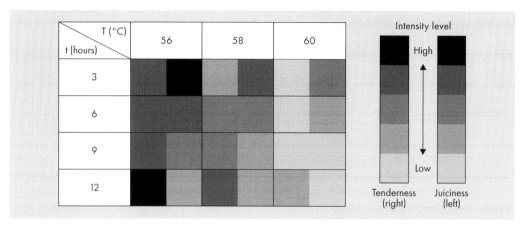

Figure 9.6. Assessment of tenderness and juiciness of sous-vide cooked beef eye of round as a function of temperature (T) and time (t) (modified from Mortensen *et al.*, 2012).

roasting

frying

But since no browning occurs, plain or sous-vide cooking leaves the meat without the attractive roast flavour. On the other hand, roasting and frying are distinct from water methods in providing colour and aroma. During frying or roasting, the meat is subjected to a much higher temperature than when boiling in water. When we place the meat on the hot pan or in the oven, the meat surface will reach temperatures well above 100 °C, resulting in water evaporation. Once the surface water has evaporated, the temperature rises and above 140 °C the Maillard reaction (a series of chemical reactions) occurs between the amino acids of the proteins and sugars from glycogen. It is the Maillard reaction that produces the molecules responsible for the nutty and meaty taste and flavour as well as the brown colour. Many of the desirable flavours are actually only formed at these high temperatures. Specifically, one important flavour molecule that is responsible for the meaty flavour in chicken and beef is 2-methyl-3-furanthiol formed by the reaction between pentose sugars and the amino acid cysteine (Melton, 1999). Keeping the high temperature through roasting or frying is necessary for the surface dehydration and the continuous development of compounds giving the desirable meaty flavour securing the tasty, delicious and crispy Maillard-crust desirable for foods in general. Knowledge about making high-quality and palatable meat dishes with intense savoury flavours may promote a meal design and eating behaviour that helps in downsizing the meat portion of a dish, and in general the consumption of meat in line with the green transition.

9.2.7 Sauce

emulsion

Some type of sauce usually accompanies the meat, seafood, or vegetable dish. Its purpose is to complement the dish in addition to lubricating the bolus and make the chewed food slide down the throat upon swallowing. Scientifically, a sauce is an emulsion. An emulsion contains two immiscible liquids that are mixed/emulsified together. One phase is dispersed (as droplets, also called the dispersed, internal, or discontinuous phase) within the other phase, the external or continuous phase. Food emulsions are normally made up of lipid and water leading to either oil-in-water (O/W) or water-in-oil (W/O) emulsions. Well-known examples are mayonnaise (O/W) and butter (W/O), Figure 9.7. Many different food emulsions exist, since they can be cold or hot, smooth or thick, spicy to sweet, and anything in between. All emulsions are invented to enhance the taste of the dish, bringing (another level of) flavour to the food. A sauce is an O/W emulsion and combining oil and water results in a coating property in the mouth, allowing time to release the important flavours and aromas before the food is swallowed.

vinaigrette

The quality of an emulsion spans from unstable/temporary to stable/permanent. Vinaigrette is the cold sauce or dressing typically made to match a salad and is a simple and highly unstable emulsion. The oil is whisked into the vinegar thereby forming small oil droplets, and finally the two phases come together as a unified liquid. During oil addition and agitation, the dressing becomes a

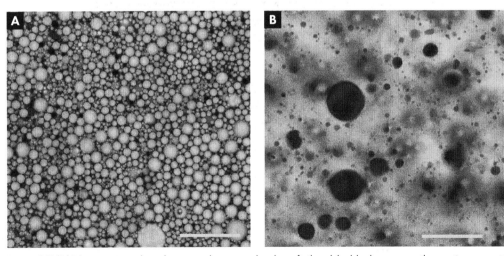

Figure 9.7. (A) Mayonnaise, where the grey spheres are droplets of oil and the black areas are the continuous water phase (CARS microscopy image, scale bar 10 μm). (B) Butter, where the grey solid is fat and the black droplets are liquid water (CARS microscopy, scale bar 10 μm) (photo courtesy of Dr Mathias P. Clausen and Dr Morten Christensen.

thicker liquid and the colour will also change into something between the two ingredients. However, if the vinaigrette stands for a while, the oil and vinegar will separate into the two phases again. Fortunately, a shake or whisk will emulsify them, ready for use again.

emulsion An emulsion is inherently thermodynamically unstable owing to the interfacial tension existing in the interface between two phases. When the interfacial tension increases with an enhanced area of contact, the stability of the emulsion will be broken down more easily. In the vinaigrette example, the oil droplets will spontaneously form aggregates and further coalescence, i.e. merge into larger droplets, and this continues until only 'one big oil droplet' is formed, i.e. the oil phase has separated from the water phase. The truly interesting science behind emulsions is concerned with how the liquids blend and how to stabilise them once they are mixed.

béarnaise sauce Béarnaise sauce is an O/W emulsified sauce, which complement a beef dish very well. Basically, the sauce is made by mixing the butter fat (often clarified butter) with another type of continuous aqueous phase namely egg yolk. The sauce becomes a béarnaise sauce upon flavouring with the specific herbs, parsley, chervil, and estragon. The sauce is made by slowly adding the oil to the egg yolk while stirring. A common error is adding the oil too quickly, causing the sauce to separate. A similar cold O/W emulsion is the mayonnaise. It is made by slowly, but roughly whisking a vegetable oil into egg yolks. A fine/smooth mayonnaise is characterised by the small oil drops being so close together that they give the emulsion a firmness and a slightly elastic mouthfeel. Once the mayonnaise emulsion is formed, it is typically stable for a long time. Thus, mayonnaise is an example of a stable/permanent emulsion. The difference between the unstable vinaigrette emulsion and the stable mayonnaise emulsion is the continuous water phase. The lecithin in the egg yolks does the magic and acts as an emulsifier. Lecithin is an interfacially active lipid that solubilises in both lipid and water, hence combining the egg yolk and the oil, essentially holding the two liquids together permanently.

emulsifiers Certain molecules act as emulsifiers, which means they help to mix two immiscible liquids together and stay together. An emulsifier's activity is based on the special molecular structure. It has a lipophilic (also called hydrophobic) part that is soluble in the oil (lipid) phase, and a hydrophilic (also called polar) part that is soluble in the aqueous phase. Thus, when the emulsifier is mixed into the immiscible oil-water system, it will locate and orient itself on the oil droplet surface (the oil/water interface) with the lipophilic part in the oil and the hydrophilic part in the water. Thereby it decreases interfacial tension and stabilises the emulsion. Many different types of emulsifiers exist, and in industry they are produced and optimised for each of their specific applications.

Generally, emulsifiers facilitate the distribution and stabilisation of one phase within the other at low concentrations. However, it is emphasised that each emulsifier can only disperse a limited amount of the discontinuous phase, i.e. it has a maximum emulsification capacity.

9.2.8 Dessert

No meal is complete without a dessert. A typical dessert is either a gel or a foam type. From the food science perspective, these two product types have special structures and textures that are produced by considering food as a material that is designed to provide specific properties. Therefore, the aim is to restructure and assemble the food molecules into gel and foam structures that are pleasant to eat. However, this is easier said than done. Most people have had the experience of over-whipping when making whipped cream – the result is definitely not pleasant.

whipped cream

Many professional chefs and amateurs have cooked and experienced with various structures and textures, with terrific outcomes but also learning from the failures, like knowing when to stop whipping and not get an over-whipped cream. Producing a gel means transforming a liquid into a solid-like texture; thus, a gel is in-between liquid and solid, possessing both flow (liquid) and elastic (solid) characteristics. Yoghurt, pudding, and ice cream are examples of such semi-solids. The main textural difference between a liquid and a gel is that the gel requires some kind of oral treatment to transform the gel into a bolus capable of being swallowed. The reason for this necessary oral manipulation is that a gel is a three-dimensional polymeric network that possesses a mechanical rigidity, meaning it resists flow and retains its structural shape. Overall, a gel can be one of two types: a continuous network of interconnected particles or assorted macromolecules dispersed in a continuous liquid phase, typically water. The general gelation process, that is the conversion from the dissolved macromolecules (the liquid) to the gel (semisolid), involves the association or crosslinking of the macromolecular chains to form a three-dimensional network that both traps water within and forms a rigid structure.

gel

milk gel

Let us take milk gel as an example, for which much knowledge is available, yet the mechanisms of gel structuring are still not fully understood. Gel formation by milk proteins is the foundation in producing cheese and yoghurt and many other dairy-based products. The main gel-forming macromolecules in cheese are the caseins, while both the caseins and whey proteins are necessary for producing yoghurt and other dairy gel products. Yoghurt is a good example of how to transform the raw material, milk, into another type of product, yoghurt, by means of controlling molecular modification.

gel formation Gel formation is a spontaneous process from a protein solution under controllable conditions of temperature or solution composition. However, the gelation of proteins requires a driving force to first unfold the native protein structure, followed by an aggregation step resulting in an organised network of aggregates of particles or strands of molecules cross-linked by non-covalent bonds or covalent bonds spanning the entire volume. The type of macromolecules and gelation formation leads to different gel microstructures; homogeneous, particulate, protein continuous, bicontinuous, serum continuous, or coarse stranded (Foegeding *et al.*, 2017).

yoghurt production Yoghurt production, which is the gelation of milk proteins, can be induced chemically by enzymatic or acidic reactions, fermentation and/or physical processes like heat or high-pressure treatment. For yoghurt and other milk-based gels, the microstructure and rheological properties together with the overall visual appearance are important physical attributes, which contribute to the sensory perception of such products. The physical structure of the gel (its geometry) and the way the gel behaves (its mechanical property) and feels in the mouth (its surface properties) are important gel texture characteristics. The behaviour of the gel in the mouth upon chewing depends mostly on the mechanical property and is typically assessed instrumentally by the parameters: hardness, cohesiveness, elasticity, viscosity, and adhesiveness. In addition, particle size and size distribution contribute to the perception of the gel's texture attributes, such as powdery, grainy, and spreadable. Most of us have experienced the unpleasant feeling of powdery and grainy upon eating a 'wrong' gel and this is directly related to particle size. Figure 9.8 shows two different types of milk gels produced in different ways, the well-known soft yoghurt and a hard high-pressure gel.

Figure 9.8. (A) Traditional (fermented) yoghurt with a touch of sweetness (homemade pickled plums). (B) An innovative milk gel produced by high pressure treatment (6,000 atm. for 10 min at 20 °C) of a 10% whole milk solution fortified with 10% whey protein powder. (C) The difference in gel hardness between high-pressure and heat-treated milk samples (Orlien *et al.*, 2006).

texture

Hence, the method of producing a milk gel determines its structure and, thereby, its texture. The first processing step for yogurt manufacture is a heat treatment of milk. Heat treatment of milk above 70 °C causes destabilisation of the native whey proteins, of which especially β-lactoglobulin is very important for gelation. Destabilising of the native casein micelles augments their possibilities of interaction to a level that ultimately causes the formation of a stable network and gelation. It is this destabilisation that is induced chemically (addition of acids, starter cultures, or enzymes) or physically (heating, cooling, or high pressure) which results in partial or total unfolding (denaturation), thereby leaving the proteins prone to interactions resulting in protein aggregation and finally gel formation. Milk protein gelation is the result of cross-links by both physical (electrostatic, hydrophobic, and hydrogen bonds) interactions and chemical (disulphide) bonds between the protein molecules. Specifically, unfolding the tertiary structure of β-lactoglobulin exposes the buried cysteine groups and the sulfhydryl-group (SH) becomes chemically reactive. The denatured β-lactoglobulins will form cross-links with each other and with casein micelles via hydrophobic interactions and intermolecular disulphide bonds. This aggregation process gives the three-dimensional organised network of aggregates or strands of cross-linked molecules, in other words, a gel is formed (Orlien, 2021). However, the technology used to produce the network will greatly affect the molecular structure and thereby the gel texture. For example, a huge difference in the gel strength is clearly seen between milk solutions subjected to high pressure or heat treatment (Figure 9.8C). Pure protein gels are generally homogeneous or particulate. However, even small variations in the gelation process may cause textural defects of the resultant gel, such as the gel being too weak, wheying-off (liquid separation from the gel) or being lumpy. Such defects definitely have a seriously negative impact on the sensory perception of the yoghurt. Therefore, several gel forming or thickening ingredients can be used to produce gels with the desired textural properties thereby providing acceptable attributes. In particular, hydrocolloids derived from plant sources and proteins can perform a number of functions including thickening and gelling.

globulins

Globulins are the main protein in plant materials like legumes and cereals. These plant globulins may possess the same functional properties as milk proteins; thus, they can form a gel under the right processing conditions. A well-known example is tofu, which is produced by heating and/or fermenting soybean milk. In general, the aggregation and gelation process of plant proteins largely resembles the gel formation of milk proteins. However, in contrast to milk, which is a system of soluble proteins already by nature, the plant protein gelation is greatly affected by the method of protein extraction. The challenge is that plant proteins are embedded in rather rigid cellular structures, so an extraction procedure is necessary before obtaining the protein ingredient.

Different protein extraction methods are used and have a notable effect on the composition and structure of the proteins. Proteins may be native or partially/fully denatured as a consequence of heating or adding salt to improve the protein solubility prior to extraction. During the extraction, the proteins may aggregate due to this destabilisation and in effect change the ability to form gels. In addition, non-protein components, such as starch and fibre, may be present in protein extracts and affect the gelation properties of the ingredient (Sharan *et al.*, 2021). Gelation with other plant proteins has been investigated but it has been concluded that the gelation depends on both the plant source and extraction methods (Nicolai and Chassenieux, 2019). A scientific and technical understanding of their composition and structure, insight into the relationship with physico-chemical and functional properties, as well as knowledge of the effect on eating quality are of major importance in order to support the green transition.

foam

A popular dessert type is foam, and whipped cream is probably the most famous foam. Foams can be very different; soft or hard, fine or coarse, frozen, cold or hot, light or dense, and examples include mousse, meringue, and ice cream. Foam is a very complex food system, as it consists of mixing two immiscible phases, air (the dispersed phase) and protein solution (the continuous, liquid phase). It is quite easy to produce, but challenging to stabilise for longer times, since such a mixture is unstable by nature. Foam is formed by whipping/beating air into the liquid/continuous phase. To form and stabilise the air bubbles, three stages are important: (1) proteins diffuse to the air/water interface, (2) the proteins unfold and orient themselves, and (3) proteins interact and form a cohesive film around the air bubbles. The proteins act as amphiphilic molecules with the hydrophilic groups located in the liquid phase and the hydrophobic groups located in the air phase. The ability of proteins to foam and form a stabile film around the air bubbles depends on factors such as intramolecular bonds, surface hydrophobicity, and molecular weight. The strength of the film depends mostly on the ability of the proteins to interact with each other. The size and distribution of the air bubbles have a direct impact on the foam's appearance and textural properties and are crucial for the foam's stability. Figure 9.9. shows a microscopy image of a foam with small (0.01-0.1 mm in diameter), uniformly distributed bubbles that give a smooth and light product, while larger (>1 mm) bubbles will feel

foaming stabilisers

like a coarse foam. To stabilise a foam, various foaming stabilisers, or surfactants may be used. On the one hand, egg white and whey proteins are excellent foaming agents. On the other hand, hydrocolloids will thicken the liquid that surrounds the air bubbles, thereby effectively slowing down drainage from the bubbles, thus stabilising the foam.

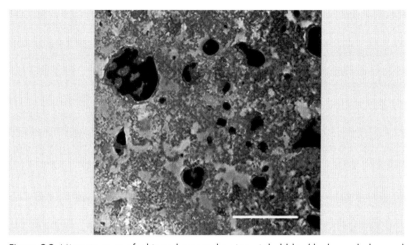

Figure 9.9. Microstructure of whipped cream showing air bubbles (the large dark areas) and the walls between them are primarily of fats interspersed with water (CARS microscopy, scale bar 200 μm) (photo courtesy of Dr Mathias P. Clausen and Dr Morten Christensen).

whipped egg whites Whipped egg whites form a soft foam stabilised by the egg white proteins and are the basis for mousses and cream puffs. Adding sugar during whipping both stabilises the foam and makes it sweet and delicious. While the egg-white proteins cover and stabilise the air bubbles at the molecular level, the sugar makes the continuous solution (phase) thicker and, thereby, stabilises the air bubble/water system at the macroscopic level. To get a meringue we bake the egg white/sugar foam. By playing with the amount of sugar and baking conditions (temperature and time), you can obtain different textures of your meringue, from soft and chewy to crisp and hard.

ice cream Structurally, ice cream is a partly frozen foam. The continuous phase is a concentrated, unfrozen solution containing proteins and sugar. Ice crystals exist as a coarsely dispersed phase and occupy a major portion of the space between the air bubbles. The ice crystal size has a major impact on the texture of ice cream; crystals larger than 40-50 μm result in a coarse/sandy texture, whereas a smooth texture is obtained with crystals smaller than 20 μm. The texture of ice cream depends mostly on the physical properties of the crystals. However, all constituents make an important contribution to the structural and sensory properties of ice cream.

Egg white and whey proteins are, due to their superior foamability, traditionally used foaming agents in processing bakery and confectionary foods. Plant-based proteins are a much more sustainable protein source with the potential to replace the animal-derived proteins. Thus, proteins from various legumes, e.g. fava bean, chickpea, cereals like oat, and new plant sources such as green leaves and grasses, are attracting increasing interest from consumers, food industry, and researchers. A direct substitution of the animal-based ingredients by plant-based ingredients challenges both sensory and foaming properties. Aquafaba, wastewater from cooking of chickpeas or other pulses, was discovered to have good foaming properties, and by the addition of sugars and flavour compounds to mask the typically beany flavour, plant-based cream puffs and marshmallows became available on the market. Thus, the knowledge of foam and gel formation opens new initiatives to use side-streams from plant-based food production and substitute the 'traditional' animal-based ingredients, such as eggs, in a sustainable production of food products.

9.3 Conclusions

We have come a long way from the early hominins' use of fire to prepare food, over twelve millennia of home cooking after the advent of agriculture, to the current era's use of massively industrialised food production and food processing technologies, exemplified by our hypothetical menu. During that time the World's population has increased from 4 million to 7.9 billion. Our early ancestors were driven by a desire for delicious food (Dunn and Sanchez, 2021), and we are no different. Food processing holds part of the key to make it possible for an increasing population on a finite planet to convert our food system into something sustainable while still fulfilling our need for nutritious, healthy, and delicious food (Schmidt and Mouritsen, 2020).

The question is how to move from molecules to structure, texture, and aroma in a rational manner. The key is science-based understanding of food material formation in order to create foods with specific properties that relate to stability and enjoyment. It is worth noting that empirical-based knowledge may consume resources and, worst of all, waste raw materials. A fundamental understanding of all the processing steps is required for efficient and effective food production, especially when moving to a more plant-based diet that has the basic tastes we crave without being ultra-processed and without putting stresses on both human health and the planet.

References

Bolumar, T., Orlien, V., Sikes, A., Aganovic, K., Bak, K.H., Guyon, C., Stübler, A.-S., de Lamballerie, M., Hertel, C. and Brüggemann, D.A., 2021. High pressure processing of meat: molecular impacts and industrial applications. Comprehensive Reviews in Food Science and Food Safety 20: 332-368. https://doi.org/10.1111/1541-4337.12670.

Burke, R., Kelly, A., Lavelle, C. and This, H. (eds), 2021. Handbook of molecular gastronomy: scientific foundations and culinary applications. CRC Press, Boca Raton, CA, USA.

Christensen, M., Purslow, P.P. and Larsen, L.M., 2000. The effect of cooking temperature on mechanical properties of whole meat, single muscle fibres and perimysial connective tissue. Meat Science 55: 301-307. https://doi.org/10.1016/S0309-1740(99)00157-6

Clark, S., Jung, S. and Lamsal, B., 2014. Food processing: principles and applications. 2nd ed., Wiley, New York, NY, USA.

Clausen, M.P., Christensen, M. and Mouritsen, O.G., 2021. Imaging foodstuff and products of culinary transformations. In: Burke, R., Kelly, A., Lavelle, C. and This vo Kientza, H. (eds) Handbook of molecular gastronomy: scientific foundations and culinary applications. CRC Press, Boca Raton, CA, USA, pp. 404-409.

Comody, R.N., Weintraub, G.S. and Wrangham, R.W., 2011. Energetic consequences of thermal and nonthermal food processing. Proceedings of the National Academy of Sciences of the USA 108: 19199-19203.

Crawford, M.A., 2010. Long-chain polyunsaturated fatty acids in human brain evolution. In: Cunnane, S.C. and Stewart, K.M. (eds) Human brain evolution: the influence of freshwater and marine food resources. John Wiley & Sons, Inc., Hoboken, NJ, USA, pp. 13-32. https://doi.org/10.1002/9780470609880.ch2

Cunnane, S.C. and Crawford, M.A., 2014. Energetic and nutritional constraints on infant brain development: implications for brain expansion during human evolution. Journal of Human Evolution 77: 88-98. https://doi.org/10.1016/j.jhevol.2014.05.001

Dunn, R. and Sanchez, M., 2021. Delicious: the evolution of flavor and how it made us human. Princeton University Press, Princeton, NJ, USA.

Faxholm, P.L., Schmidt, C.V., Brønnum, L.B., Sun, Y-T., Clausen, M.P., Flore, R., Olsen, K. and Mouritsen, O.G., 2018. Squids of the North: gastronomy and gastrophysics of Danish squid. International Journal of Gastronomy and Food Science 14: 66-76. https://doi.org/10.1016/j.ijgfs.2018.11.002

Foegeding, E.A., Stieger, M. and Van de Velde, F., 2017. Moving from molecules, to structure, to texture perception. Food Hydrocolloids 68: 31-42. https://doi.org/10.1016/j.foodhyd.2016.11.009

Fellows, P.J., 2011. Food processing technology. Woodhead Publishing Series in Food Science, Technology and Nutrition, 3rd ed. Elsevier, Amsterdam, the Netherlands.

Lieberman, D.E., 2011. The evolution of the human head. Harvard University Press, Cambridge, MA, USA.

Melton, S.L., 1999. Current status of meat flavour. In: Xiong, Y.L., Ho, C.T. and Shahidi, F. (eds) Quality attributes of muscle foods. Kluwer Academic/ Plenum Publishers, New York, NY, USA, pp. 115-130.

McGee, H., 2004. On food and cooking: the science and lore of the kitchen. Scribner, New York, NY, USA.

Mortensen, L.M., Frøst, M.B., Skibsted, L.H. and Risbo, J., 2012. Effect of time and temperature on sensory properties in low-temperature long-time sous-vide cooking of beef. Journal of Culinary Science and Technology 10: 75-90. https://doi.org/10.1080/15428052.2012.651024

Mouritsen, O.G., 2016. Deliciousness of food and a proper balance in fatty-acid composition as means to improve human health and regulate food intake. Flavour 5: 1.

Mouritsen, O.G. and Styrbæk, K., 2014. Umami: unlocking the secrets of the fifth taste. Columbia University Press, New York, NY, USA.

Mouritsen, O.G. and Styrbæk, K., 2017. Mouthfeel: how texture makes taste. Columbia University Press, New York, NY, USA.

Mouritsen, O.G. and Schmidt, C.V., 2020. A role for macroalgae and cephalopods in sustainable eating. Frontiers in Psychology 11: 1402. https://doi.org/10.3389/fpsyg.2020.01402

Mouritsen, O.G. and Styrbæk, K., 2020. Design and 'umamification' of vegetables for sustainable eating. International Journal of Food Design 5: 9-42.

Mouritsen, O.G. and Styrbæk, K., 2021a. Tsukemono: decoding the art and science of Japanese pickling. Springer Nature, Cham, Switzerland.

Mouritsen, O.G. and Styrbæk, K., 2021b. Octopuses, squid & cuttlefish: seafood for today and for the future. Springer Nature, Cham, Switzerland.

Mouritsen, O.G., Duelund, L., Petersen, M.A., Hartmann, A.L. and Frøst, M.B., 2019. Umami taste, free amino acid composition, and volatile compounds of brown seaweeds. Journal of Applied Phycology 31: 1213-1232. https://doi.org/10.1007/s10811-018-1632-x

Nicolai, T. and Chassenieux, C., 2019. Heat-induced gelation of plant globulins. Current Opinion in Food Science 27: 18-22. https://doi.org/10.1016/j.cofs.2019.04.005

Nishimura, T. and Kuroda, M. (eds), 2019. Koku in food science and physiology. Springer, Cham, Switzerland.

Organ, C., Nunn, C.L., Machanda, Z. and Wrangham., R.W., 2011. Phylogenetic rate shifts in feeding time during the evolution of Homo. Proceedings of the National Academy of Sciences of the USA 108: 14555-14559.

Orlien, V., 2021. Structural changes induced in foods by HPP. In: Knoerzer, K. and Muthukumarappan, K. (eds) Innovative food processing technologies. Elsevier, New York, NY, USA, pp. 112-129. https://doi.org/10.1016/B978-0-08-100596-5.22685-1

Orlien, V., Pedersen, H.B., Knudsen, J.C. and Skibsted, L.H., 2006. Whey protein isolate as functional ingredient in high-pressure induced milk gels. Milchwissenschaft 61: 3-6.

Rico-Campà, A., Martínez-González, M. A., Alvarez-Alvarez, I., de Deus Mendonça, R., de la Fuente-Arrillaga, C., Gómez-Donoso, C. and Bes-Rastrollo, M., 2019. Association between consumption of ultraprocessed foods and all-cause mortality: SUN prospective cohort study. British Medical Journal 365: 1949. https://doi.org/10.1136/bmj.l1949

Schatzker, M., 2015. The dorito effect. Simon & Schuster, New York, NY, USA.

Sharan, S., Zotzel, J., Stadtmüller, J., Bonerz, D., Aschoff, J., Saint-Eve, A., Maillard, M.-N., Olsen, K., Rinnan, Å. and Orlien, V., 2021. Two statistical tools for assessing functionality and protein characteristics of different fava bean (*Vicia faba* L.) ingredients. Foods 10: 2489. https://doi.org/10.3390/foods10102489

Snitkjær, P., Frøst, M.B., Skibsted, L.H. and Risbo, J. 2010. Flavour development during beef stock reduction. Food Chemistry 122: 645-655. https://doi.org/10.1016/j.foodchem.2010.03.025

Sproesser, G., Ruby, M.B., Arbit, N., Akotia, C.S., Alvarenga, M.D.S., Bhangaokar, R., Furumitsu, I., Hu, X., Imada, S., Kaptan, G., Kaufer-Horwitz, M., Menon, U., Fischler, C., Rozin, P., Schupp, H.T. and Renner, B., 2019. Understanding traditional and modern eating: the TEP10 framework. BMC Public Health 19: 1606. https://doi.org/10.1186/s12889-019-7844-4

Srour, B., Fezeu, L. K., Kesse-Guyot, E., Allès, B., Méjean, C., Andrianasolo, R. M., Chazelas, E., Deschasaux, M., Hercberg, S., Galan, P., Monteiro, C.A., Julia, C. and Touvier, M., 2019. Ultra-processed food intake and risk of cardiovascular disease: Prospective cohort study (NutriNet-Santé). British Medical Journal 365: l1451. https://doi.org/10.1136/bmj.l1451

Searchinger, T., 2019. World resources report: creating a sustainable food future. A menu of solutions to feed nearly 10 billion people by 2050. World Resources Institute, Washington DC, USA.

Schmidt C.V. and Mouritsen, O.G., 2020. The solution to sustainable eating is not a one-way street. Frontiers in Psychology 11: 531. https://doi.org/10.3389/fpsyg.2020.00531

Schmidt, C.V., Poojary, M.M., Mouritsen, O.G. and Olsen, K., 2020. Umami potential of Nordic squid (*Loligo forbesii*). International Journal of Gastronomy and Food Science 22: 100275. https://doi.org/10.1016/j.ijgfs.2020.100275

Schmidt, C.V., Plankensteiner, L., Clausen, M.P., Walhter, A.R., Kirkensgaard, J.J.K., Olsen, K., Mouritsen, O.G., 2021a. Gastrophysical and chemical characterization of structural changes in cooked squid mantle. Journal of Food Science 86: 4811-4827. https://doi.org/10.1111/1750-3841.15936

Schmidt, C.V., Plankensteiner, L., Faxholm, P.L., Olsen, K., Mouritsen, O.G. and Frøst, M.B., 2021b. Physicochemical characterisation of sous vide cooked squid (*Loligo forbesii* and *Loligo vulgaris*) and the relationship to selected sensory properties and hedonic response. International Journal of Gastronomy and Food Science 23: 100298. https://doi.org/10.1016/j.ijgfs.2020.100298

Varzakas, T. and Tzia, C. (eds), 2016. Handbook of food processing. CRC Press, Boca Raton, CA, USA.

Willett, W., Rockström, J., Loken, B., Springmann, M., Lang, T., Vermeulen, S., Garnett, T., Tilman, D., DeClerck, F., Wood, A., Jonell, M., Clark, M., Gordon, L.J., Fanzo, J., Hawkes, C., Zurayk, R., Rivera, J.A., De Vries, W., Majele Sibanda, L., Afshin, A., Chaudhary, A., Herrero, M., Agustina, R., Branca, F., Lartey, A., Fan, S., Crona, B., Fox, E., Bignet, V., Troell, M., Lindahl, T., Singh, S., Cornell, S.E., Srinath Reddy,

K., Narain, S., Nishtar, S. and Murray, C.J.L., 2019. Food in the Anthropocene: the Lancet Commission on healthy diets from sustainable food systems. The Lancet 393: 447-492. https://doi.org/10.1016/S0140-6736(18)31788-4

Wrangham, R.W., 2009. Catching fire: how cooking made us human. Basic Books, New York, NY, USA.

Wrangham, R.W. and Conklin-Brittain, N., 2003. Cooking as a biological trait. Comparative Biochemistry and Physiology A 136: 35-46. https://doi.org/10.1016/s1095-6433(03)00020-5

10. Hazard analysis and critical control points (HACCP)

This is what you need to know about food safety management systems based on HACCP principles

Iain M. Ferris

Engineering and Physical Sciences, School of Chemical Engineering, University of Birmingham, Edgbaston, Birmingham B15 2TT, United Kingdom; i.ferris@bham.ac.uk

Abstract

There are many hazards involved in the production of food. Pathogenic bacteria, viruses and parasites pose biological risks (Chapter 12; Vaskoska, 2022), whilst the use of chemicals and pesticides can also render foods unsafe for consumption (Chapter 13; Bast, 2022). Furthermore, the environment in which foods are produced has the potential to contaminate foods with foreign bodies, such as dirt, stones, glass and metal. To limit the risk posed by these food safety threats food businesses need to put in place procedures and controls. A food safety management system based on the principles of hazard analysis and critical control points (HACCP) is a common method used to help businesses identify and control the broad spectrum of food safety hazards. In many countries it is also a legal requirement to implement such a system. Therefore, having a sound grasp of the principles involved is beneficial for all food businesses or food safety professionals.

Key concepts

- *Codex Alimentarius*: Latin for 'Food Code', is a collection of internationally agreed food standards and related texts presented in a uniform manner. These standards aim to protect consumers' health and ensure fair practices in the food trade. The publications of the *Codex Alimentarius* are intended to guide and promote the elaboration and

Bart Wernaart and Bernd van der Meulen (eds)
Applied food science
DOI: 10.3920/978-90-8686-933-6_10, © Iain M. Ferris 2022

187

establishment of definitions and requirements for foods to assist in their harmonisation and in doing so to facilitate international trade. *Codex Alimentarius'* 'General Principles of Food Hygiene' is recognised as the primary source of information on the application of HACCP. See http://www.fao.org/fao-who-codexalimentarius/en/ for more information.

► Critical Control Point (CCP): A step at which a control measure or control measures, essential to control a significant hazard, is/are applied in an HACCP system.

► Critical limit: A criterion, observable or measurable, relating to a control measure at a CCP which separates acceptability from unacceptability of the food.

► Food Safety Management System: The combination of the prerequisite programmes and the HACCP system which will produce safe food.

► Good Hygiene Practices (GHPs), Good Manufacturing Practice (GMP), Good Agricultural Practice (GAP): Fundamental measures and conditions applied at steps within the food chain to provide safe and suitable food.

► Hazard: A biological, chemical or physical agent in food with the potential to cause an adverse health effect.

► Hazard analysis: The process of collecting and evaluating information on hazards identified in raw materials and other ingredients, the environment, in the process or in the food, and conditions leading to their presence, to decide whether or not these are significant hazards.

► Hazard Analysis and Critical Control Points (HACCP): an approach to the assessment of food safety hazards and establishing control systems that focus on control measures for significant hazards along the food chain, rather than relying on end-product testing. It consists of seven key principles to guide the HACCP team in its deliberations. Before it is completed the prerequisite programmes must be in place and five preliminary steps carried out.

► Prerequisite programmes (PRP): Programmes, practices and procedures such as training, pest control, premises design, cleaning and traceability, that establish the basic environmental and operating conditions that set the foundation for implementation of an HACCP system.

► Risk: A function of the likelihood of a food safety hazard occurring and the severity of the adverse effect that may occur.

► Safer Food Better Business: A food safety management system produced by the UK's Food Standards Agency to help small catering businesses with their legal obligation to implement food safety procedures.

► Validation: Evidence that the plan if implemented correctly should result in safe food.

► Verification: Checks that your plan is being implemented and working as intended.

10.1 The development of HACCP

Dr Howard Bauman of the Pillsbury Company (Case 10.1) first presented HACCP to the food industry in 1971, but the early attempts by food companies barely resemble today's detailed systems. In the 1980s organisations including the Codex Committee on Food Hygiene (CCFH) produced guidance that helped businesses implement the HACCP principles (Bernard, 1998). In 1997 the CCFH incorporated HACCP requirements into its international code of practice called the 'General Principles of Food Hygiene'. This document is now considered a primary source of information on food safety requirements and in particular HACCP. Up until the 1990s the implementation of HACCP was mostly voluntary, but many countries have since introduced legislative requirements that make it mandatory for food businesses. Within the UK and European Union, Regulation 852/2004 on the hygiene of foodstuffs largely resembles the Codex document and article 5 makes it a legal requirement to 'put in place, implement and maintain a permanent procedure or procedures based on the HACCP principles' (EC, 2004). Regulation 852/2004 also states that the HACCP requirements should take account of the principles contained within Codex.

Case 10.1. HACCP from out of space.

HACCP as a method of ensuring food safety is credited to the American space programme. Working with the National Aeronautics and Space Administration (NASA) in the 1960s, the Pillsbury Company were tasked with the challenge of producing foods that could be guaranteed safe to eat but also palatable. Previously, foods that required such high-level assurance were usually sterilised, involving severe thermal processes that often affected the quality. Prior to the introduction of HACCP, end-product sampling was commonly used as a method of verifying the microbiological safety of the foods. However, due to the non-homogenous nature of contamination it is difficult to provide the level of confidence required that the food is free from pathogenic bacteria without sampling large quantities of the product. Therefore, a system was developed that examined the whole process to identify where and how contamination may occur and the critical points for assuring the safety of the food. This systematic approach to the complete production when combined with appropriate checks and records should provide assurance that the food is safe without the reliance on end-product sampling (Lytton, 2019: 65-66).

10.2 Food safety management system based on HACCP principles

In its most basic form HACCP is about hazards and controls. *Codex Alimentarius* (2020) describes a hazard as 'a biological, chemical or physical agent in food with the potential to cause an adverse health effect' and a control measure as 'any action or activity that can be used to prevent or eliminate a hazard or reduce it to an acceptable level'. Essentially, if the hazards are well understood and measures are put in place to control them, then the objective of a HACCP based system will be achieved. The seven HACCP principles help food professionals to ask the right questions, which aids understanding of the hazards and which controls are needed. These principles are detailed in Figure 10.1 and will be discussed later in this chapter.

10.2.1 Preliminary activities

Codex Alimentarius' (2020) General Principles of Food Hygiene identifies five actions that are necessary before beginning the plan. These activities ensure there is sufficient expertise and information about the product and how it is processed in order to be able to assess the hazards and ensure the controls are adequate. The preliminary activities are as follows: (1) Assemble an HACCP team; (2) describe the product; (3) identify the intended use; (4)construct a flow diagram; and (5) on-site confirmation of flow diagram.

assemble HACCP team and identify scope

To be able to create a robust plan will require knowledge across a range of subjects in addition to the HACCP principles. Identifying and analysing hazards requires knowledge of the food, its ingredients and the risk of different biological, chemical or physical contaminants that may be present. Additionally, site specific knowledge, and an understanding of the equipment and production

PRINCIPLE 1 Conduct a hazard analysis and identify control measures.
PRINCIPLE 2 Determine the Critical Control Points (CCPs).
PRINCIPLE 3 Establish validated critical limits.
PRINCIPLE 4 Establish a system to monitor control of CCPs.
PRINCIPLE 5 Establish the corrective actions to be taken when monitoring indicates that a deviation from a critical limit at a CCP has occurred.
PRINCIPLE 6 Validate the HACCP plan and then establish procedures for verification to confirm that the HACCP system is working as intended.
PRINCIPLE 7 Establish documentation concerning all procedures and records appropriate to these principles and their application.

Figure 10.1 The seven HACCP principles.

methods used are also important in order to establish whether the hazards identified are being controlled. Assembling a multidisciplinary team is therefore beneficial to provide sufficient knowledge to create an effective HACCP plan.

Before establishing an HACCP plan the team needs to agree the scope. This should detail the range of activities considered within the plan and the types of hazards identified. The scope of the plan should cover all steps from raw ingredient purchasing through production and supply to the customer. It should also establish the nature of the biological, chemical and physical hazards considered. Where a business may not have the expertise already available internally, advice may need to be obtained from other sources, such as industry or regulatory bodies, independent experts, and HACCP literature.

Sara Mortimer and Carol Wallace (1998: 34-36) suggest that the team requires several other attributes in addition to their expertise. They need research skills and to be able to solve problems and evaluate data, but also leadership and communication skills to ensure that the plan is implemented.

Although in many organisations the HACCP team will be comprised of a group with a variety of expertise, *Codex* suggests that it is feasible that a well-trained individual with access to adequate guidance could establish an HACCP plan. It has however been acknowledged that small and less developed businesses can have difficulty producing an HACCP plan due to a lack of expertise, and governments have been encouraged to produce strategies to address this problem (FAO/WHO, n.d.). The European Commission (2016) acknowledges this problem and has advised on the flexibility that can be applied with regards to the implementation of HACCP-based procedures. The guidance suggests some ways in which the principles can be complied with in a manner that requires less in-depth knowledge. The degree of flexibility, however, should relate to both the nature and the scale of activities, and it should not compromise food safety. One such suggestion is the use of generic guides where many businesses have common hazards and controls. In the UK, for example, the Food Standards Agency has produced a food safety management system for small catering businesses called 'Safer Food Better Business' that addresses the common hazards associated with this type of business. The expertise is thus provided by the authority and the food business is guided through a basic assessment of their controls without the need for much knowledge of HACCP or other subjects such as microbiology. (Food Standards Agency 2020)

To carry out an analysis of the hazards the food business needs to provide a full description of the product that includes the ingredients and both intrinsic and extrinsic characteristics. This information will be useful in determining which hazards are likely to be present and whether the food could be rendered unsafe.

describe product The ingredients used may introduce biological, chemical and physical hazards, some of which may be specific to particular ingredients. Raw poultry meat, for example, is commonly associated with pathogenic bacteria such as *Salmonella* and *Campylobacter,* and scombrotoxin is a hazard in oily fish, such as mackerel and tuna. Pesticides may be a chemical concern for fruit and vegetable ingredients, and certain food additives are restricted for safety reasons. The nature of the ingredients may also determine what physical contaminants may be present, including stones, dirt and insects in vegetables. A full list of the ingredients used in different products will therefore allow the HACCP team to identify many of the hazards that have to be controlled.

In addition, the description should include the intrinsic characteristics of the food as these will help in the assessment of the risk posed by certain hazards (see also Case 10.2). Biological hazards such as food poisoning bacteria will only grow in foods that provide a suitable environment. Foods that are acidic and have a low pH, are dry and have a low water activity (a_w) or contain significant amounts of sugar or salt are not conducive to bacterial growth. Therefore, characteristics, such as the pH and a_w must be identified.

Case 10.2. Botulism in hazelnut puree yoghurt.

The UK's largest outbreak of botulism food poisoning that affected 27 people and killed 1 person serves as a lesson in the importance of understanding the intrinsic characteristics of food and how the process renders it safe. The outbreak in June 1989 was linked to the consumption of hazelnut yoghurt and subsequent investigations found the presence of *Clostridium botulinum* toxin in unopened containers of both the puree and the yoghurt. An investigation of the process identified that the heat process used to produce the hazelnut puree was insufficient to eliminate *C. botulinum* spores and the pH not low enough to prevent the growth and toxin formation. For other, more acidic fruit purees produced by the manufacturer the heat process would have been sufficient as the pH produces conditions in which this pathogen will not grow. The hazelnut purees were also part of a batch that had used a sweetener as a replacement for sugar within the product that may have also previously controlled the growth of this deadly pathogen (O'Mahony *et al.* 1990).

Extrinsic factors should also be described and would include how the product is packaged and stored, as these will also have an impact on the safety of the product. Canned and vacuum-packed foods exclude oxygen from the product which provides ideal conditions for pathogens such as *Clostridium botulinum* to grow and thus would need to be considered as a hazard within products packaged this way (see Case 10.2, for example). The temperature at which foods are stored will also determine which bacteria will or won't grow. The growth of *Listeria monocytogenes*, for example, would pose a hazard in ready-to-eat foods even at chilled temperatures.

Any treatments that the product has been subjected to should also be included within the description, as these may determine any remaining hazards in the final product. For example, if the product has been subjected to a heat treatment, such as pasteurisation, many pathogenic bacteria will be eliminated, although more heat-resistant spore formers may remain.

To ensure an effective plan the HACCP team needs to understand the food being produced and the hazards associated with its production. Where foods are produced that are sufficiently similar then it may not be necessary to produce a plan for each product, although careful consideration should be given to any significant differences.

identify intended use How the product is intended to be used may have an impact on the assessment of the safety of the product. One of the most significant points is whether the product needs to be cooked or is ready to eat. Care should be taken here, however, as in many countries the law requires that some foods intended to be cooked must still be free from certain pathogens even though the cooking should render the product safe. For example, the presence of *Salmonella* in poultry products intended to be eaten cooked (Commission Regulation (EC) No 2073/2005 (EC, 2005)) and verotoxigenic *Escherichia coli* in ground beef may be considered unsafe (Centers for Disease Control and Prevention, 2019) and it may not be appropriate to rely on cooking by the consumer to render the product safe.

Who the product is intended for may also affect the plan. If the product is intended for a particular group of consumers such as infants or people with allergies it will impact the analysis of the hazards. For instance, pesticide levels in infant foods are a greater risk than in foods intended for adults and thus may require tighter controls.

construct flow diagram The HACCP team should construct a flow diagram that details all of the significant steps in the operation for a specific product. This is useful as it breaks down the process and allows the HACCP team to consider which hazards may occur at different points in the process and should therefore contain sufficient technical data.

Where products use similar steps, it is possible to use the same flow diagram. It is important that the team also considers activities before and after the processes within the operation. What may have happened to the food before it reaches the food business and what happens when it has left the business should form part of the HACCP team's deliberations. In addition, any reworking of product and waste should also be included within the process flow diagram.

on site confirmation flow diagram In order to verify that the steps included within the plan are accurate, the HACCP team should check the processing operation against the flow diagram. Consideration should also be given to any changes that may occur during different stages and hours of operation. Where discrepancies are identified between the flow diagram and what is observed, then the diagram should be amended.

10.2.2 Prerequisite programmes (PRPs) and HACCP

In order to successfully implement a food safety management system based on the HACCP principles, a food business must first establish environmental and operational conditions conducive to the success of the plan. Food business operators should be complying with good hygiene practice (GHP), good agricultural practices (GAP) and good manufacturing practice (GMP). These practices are considered prerequisites and are detailed in chapter 1 of the *Codex Alimentarius* (2020) 'General Principles of Food Hygiene'. To achieve the objectives of these prerequisites involves establishing programmes before and during the implementation of HACCP procedures. These are commonly referred to as prerequisite programmes or PRP.

A comprehensive list of prerequisites can be found in chapter 1 of the *Codex* (2020) 'General Principles of Food Hygiene' and includes the following:
- ► premises design & facilities;
- ► pest control;
- ► maintenance;
- ► cleaning;
- ► supplier management;
- ► allergen management;
- ► personal hygiene;
- ► training;
- ► operational controls.

Prerequisite programmes are essential as they provide the foundations for the implementation of the HACCP plan. PRPs apply generally rather than to a specific hazard, however they will often contribute to the management of many hazards. For example, the training programme, cleaning and premises design will all contribute to minimise the risk of cross-contamination.

PRPs are likely to include records such as standard operating procedures (SOPs), training plans, cleaning instructions, pest control checks and internal audits.

The relationship between PRPs and HACCP is often presented as a pyramid (Figure 10.2) with the PRPs providing the foundation upon which the HACCP plan is built. Whilst they are often contemplated separately, the true relationship between them is more interdependent. For instance, in the first HACCP step the effectiveness of the PRPs are considered within the analysis of the hazards. If the PRP adequately controls the hazard, then no further consideration is required within the HACCP plan. However, *Codex Alimentarius* (2020) states that there will inevitably be situations where some of the specific prerequisites contained in the General Principles of Food Hygiene are not applicable. *Codex* identifies this by using phrases such as 'where necessary' and 'where appropriate'. In practice, this means that, although the requirement is generally appropriate and reasonable, there may be some situations where it is not necessary on the grounds of food safety. What is required within a PRP will require an assessment of the potential hazards involved and the nature of the foods produced, which is in line with the HACCP approach. For example, the adequacy of handwashing provisions will depend largely on an assessment of the hazards posed by a food business (Case 10.3).

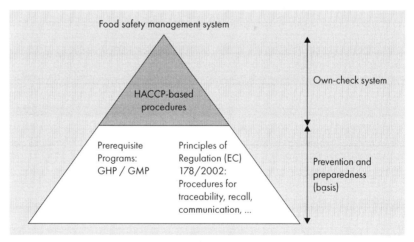

Figure 10.2. Food safety management system (EC, 2016).

Case 10.3. Handwash basin provisions.

The European court has made rulings that support this approach to the prerequisites. In 'Astrid Preissl KEG v Landeshauptmann von Wien' (European Court, 2011), the court was asked to rule on the requirements of paragraph 4 of Chapter I of annex II which states:

> An adequate number of washbasins is to be available, suitably located and designated for cleaning hands. Washbasins for cleaning hands are to be provided with hot and cold running water, materials for cleaning hands and for hygienic drying. Where necessary, the facilities for washing food are to be separate from the hand-washing facility.

In this case, the competent authority ordered a food business to install a washbasin with hot and cold running water, a soap dispenser and a paper towel dispenser in the staff toilet of the establishment managed by it. It was also laid down that the taps could not be hand-operated.

The establishment in question was a bar which serves almost no food, except toast, and there was a sink equipped with hot running water which could be used at any time for washing hands, but which was also used for washing dishes. An appeal was lodged against the authority's interpretation of the requirement. The question was asked as to whether the requirements of paragraph 4 meant that a food premises should have a basin exclusively for the use of washing of hands and that the taps should not be operated by hand.

In its ruling the court said that the law should not be interpreted such that a food premises should have a basin exclusively for the use of washing of hands and that the taps should not be operated by hand, but the particular circumstances of a case should be considered when interpreting the requirements, rather than in a general way and irrespective of those circumstances. Essentially, what the ruling is saying is that consideration should be given to whether it was necessary based on risk to food safety to have sinks dedicated solely for the washing of hands.

As mentioned earlier, EU Regulation 852/2004 largely reflects the *Codex* document. Article 4 of 852/2004 requires food business operators to comply with annex I or II that contains much of the prerequisites (EC, 2004). The language used in the annexes also provides the flexibility to enable the requirements to apply appropriately to any food business.

What is required in creating a prerequisite programme will therefore depend on the nature of the business and type of foods handled. Consider, for example, a premises that is merely retailing prepackaged foods compared with a butcher who is handling both raw and cooked meats. PRPs such as training and cleaning will be very different in these different businesses. In Case 10.3 we describe a court ruling that demonstrates how the context of the particular food business determines the required hygiene requirement.

responsibility of the food business operator

In the first instance it is the responsibility of the food business operator to put in place adequate programmes, although governments and organisations are encouraged to produce guidance. Industry-specific guides to good hygiene practice are a valuable source of information for which prerequisite procedures should be in place and can provide some of the expertise needed to implement a food safety management system. *Codex Alimentarius*, in addition to providing the General Principles of Food Hygiene, also provides specific codes of hygienic practice related to particular food industries or a particular prerequisite such as allergen management.[46]

It is important to ensure that PRPs are well considered, as inadequate programmes can have an impact on the overall food safety management system. Poorly implemented PRPs are likely to introduce hazards not controlled within the HACCP plan, e.g. if staff failed to wear the appropriate protective clothing, it would increase the risk of physical contaminants, such as hair. Failure of the PRP in most cases will not render the food unsafe for consumption, although finding a hair in your food is not very pleasant. Some PRPs, however, may be of greater importance, such as the cleaning of food contact surfaces in the ready-to-eat production area. This will mean that some programmes will need more stringent controls depending on the nature of the business. These are often referred to as operational PRPs or oPRPs.

10.2.3 The 7 HACCP principles

The objective of the HACCP system is to examine the various hazards posed in food production and to strengthen controls in the areas that are critical to food safety. Whereas the PRPs relate to the general environment, in HACCP the hazards will be much more related to the specific product and the process. The seven HACCP principles provide a systematic method of conducting the study and they are as follows.

---

[46] Code of Practice on Food Allergen Management for Food Business Operators (CXC 80-2020) available at http://www.fao.org/fao-who-codexalimentarius/codex-texts/codes-of-practice/en/

Principle 1: conduct a hazard analysis and identify control measures

This first principle will inevitably take the longest and calls upon the expertise of the HACCP team. In the first instance the team should identify all major hazards that may occur during the different steps throughout the process. The preliminary steps, if completed thoroughly, should provide some of the information needed to carry out the assessment. The product description should include the raw ingredients used within the product that are likely to be one of the major hazard sources. Many of these hazards are likely to be specific to a food and the team may need to conduct research to identify and understand them. Having a full description of the process will also allow the team to assess which hazards may occur within the operation but also some of the controls that are present.

significant hazards

The HACCP team then has to identify those hazards that must be controlled to ensure the safety of the end product. These are often referred to as the 'significant hazards' and defined in *Codex Alimentarius* (2020) as hazards 'identified by a hazard analysis, as reasonably likely to occur at an unacceptable level in the absence of control, and for which control is essential given the intended use of the food.' A worksheet (Figure 10.3) is often used to help organise the identification and assessment of the different hazards at the various steps in the process and to record the team's assessment. When describing the potential hazards it is important to be specific to enable an informed assessment. So rather than 'glass' or '*Salmonella*' being identified as a hazard, the plan would detail the source or reason, e.g. 'glass fragments from breakages during filling' or 'Insufficient cooking resulting in the survival of *Salmonella*'.

During the deliberations the team would consider whether the particular hazard in question is already addressed partly or in full by the prerequisite programmes. Glass breakages could potentially be covered by a glass policy with standard operating procedures and thus it doesn't have to be addressed by the HACCP plan. Where the hazard is deemed to be controlled by the PRP, the team should satisfy themselves that the programme is adequate. In contrast, the risk that *Salmonella* would survive the cooking process would not be covered by the PRPs. In this instance the hazard would need to be included within the plan and details of the measures required should be provided, such as the cooking time and temperature combination that may be needed.

Guidance issued by the European Commission (2016) suggests a number of areas for consideration in the hazard analysis stage. Firstly, the team should consider the risk that the defined hazard will cause a problem in the end-product. The degree of risk is a product of the likelihood of occurrence of hazards and the severity of their adverse health effects. When assessing the

(1) Step*	(2) Identify potential hazards introduced controlled or enhanced at this step B = biological C = chemical P = physical	(3) Does this potential hazard need to be addressed in the HACCP plan?		(4) Justify your decision for column 3	(5) What measure(s) can be applied to prevent or eliminate the hazard or reduce it to an acceptable level?
		Yes	No		
	B				
	C				
	P				
	B				
	C				
	P				
	B				
	C				
	P				

*A hazard analysis should be conducted on each ingredient used in the food; this is often done at a 'receiving' step for the ingredient. Another approach is to do a separate hazard analysis on ingredients and one on the processing steps.

Figure 10.3. Hazard analysis worksheet (*Codex Alimentarius* General Principles of Food Hygiene, 2020).

risk of a particular hazard, thought should be given to any subsequent steps in the process that may eliminate the hazard. So, for instance, the risk posed by the presence of *Salmonella* in raw ingredients at the goods received stage is low if the ingredients will be cooked later, but may be considered a greater risk if the food is not going to be cooked. Again, the role of the PRPs should also be contemplated when considering the risk and whether they will be effective in eliminating the hazard. If so, then the risk posed is low.

The analysis of the hazards requires the HACCP team to have sufficient understanding of them and highlights the importance of assembling a team with sufficient expertise. Different hazards cause the food to become unsafe in a variety of ways and similarly the methods of control can vary. The analysis should include a qualitative and quantitative evaluation of the presence of hazards and thus would include a review of the mechanism by which the hazard affects the food and also any numerical information such as legal limits or amounts needed to cause illness.

Many of the hazards have diverse characteristics that affect the likelihood of them rendering the food unsafe. Pathogenic bacteria, for instance, will multiply only where the conditions are right and different species have varying ability to grow in foods with low water activity, pH and oxygen levels or when stored under refrigeration. Some may cause illness by producing a toxin within the food and others infect the host. Some pathogens have a low infective dose and therefore do not need to grow but others require large numbers before they cause harm. Additionally, their differing ability to resist heat treatments means the process may not be sufficient to eliminate all strains of bacteria or their toxins that may be present.

The preliminary activities of describing the product and the process will provide valuable information to assess the risk posed. Furthermore, with sufficient understanding of the hazard the team can identify controls that eliminate the risk or reduce it to an acceptable level. In some cases the controls may be a combination of factors that control the hazard rather than a single measure. Parameters such as time, temperature, water activity, pH and humidity may combine to ensure the safety of food. This is often referred to as 'hurdle' technology (Mukhopadhyay and Gorris, 2014: 221-227).

barriers to implementation One of the most common barriers to implementing HACCP in small and less developed businesses, however, is their lack of knowledge (Walker *et al.*, 2003). This can be partly overcome by using generic HACCP-based procedures provided by industry bodies or competent authorities. Hence the expertise is provided externally and as long as the business implements the recommended controls they are likely to produce safe food. Plans such as 'Safer Food Better Business' provided for small catering businesses by the UK's Food Standards Agency (2020) remove much of the technical information related to the hazards and require less expertise to implement. The business still has to ensure, however, that the plan fully covers all of its activities.

Principle 2: determine the critical control points

Once the HACCP team has established which hazards are significant, the next step is to determine the critical control points (CCPs). *Codex Alimentarius* (2020) defines a CCP as 'a step at which a control measure or control measures, essential to control a significant hazard, is/are applied in a HACCP system' and the European Commission (2016) advises that they are the 'points in a production process where a continuous/batch wise control via a specific control measure is required to eliminate or to reduce the hazard to an acceptable level'.

In order to identify whether the point in the process is critical to the safety of the food, and therefore a CCP, a series of questions or a decision tree is sometimes used (see, for instance, Figure 10.4). The purpose of the questions is to consider whether the step in the process is essential to ensure that a particular hazard is controlled or whether there are others controls later in the process that would do so. If the point in the process does not control the hazard, or there is another point later in the process that will eliminate the hazard, then the point is not a CCP for that hazard.

It is important to consider each hazard at each step individually as the step may be critical to control one hazard but not others. For instance, pasteurisation is likely to be a CCP for some pathogenic bacteria, e.g. *Listeria*, but others will not be controlled by these temperatures as they are more heat tolerant, e.g. *Clostridium perfringens*. The cooling step may subsequently be a CCP for these heat-resistant pathogens to prevent their growth but the elimination of other pathogens at the previous step will mean the cooling is not significant for their control. It is possible to group some hazards together where the control would eliminate all of them. In the case of heat processing, temperatures used to eliminate *Listeria*, for example, would also control other pathogens such as *Salmonella*.

Although traditionally a decision tree may be used to identify CCPs, *Codex* does not prescribe that they have to be used. The expertise of the HACCP team may be sufficient to determine what is critical to the safety of the food.

When identifying critical control points the team must be able to identify what the preventative measures are for each of the given hazards. If the step being examined is not the critical control point, then the team should be able to

---

Question 1: Do preventative measure(s) exist within the system that will control the hazard?
If yes, go to Question 2.
If no, is control at this step necessary for safety?
If yes, modify step, process, or product.
If no, not a CCP. Stop and proceed to the next identified hazard in the described process.

Question 2: Is the step in question specifically designed to eliminate or reduce the likely occurrence of a hazard to an acceptable level?
If yes, it's a Critical Control Point.
If no, go to Question 3.

Question 3: Could contamination with identified hazards occur in excess of acceptable level(s) or could these increase to unacceptable levels?
If yes, go to Question 4.
If no, not a CCP. Stop and proceed to the next identified hazard in the described process.

Question 4: Will a subsequent step eliminate identified hazard(s) or reduce likely occurrence to an acceptable level?
If yes, it's not a CCP. Stop and proceed to the next identified hazard in the described process.
If no, it's a Critical Control Point.

Figure 10.4. Example of critical control points decision tree.

identify how the hazard is controlled. In some instances there may be different methods of controlling specific hazards that could be used, for example, in the control of *E. coli* O157 in beef burgers (discussed in Case 10.4).

Case 10.4. Burgers less than thoroughly cooked.

Raw beef is hazardous and is commonly associated with *E. coli* O157. This particular pathogen is found in the intestine of some cattle and during the slaughtering process the surface of meat can sometimes become contaminated. Since *E. coli* is not particularly resistant to heat treatment, a few seconds on a hot grill to sear each side of a steak should eliminate this hazard since the contamination is found on the surface of the meat. However, when the beef is minced and made into burgers any contamination on the surface of the meat will be transferred to the centre, and therefore cooking the burger thoroughly is required to control the risk. It is suggested that the centre of the burger is cooked to a temperature of 70 °C for 2 minutes or equivalent (Food Standards Agency, 2016). This would be considered the CCP. Some restaurants, however, prefer to cook their burgers less thoroughly where the meat remains pink in the middle. The restaurant thus no longer has the control of thorough cooking for the hazard and should identify how they reduce the risk.

The UK's Food Standards Agency provides a number of suggestions of alternative methods that could be used to reduce the risk. One suggestion is that the burgers are produced using a 'sear and shave' method where the whole meat cuts are heated on the outside first and then the meat is trimmed. The trimmed meat can then be used to make burgers that do not require to be thoroughly cooked since the searing has eliminated the hazard. The searing of the meat would become the CCP. Alternatively, the restaurant could source mince or burgers from suppliers whose controls are sufficient to reduce the likelihood of the meat being contaminated with the pathogen in the first place. In this case the restaurant will need to establish that the supplier has sufficient hygiene practices and/or measures to control pathogens, such as *E. coli* O157.

During the analysis of the hazards and determination of CCPs it may become apparent that some PRPs play a more significant role in the control of certain hazards. In the case study above, clearly supplier management for the restaurant will be extremely important as a method of controlling the risk posed by pathogens, such as *E. coli* O157 when cooking is less than thorough (Case 10.4). Where this occurs, these parts of the prerequisite programme should be highlighted for their importance. These areas are often identified as 'operational prerequisite programmes' or 'oPRPs' and are described as 'PRPs that are typically linked to the production process and are identified by the hazard analysis as essential, in order to control the likelihood of the introduction, survival and/or proliferation of food safety hazards in the product(s) or in the processing environment' (European Commission, 2016).

Further examples of oPRPs include the increased cleaning required within the ready-to-eat areas of production or where different allergenic ingredients are processed. Supplier management may also be an oPRP where certain hazards may be present within the raw ingredients that are difficult to process out such as pesticides. Additionally, the establishment of an appropriate shelf life for foods that support the growth of pathogenic bacteria would be an oPRP. Where PRPs are identified as essential to control a significant hazard, then the programme should ensure that there is sufficient monitoring in place.

Principle 3: establish validated critical limits

Codex Alimentarius (2020) defines a 'critical limit' as 'a criterion, observable or measurable, relating to a control measure at a CCP which separates acceptability from unacceptability of the food'. Essentially these are criteria beyond which the food should be treated as unsafe. Typical measurements would include parameters such as time, temperature, water activity, pH, salt content and humidity. In some cases observations such as how a food looks, for example, when thoroughly cooked may also be acceptable.

These limits must be proven to control the hazard and can be based on generally accepted scientific advice, guides to good practice or legal standards. Alternatively, if this is not available, then scientific studies should be carried out to establish the critical limits and/or demonstrate that a particular control will actually work. The proof that the critical limits will control the hazard is part of principle 6 described below.

A business may also establish target measurements that may be stricter than the critical limits, to factor in a level of uncertainty or variability within the process.

Principle 4: establish a system to monitor control of CCPs

Where a point has been identified that is critical to the safety of the food, the HACCP team must plan a sequence of observations or measurements of control parameters to assess whether the control measure is achieved. The monitoring needs to check that the production process does not breach any of the critical limits, and the timing and frequency of the checks should ensure that where the food has exceeded a critical limit it can be identified and isolated (*Codex Alimentarius*, 2020).

In some cases, the monitoring may be continuous, such as the cooking temperatures or the speed of a conveyor belt, but in other cases it may be intermittently such as checking the pH of products. There may be alternative ways to monitor the CCP such as measuring weight loss as an alternative to checking the reduced water activity in some dried foods. Additionally, some monitoring may not consist of a measurement but may be an observation, such as checking the settings on machinery or that a liquid is boiling. However, the team decides to monitor the CCP, it must ensure that it is timed to be able to take corrective action and prevent unsafe product from reaching the consumer.

In smaller businesses, such as catering outlets, the European Commission (2016) suggests that monitoring may not need to be carried out on products when following established procedures that have been shown to produce safe food. An example of this may be where the business has checked that a certain time in the microwave on a particular setting will ensure the food is heated thoroughly. The business need not probe every food as long as staff follow the correct procedure and the business verifies that the microwave is working correctly, which may involve periodic checking of foods.

Principle 5: establish the corrective actions to be taken when monitoring indicates that a deviation from a critical limit at a CCP has occurred

If a critical limit is breached, it is important that the person carrying out the checks knows what action to take. This should be considered beforehand so that the intervention can be taken without delay.

The actions required will depend on the hazard posed. A product may be able to be reworked or subjected to further treatment, such as heating, but in other cases this may not be possible, for example where a heat stable toxin could have formed within the food, in which case the food would have to be discarded.

The actions should also involve an investigation into the root cause of the deviation and a periodic review that identifies any regular breaches of critical limits. This enables the business to address problems that may be the cause of the deviation.

Principle 6: validate the HACCP plan and then establish procedures for verification to confirm that the HACCP system is working as intended

The plan must be effective in controlling hazards associated with the food, i.e. that the controls are valid. Validation is the action of 'obtaining evidence that a control measure or combination of control measures, if properly implemented, is capable of controlling the hazard to a specified outcome' (*Codex Alimentarius,* 2020). Evidence may take different forms, but the plan should be validated before it is implemented.

validation

Firstly, validation will usually involve a review of established scientific data or guidance issued by competent authorities. In some cases, this may be sufficient to validate that the control of some hazards is adequate; for example, the times and temperatures required to destroy certain pathogenic bacteria is well established. However, the food business should ensure that the information is properly applied and any conditions relating to the data are met. For instance, guidance on commercial sterilisation processes such as canning has different temperature requirements based on the acidity of the product. As part of the validation process, in addition to establishing the pH, the business would also need to demonstrate that the process used would achieve the necessary temperatures. Where generic systems have been created by a competent authority, e.g. a guide for caterers, the validation should have been conducted by that authority and following the guidance should result in safe food.

Other methods of validation include mathematical modelling. This is often used to predict the growth of bacteria under specified conditions such the pH, water activity and temperature using computer programmes. Another alternative is to carry out tests such as 'challenge studies', to determine whether food pathogens would grow or are reduced. In these cases, foods are often deliberately inoculated with a particular bacterium and the food is monitored to see if the bacteria grow. In some cases, validation studies may have been carried out already, for example cleaning chemicals are often certified to a standard such as BS EN 1276/2019. This standard certifies that the chemical can achieve at least a 5 $\log_{10}$ reduction of specified strains of bacteria within 5 minutes of application (British Standards Institute, 2020). Thus, provided the business follows the manufacturer's instructions, the method of disinfecting surfaces is validated. This may be sufficient validation in small catering businesses, but larger food manufacturers may have to initially conduct studies to determine if the disinfection process works.

Where there are changes to the product or process, the HACCP team should review what effect this will have and revalidate the process. In Case 10.2 earlier, the substitution of sugar as an ingredient with a sweetener should have resulted in a review and revalidation of the process.

verification

Whereas validation is about showing that the controls will work, it relies on them being implemented fully. Verification is required to show that this is the case. *Codex Alimentarius* (2020) describes verification as 'the application of methods, procedures, tests and other evaluations, in addition to monitoring, to determine whether a control measure is or has been operating as intended'. Verification could, for example, include checks to ensure that staff are carrying out the monitoring of CCPs as required, but it could also include sampling to determine if the control is working as expected.

Verification is also important in the management of PRPs as they play a significant role in the overall food safety management system. If PRPs are not being implemented, they may result in the introduction of hazards not considered as part of the HACCP plan. Examples of verification of PRPs could include testing that staff understand the training received, reviewing pest control reports, checking cleaning standards and testing that ingredients meet product specifications.

Validation and verification are quite easily confused (not least because they sound alike). Figure 10.5 summarises these terms.

Principle 7: establish documentation concerning all procedures and records appropriate to these principles and their application

The final HACCP principle requires the team to create documents and keep records to support the system. *Codex Alimentarius* (2020) states that the 'documentation and record keeping should be appropriate to the nature and size of the operation and sufficient to assist the business to verify that the HACCP controls are in place and being maintained'. Therefore, larger and more complex businesses should generate more documents and records to support the system.

food safety management system The food safety management system should firstly contain details of the plan including who was involved in its creation, details of the products and processes, and the decisions that were made. The *Codex* guidance provides some example tables to summarise the conclusions during the creation of the plan. Figure 10.3 details the decisions during the hazard analysis step and Figure 10.6 summarises the critical control points. Where necessary, the plan should also contain documents that support and validate the plan; for example, copies of codes of practice, scientific papers or relevant regulations that were referred to during its construction. Further documents may include standard

VALIDATION = Evidence that the plan should result in safe food.
VERIFICATION = Checks that the plan is being implemented and working as intended.

Figure. 10.5. The difference between validation and verification.

Critical control points (CCPs)	Significant hazard(s)	Critical limits	Monitoring				Corrective actions	Verification activities	Records
			What	How	When (Frequency)	Who			

Figure 10.6. Critical control points.

operating procedures and policies for the PRPs detailing how particular tasks are carried. Essentially the documentation should detail how the different programmes and controls operate and manage the hazards.

records

Records should also be maintained that verify that controls are being implemented. These would include supporting evidence that the PRPs are in place and that critical control points are being monitored. Typical records that could be found within the PRPs include cleaning checks, delivery invoices, product specifications, training records, pest control reports and results from environmental swabs and product sampling. The monitoring of the CCPs should also be evidenced by adequate records that detail the checks carried out, e.g. temperature checks and any corrective actions that need to be taken. In addition to the records for specific areas, the plan is likely to contain records of audits of the overall system verifying that the whole system is operating according to the plan.

10.2.4 Management commitment and food safety culture

Creating a food safety management system relies heavily on an understanding of food science, but its success is dependent on people implementing it. Therefore, it has been suggested that more focus should be placed on the behavioural sciences and the food safety culture. As Frank Yiannas (2008) puts it,

You can have the best documented food safety processes and standards in the world, but if they are not consistently put into practice by people, they are useless. Accordingly, our system has to address both the science of food safety and the dimensions of organisational culture and human behaviour.

As far back as the 1990s some academics were highlighting the role that behavioural science can play within food safety and why people do not always implement food safety practices despite having received training (Griffith *et al.,* 1995). In 2009, food safety culture was identified as a contributing factor that led to an outbreak of *E. coli* O157 that affected 157 people, including 127 under the age of 18. Within the public inquiry Professor Griffith described food safety organisational culture 'as a manifestation of the values and beliefs and attitudes within a workforce. Its formation is dependent upon the knowledge, standards, motivation and leadership of the person in charge, how they communicate with, and are trusted by, the staff' (Pennington, 2009).

organisational culture

Whilst *Codex Alimentarius'* 'general principles of food hygiene' has always advised that management commitment was necessary in order to ensure the effectiveness of the food safety management system, there was no mention of food safety culture until the 2020 edition. It now states that:

> fundamental to the successful functioning of any food hygiene system is the establishment and maintenance of a positive food safety culture acknowledging the importance of human behaviour in providing safe and suitable food.

Codex also suggests some elements that contribute to the establishment of a positive food safety culture:

- ► commitment of the management and all personnel to the production and handling of safe food;
- ► leadership to set the right direction and to engage all personnel in food safety practices;
- ► awareness of the importance of food hygiene by all personnel in the food business;
- ► open and clear communication among all personnel in the food business, including communication of deviations and expectations; and
- ► the availability of sufficient resources to ensure the effective functioning of the food hygiene system.

It also advises that management must also ensure the effectiveness of the food safety management system by:

- ► ensuring that roles, responsibilities, and authorities are clearly communicated in the food business;
- ► maintaining the integrity of the food hygiene system when changes are planned and implemented;
- ► verifying that controls are carried out and working and that documentation is up to date;

▶ ensuring that the appropriate training and supervision are in place for personnel;

▶ ensuring compliance with relevant regulatory requirements; and

▶ encouraging continual improvement, where appropriate, taking into account developments in science, technology and best practice.

How food safety culture is measured, assessed and influenced has attracted much interest in the early part of the 21st century, with a number of academics and organisations putting forward models and theories (Global Food Safety Initiative, 2018; Jespersen *et al.*, 2017; Wilson *et al.*, 2010). Since the Pennington Inquiry (2009), the UK's Food Standards Agency in particular has commissioned a number of studies to investigate food safety culture and its impact on the provision of safe food (Bolanos, n.d.). From an enforcement perspective it is difficult to assess values, beliefs and attitudes and it is a challenge for everyone involved in the food industry as to how a positive food safety culture can be accomplished. In 2021, however, the European Union introduced Regulation 2021/382 (EC, 2021) that amended the 'Hygiene of Foodstuffs' legislation to incorporate a requirement to 'establish, maintain and provide evidence of an appropriate food safety culture' based on the *Codex* requirements.

Organisational culture and behavioural science are vast subjects and this short section can only provide a glimpse of what is involved. It is clear, however, that the role of food safety culture is likely to form part of future food safety programmes (and possibly warrant a whole chapter in textbooks such as this).

10.3 Threat/vulnerability assessment and critical control points

Whereas HACCP is focused on hazards associated with the food and the process, there are a number of other risks that are not considered within its scope, such as the risk from food adulteration or deliberate contamination from disgruntled members of staff. The main principles used in HACCP have also been adopted to address some of these other threats to businesses such as fraud, deliberate contamination and cybercrime. Threat/Vulnerability Assessment and Critical Control Points (TACCP/VACCP) requires the business to analyse these threats and put in place controls to mitigate the risk, using similar principles to those applied in HACCP.

Rather than physical, chemical and biological hazards, the British Standards Institution's (2014) 'Guide to protecting and defending food and drink from deliberate attack' identifies people as the main threat to the business. It requires the business to identify the threats posed by persons both inside and outside of the business. Whilst some of the threats may pose a risk to food safety, e.g. deliberate contamination with glass, others can affect the brand image or have a financial impact, e.g. food fraud and cybercrime.

The team or individual tasked with producing the TACCP plan are likely to require different expertise in areas such as human resources and information technology to enable them to assess the risks to the business and ensure that adequate controls are put in place to mitigate the threats. The British Standards Institution's (2014) guide suggests that the plan should address the following four questions:

1. Who might want to attack us?
2. How might they do it?
3. Where are we vulnerable?
4. How can we stop them?

10.4 Conclusions

Designing a food safety management system can be quite a daunting task. It may be difficult in the beginning to identify what should be included within the system and what is appropriate for the business in question. Following guidelines such as those provided by *Codex Alimentarius* (2020) should help the business to put in place appropriate procedures and controls. Part of the success of HACCP is its flexibility to allow application in all businesses, although it has been acknowledged that small and less developed businesses may experience more difficulties, particularly due to the lack of expertise in HACCP and the hazards involved in food production.

Whilst a food safety management system is often divided into PRPs and HACCP, they are inextricably linked. Failure to implement adequate PRPs can introduce hazards not considered and controlled by the HACCP plan. However, some of the principles involved in HACCP are needed to determine appropriate PRP programmes. This is particularly true of the hazard analysis principle, as what is required within a PRP depends largely on the type of food being produced and the potential hazards. The importance of some PRPs, e.g. cleaning, may require them to be validated and there should be some verification activities for most PRPs.

There will often be occasions when the HACCP team may be unsure how to apply one of the principles. This often happens around the area of identifying whether a point is a CCP, a PRP or an oPRP. For example, there are different schools of thought on whether metal detection is a CCP or a form of verification that prerequisite programmes are working. Whether it is critical to the safety of the food may depend on the likelihood of metal being present in the final product without the detector in place. If the detector is frequently rejecting product and it is relied upon as the sole control for metal contamination, it may point more towards it being a CCP. Furthermore, some customers may require the business to identify it as a CCP. Whether it is called a PRP, oPRP or CCP is a matter of judgement and is unlikely to invalidate the food safety management system. What is more important is that the hazard of metal contamination has been analysed and sufficient controls are in place.

It is also important to remember that the objective of a food safety management system is to produce safe food and that the HACCP principles help in this process. Thus, assuring food safety is more important than following the principles to the letter. On this matter the European Commission (2016) advises:

> The seven HACCP principles are a practical model for identifying and controlling significant hazards on a permanent basis. This implies that where that objective can be achieved by equivalent means that substitute in a simplified but effective way some of the seven principles, it must be considered that the obligation laid down in Article 5, paragraph 1 of Regulation (EC) No 852/2004 is fulfilled.

Finally, it is important not to overlook the role that behaviour plays within the implementation of an effective food safety management system. The shared attitudes and beliefs that form the food safety culture of the business and its commitment to ensuring safe food are essential to ensure the success of the HACCP plan.

References

Bast, A., 2022. Food toxicology. In: Wernaart, B.F.W. and Van der Meulen, B.M.J. (eds) Applied Food Science. Wageningen Academic Publishers, Wageningen, the Netherlands, pp. 267-288.

Bernard, D., 1998. Developing and implementing HACCP in the USA. Food Control 9: 91-95. https://doi.org/10.1016/S0956-7135(97)00056-X

Bolanos, J.A., n.d. Organisations, culture & food safety: a rapid comparative overview of organisational culture frameworks in the food sector. Available at: https://www.food.gov.uk/sites/default/files/media/document/organisations-culture-food-safety.pdf.

British Standards Institute, 2014. Guide to protecting and defending food and drink from deliberate attack PAS 96/2014.

British Standards Institute, 2020. Chemical disinfectants and antiseptics. Quantitative suspension test for the evaluation of bactericidal activity of chemical disinfectants and antiseptics used in food, industrial, domestic and institutional areas. Test method and requirements (phase 2, step 1).

Centers for Disease Control and Prevention, 2019. Outbreak of *E. coli* infections linked to ground beef. Available at: https://www.cdc.gov/ecoli/2019/o103-04-19/index. html

Codex Alimentarius, 2020. General principles of food hygiene. Available at: http:// www.fao.org/fao-who-codexalimentarius/sh-proxy/en/?lnk=1&url=https%253A %252F%252Fworkspace.fao.org%252Fsites%252Fcodex%252FStandards%252FC XC%2B1-1969%252FCXC_001e.pdf

European Commission (EC), 2004. Regulation (EC) No 852/2004 of the European Parliament and of the Council of 29 April 2004 on the hygiene of foodstuffs. Official Journal of the European Union L 139, 30.4.2004: 1-54.

European Commission (EC), 2005. Commission Regulation (EC) No 2073/2005 (2005) on microbiological criteria for foodstuffs. Official Journal of the European Union L 338, 22.12.2005: 1-26. Available at https://eur-lex.europa.eu/legal-content/EN/ TXT/?uri=CELEX%3A32005R2073

European Commission (EC), 2016. Commission Notice on the implementation of food safety management systems covering prerequisite programmes (PRPs) and procedures based on the HACCP principles, including the facilitation/flexibility of the implementation in certain food businesses C/2016/4608. Official Journal of the European Union C 278, 30.7.2016: 1-32. Available at https://eur-lex.europa.eu/ legal-content/EN/TXT/?uri=CELEX%3A52016XC0730%2801%29

European Commission (EC), 2021. Commission Regulation (EU) 2021/382 of 3 March 2021 amending the Annexes to Regulation (EC) No 852/2004 of the European Parliament and of the Council on the hygiene of foodstuffs as regards food allergen management, redistribution of food and food safety culture. Official Journal of the European Union L 74, 4.3.2021: 3-6.

European Court, 2011. 'Astrid Preissl KEG v Landeshauptmann von Wien' (2011) Case C-381/10. European Court Reports 2011 I-09281. Available at https://eur-lex. europa.eu/legal-content/EN/TXT/?uri=CELEX%3A62010CJ0381

Food and Agriculture Organisation of the United Nations and World Health Organisation (FAO/WHO), n.d. FAO/WHO guidance to governments on the application of HACCP in small and/or less-developed food businesses. Available at: https://www.who.int/foodsafety/publications/fs_management/HACCP_SLDB.pdf

Food Standards Agency, 2016. The safe production of beef burgers in catering establishments: advice for food business operators and LA officers. Available at: https://www.food.gov.uk/sites/default/files/media/document/lttcupdatedguidance. pdf

Food Standards Agency, 2020. Safer Food, Better Business (SFBB). Available at: https:// www.food.gov.uk/business-guidance/safer-food-better-business-sfbb

Global Food Safety Initiative, 2018. A culture of food safety: A Position Paper from the Global Food Safety Initiative (GFSI). Global Food Safety Initiative, Paris, France.

Griffith, C.J., Mullan, B. and Price, P.E., 1995. Food safety: implications for food, medical and behavioural scientists. British Food Journal 97: 23-28. https://doi.org/10.1108/00070709510100082

Jespersen, L., Griffiths, M. and Wallace, C.A., 2017. Comparative analysis of existing food safety culture evaluation systems. Food Control 79: 371-379. https://doi.org/10.1016/j.foodcont.2017.03.037

Lytton, T.D., 2019. Outbreak: foodborne illness and the struggle for food safety. University of Chicago Press, Chicago, IL, USA.

Mortimer, S. and Wallace, C., 1998. HACCP: a practical approach (2nd ed.). Aspen Publishers Inc. (Springer), Gaithersburg, MD, USA.

Mukhopadhyay, S and Gorris, L.G.M., 2014. Hurdle technology, encyclopedia of food microbiology (2nd ed.). Academic Press, Cambridge, MA, USA. https://doi.org/10.1016/B978-0-12-384730-0.00166-X

O'Mahony, M., Mitchell, E., Gilbert, R.J., Hutchinson, D.N., Begg, N.T., Rodhouse, J.C. and Morris, JE., 1990. An outbreak of foodborne botulism associated with contaminated hazelnut yoghurt. Epidemiology & Infection 104: 389-395. https://doi.org/10.1017/s0950268800047403

Pennington, H., 2009. Report of the Public Inquiry into the September 2005 Outbreak of *E. coli* O157 in South Wales. HMSO, London, UK. Available at: https://www.reading.ac.uk/foodlaw/pdf/uk-09005-ecoli-report-summary.pdf.

Vaskoska, R., 2022. Hostile microbiology. In: Wernaart, B.F.W. and Van der Meulen, B.M.J. (eds) Applied Food Science. Wageningen Academic Publishers, Wageningen, the Netherlands, pp. 247-266.

Walker, E., Pritchard, C. and Forsythe, S., 2003. Food handlers' hygiene knowledge in small food businesses. Food Control 14: 339-343. https://doi.org/10.1016/S0956-7135(02)00101-9

Wilson, S., Tyers, C. and Wadsworth, E., 2010. Evidence review on regulation culture and behaviours. Social Science Research Unit Report 12. Food Standards Agency, London, UK.

Yiannas, F., 2008. Food safety culture: creating a behavior-based food safety management system. Springer, New York, NY, USA.

11. Food microbiology

This is what you need to know about benign microorganisms, probiotics and fermentation

Bruno Pot[*] and Marjon Wolters

Yakult Europe BV, Schutsluisweg 1, 1332 EN Almere, the Netherlands; bpot@yakult.eu

Abstract

In this chapter we would like to introduce the wonderful world of the 'good bugs'. While bacteria are still quite often associated with disease, it has now become clear that without them we may not be that healthy either. Each of us carries around 10^{14} microorganisms in and on our body, 24 hours a day, 7 days a week – as many bacteria as we have cells in our body. Our bacteria bring us about 100 times more genes than we have of our own (Gilbert *et al.*, 2018). Humans have evolved for millions of years *together* with their bacteria. The resulting 'holobiont', as it is now referred to (Baedke *et al.*, 2020), is a perfect ecosystem that cannot properly function with only one of the 'partners'. Thanks to large-scale metagenomic research, we now know that diversity of the microbiota is key to a proper functioning of this holobiont. A sufficiently diverse microbiota guarantees that the necessary metabolic pathways will always be present and functional under differing conditions, securing the health of the host. Diversity of the gut ecosystem starts to develop during and immediately after birth, fuelled by contacts with parents, nursing staff and visiting family members and friends. Later, contacts with animals, the environment and, importantly for the gut microbiota, with microbes from the diet will complete the 'collection' of essential microorganisms. In this chapter we present the earlier hypothesis that higher hygiene and reduced consumption of fermented foods have weakened the microbial diversity, generation after generation, and therefore the resilience of the gut microbiota. There is proof that this is impacting on e.g. the functionality of the holobiont, with influence on the metabolism and the immune system mainly. It is thought that these changes of the microbiota in our Western society are partly responsible for the observed increase in non-communicable diseases, such as allergy, obesity and, possibly, even with some cardiovascular and neurological diseases (Durack and Lynch, 2019). While the solution may be to educate people on the importance of a diet rich in living good bugs, the current regulatory situation is not making the communication on this very easy.

Key concepts

- ► A definition of the term microbiome was recently proposed by a large group of 40 experts (Berg *et al.*, 2020). The short description, 'a characteristic microbial community occupying a reasonable well-defined habitat which has distinct physio-chemical properties' was further substantiated with elements such as, 'The microbiome not only refers to the microorganisms involved but also encompasses their theatre of activity, which results in the formation of specific ecological niches', and 'The microbiome, which forms a dynamic and interactive micro-ecosystem prone to change in time and scale, is integrated in macro-ecosystems including eukaryotic hosts, and is crucial for their functioning and health'. This description, while covering a wider ecological range, is perfectly compatible with the human microbiome concept.
- ► In the same paper, it was said that the microbiota 'consists of the assembly of microorganisms belonging to different kingdoms (Prokaryotes [Bacteria, Archaea], Eukaryotes [e.g. Protozoa, Fungi and Algae]), while 'their theatre of activity' includes microbial structures, metabolites, mobile genetic elements (e.g. transposons, phages and viruses), and relic DNA embedded in the environmental conditions of the habitat'.
- ► Fermented foods were also quite recently (re)defined by Maria Marco and colleagues (Marco *et al.*, 2021), under the auspices of the International Scientific Association for Pro- and Prebiotics (ISAPP), as 'foods made through desired microbial growth and enzymatic conversions of food components'. This definition resembles an earlier definition from Maria Marco's team in 2017, 'foods or beverages produced through controlled microbial growth, and the conversion of food components through enzymatic action', in an article focusing on the health benefits of fermented foods.
- ► Probiotics probably have the oldest established, stable definition. Originally proposed in 2001 by expert groups created by the FAO and WHO organisations (Araya *et al.*, 2002; Gilliland *et al.*, 2001), it was grammatically revised in 2014 by ISAPP (Hill *et al.*, 2014) to 'live microorganisms which, when administered in adequate amounts, confer a health benefit on the host'.

▶ Live dietary microorganisms are microorganisms that are present in foods (fermented foods mainly), in sufficient numbers and can be isolated (are alive) from the food. They differ from probiotics in the fact that they have not undergone clinical research to formally support their health benefit. Many dietary microbes, however, may contribute to the gut microbiota diversity (see below for the importance of this).

▶ Prebiotics were defined in 2017 by ISAPP as, 'a substrate that is selectively utilized by host microorganisms conferring a health benefit' (Gibson *et al.*, 2017).

▶ Synbiotics, recently also defined by ISAPP, are 'a mixture comprising live microorganisms and substrate(s) selectively utilized by host microorganisms that confers a health benefit on the host' (Swanson *et al.*, 2020).

▶ Postbiotics is a term that is currently generating some controversy. The ISAPP definition, 'A preparation of inanimate microorganisms and/or their components that confers a health benefit on the host' (Salminen *et al.*, 2021), was disputed by a large number of researchers (Aguilar-Toala *et al.*, 2021) but defended further by Vinderola and colleagues (Vinderola *et al.*, 2022).

▶ Holobiont, introduced above, is a concept that dates back to the previous century, mostly in a wider biological context (Baedke *et al.*, 2020; Faure *et al.*, 2018). Its modern meaning, in terms of a human ecosystem being a symbiosis between the human host and his/her microbes, was nicely discussed and illustrated in Berg *et al.* (2020).

▶ Non-communicable diseases are defined by the WHO as '*chronic diseases which are not passed from person to person*'. According to WHO (Anonymous), non-communicable diseases kill 41 million people every single year, equivalent to 71% of all deaths globally. 85% of these 'premature' deaths occur in low- and middle-income countries.

11.1 Introduction: some historical context

Publications by Jean-François Bach (Bach, 2002), Bach and Chatenoud (Bach and Chatenoud, 2012) and later Okada *et al.* (2010) proposed a 'hygiene hypothesis' – a link between changed (improved) sanitary conditions and (increased) susceptibility to allergic and autoimmune diseases. Another paper, however, pointed out that the increase in diseases caused by immune dysregulation could be associated with defects in immuno*regulation* by commensal micro-organisms (Guarner *et al.*, 2006). Immune *regulation* is generally considered the playground of *regulatory* T-cells which are mainly activated by 'beneficial' microorganisms, not by pathogens. Typical organisms that in the laboratory have been shown to have this immune regulatory potential are some lactic acid bacteria (LAB), including specific strains of bifidobacteria and lactobacilli (Foligne *et al.*, 2007), but also helminths or saprophytic mycobacteria. The immunoregulatory mechanisms they induce have the potential to temper excessive immune responses such as auto-immune reactions, allergic reactions or to reduce effects of chronic inflammations such as those observed in inflammatory bowel disease (IBD) or metabolic syndrome (MS) cases. That harmless commensal microorganisms are involved in this process is probably linked to the fact that they have been part of the human holobiont ecosystem for hundreds of thousands of years, and are still provided through daily contacts with the environment and the diet.

commensal microorganisms

In the paper from 2006 (Guarner *et al.*, 2006), these commensal microorganisms have been called 'old friends' and in the meantime it has been well documented that, in our Western society, with increased hygiene and widely available antibiotics for instance, we have unfortunately lost many of these old friends. Since we are consuming significantly fewer fermented foods than a century ago, we are also not replenishing these old friends sufficiently through our diet. Warriner *et al.* (2015) discussed the possibilities of extracting information related to composition and diversity from palaeolithic sources, including coprolites or dental calculus, to support this hypothesis (paleo-microbiology).

Additional proof of the importance of the diet comes from comparing microbiota compositions of populations with more traditional diets, that are rich in fibres and fermented foods, with populations that consume a typical Western diet, rich in (sterile) processed foods (De Filippo *et al.*, 2010; Marlowe, 2005, 2010; Schnorr *et al.*, 2014; Yatsunenko *et al.*, 2012). Some concrete examples are given in Case 11.1.

Interestingly, it can even be shown that population groups who changed from a traditional to a Western diet, after moving from one continent to another for instance, have also changed the risk of developing non-communicable diseases (Holmboe-Ottesen and Wandel, 2012).

Case 11.1. The loss of gut microbiota diversity.

The gut microbiota of children from a rural village in Burkina Faso was found to be very different from an otherwise healthy 'Western' cohort in Italy (De Filippo *et al.*, 2010). The African group not only had different phylogroups, but they also showed a higher diversity, an increased representation of Bacteroidota relative to Bacillota (two dominant phyla of the human microbiota), (Bacteroidota are normally depleted in people associated with obesity (Ley *et al.*, 2005, 2006; Turnbaugh *et al.*, 2009)) and, importantly, a large increase in short chain fatty acid (SCFA) production, mainly butyrate and propionate, which coincides with nearly twice the dietary fibre intake compared to the European group.

In a study of rural groups in Malawi and Venezuela (Yatsunenko *et al.*, 2012), a greater bacterial diversity was also found, as well as differences in the microbiota composition relative to Western groups.

The interesting point is that the microbiota diversity and SCFA production from all three traditional agrarian populations (Malawi, Venezuela and Burkina Faso) resembled one another more closely than they resembled their Western counterparts.

Findings from the Hadza population, a group of about 1,000 hunter-gatherers in Tanzania who are thought to approximate pre-agricultural humans (Marlowe, 2005; Marlowe and Berbesque, 2009), showed that, when compared to 16 Italians, the microbiota of 27 Hadza had a significantly larger bacterial diversity, with relatively higher levels of the genus *Prevotella*, presence of *Treponema* species, and, strikingly, no bifidobacteria (Schnorr *et al.*, 2014).

All these observations support the view that the modern microbiota deviates substantially from the ancestral states (hunting and gathering was the lifestyle that dominated for the longest time span in our species' existence (Sonnenburg and Sonnenburg, 2014)), with a general loss in diversity and change in composition.

Multiple studies support the fact that reduced exposure to our 'old friends' at least partly explains the increase in immune disorders, including very high incidences of allergy (World Allergy Organisation statement: (Haahtela *et al.*, 2013, 2021)), obesity (Abris *et al.*, 2018; Lozupone *et al.*, 2012) and auto-immunity (Bach, 2020). Another possible factor that might be contributing to the observed depletion of old friends is the increased use of antibiotics (ABs), which obviously also has the potential to contribute to a further reduction in the 'natural' diversity, e.g. of the gut microbiota.

In the light of the above, the use of fermented foods, probiotics, prebiotics, synbiotics, able to increase the microbiota diversity with safe microorganisms, might, therefore, be considered an easy, affordable, valuable, non-medical approach to the prevention of non-communicable diseases.

The extremely strict regulation of health claims, relating to communication of possible health benefits of (fermented) foods including probiotics or prebiotics, especially in Europe, has, however, hampered the education of consumers on the importance of these 'old friends' for maintaining health and preventing the further increase in non-communicable diseases that threatens to reduce life expectancies in Western countries and increases health care costs considerably (Bloom *et al.*, 2011; European Commission, 2022).

11.2 The contemporary issues; observations, problems and (science-based) solutions

As argued above, allergies, obesity, type 2 diabetes, auto-immune diseases, inflammatory bowel disease (IBD) and irritable bowel syndrome (IBS) are all increasing in our Western society, with a speed that cannot be explained by genetic causes (Koopman *et al.*, 2016). The availability of modern research tools, such as metagenomics and metabolomics, nowadays allows us not only to identify 'who' is in our gut, but also 'what are they doing there' (Zhang *et al.*, 2019). The results of this research are fascinating. While it has become evident that each one of us has a unique microbiota, just as we have a unique fingerprint, it turns out that this 'microbiological fingerprint' can change somehow over time and that these changes may impact our health (Wilmanski *et al.*, 2021).

11.2.1 The microbiota: the importance of stability, balance and an intact intestinal barrier

Case 11.2. Livelong microbacteria.

The most obvious changes happen shortly after birth. Babies are born sterile, but will be 'contaminated' by the microbiota of the mother during delivery or by the environment after a caesarean section (Bolte *et al.*, 2022; Thomas, 2022). The colonisation of the new-born intestine is characterised by certain phases, with aerobic bacteria appearing first, consuming the oxygen present in the intestine and preparing the field for anaerobic bacteria to settle. It will take about two to three years for the microbiota to reach the expected diversity (Roswall *et al.*, 2021). During the majority of the following years the microbiota will be stable. At old age, however, it is observed that the microbiota changes again, with a reduction in diversity as well as reduction in specific bacteria, such as bifidobacteria (Ragonnaud and Biragyn, 2021; Wilmanski *et al.*, 2021).

microbiota changes
While the main changes will happen early in life and in older age (Case 11.2), microbiota changes will also result from considerable changes in the diet (Leeming *et al.*, 2019, 2021). The intake of large amounts of plant-based fibres, as compared to large amounts of proteins, will trigger a switch from a 'putrefactive' microbiota, specialised in degrading proteins and fat, towards a 'fermentative' microbiota, specialised in degrading (high molecular) fibres into

SCFA, which are known to be beneficial to health. Travelling abroad, taking antibiotics or enduring long periods of stress are also known to have a (negative) impact on the microbiota composition, and therefore on its functioning.

intestinal barrier

Another of these possible changes, other than the switch between putrefactive and fermentative digestion, is an impact on the intestinal barrier (Windey *et al.*, 2012). The intestinal barrier is the name for an important but vulnerable lining between the sterile inner part of the body and the non-sterile lumen of the intestine. In order to be able to absorb all required nutrients in a timely manner, the total surface of the small intestine, where most of the absorption takes place, has been extended to the size of a tennis court by structures called villi and microvilli (Figure 11.1).

immune capacity

Because this large surface is composed of a single cell layer of enterocytes (Figure 11.1), necessary for easy nutrient absorption, it is in need of powerful defence mechanisms, able to destroy pathogens and toxins before they can do any harm to the host. It is estimated that 70% of the immune capacity is located around the gut, ready to act whenever needed. This immune system monitors the intestinal content 24 hours a day, 7 days a week and is kept vigilant and alert by contacts with the gut microbiota. An optimal functioning of the immune system is essential to our health.

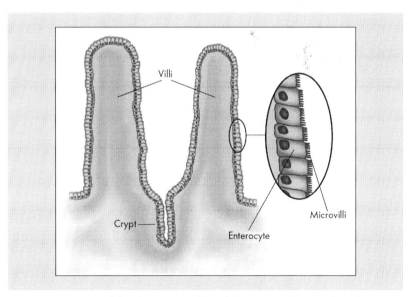

Figure 11.1. Increase of the absorption surface of the intestine through villi and microvilli. (BallenaBlanca, https://commons.wikimedia.org/w/index.php?curid=48093230)

Since the intestinal barrier is critical to our defence, it is also fortified by the presence of mucus, which is constantly renewed to eliminate microorganisms that come too close to the intestinal lining. Besides mucus, the quality of the so-called 'tight junctions' (TJs) is important too. TJs are protein structures that kind of 'glue' neighbouring epithelial cells to each other, preventing penetration of the inner body tissues by microorganisms (Figure 11.2).

It is well established that pathogens can disrupt both barriers (see Chapter 12; Vaskoska, 2022), while many benign bacteria will reinforce both mucus and TJ protection. An imbalance between both types of microorganisms might, therefore, be the basis of a deficient barrier function, leading to the chronic inflammation in different parts of the body that has been observed in most of the abovementioned non-communicable diseases.

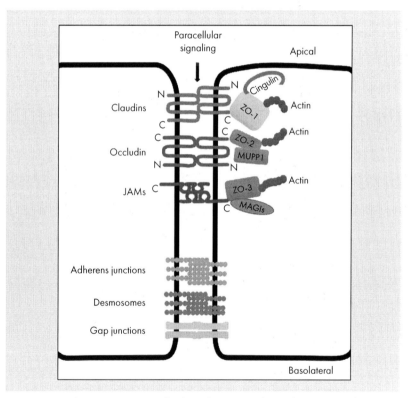

Figure 11.2. Tight Junction proteins that keep bacteria in the gut lumen, away from sterile body sites (Shi et al., 2018).

11.2.2 What the science is telling us about our microbiota composition

As mentioned before, there can be many types of 'functional combinations' of microorganisms in the gut microbiota (Yatsunenko *et al.*, 2012). Metagenomic research on large populations has revealed that there are so-called enterotypes (Arumugam *et al.*, 2011), which can be identified based on the presence of certain predominant genera of bacteria (Case 11.3).

Case 11.3. Enterotypes and diet.

Originally three enterotypes have been described (Arumugam *et al.*, 2011). The first is called the *Prevotella*-dominated type and is mainly found in people who consume a diet high in fibre, grains, pulses, vegetables and fruit, although also consuming refined carbohydrates like sweets and pastries. The *Bacteroides*-dominated enterotype is associated more with higher consumption of animal protein and fats, besides refined sugar. More recently a subtype of this enterotype was described, *Bacteroides* 2b enterotype (Vieira-Silva *et al.*, 2019), characterised by a low cell count and often associated with a diseased microbiome data set. The description was based on 106 patients with primary sclerosing cholangitis (PSC) and/or inflammatory bowel disease (IBD). It has been suggested that the *Bacteroides* type may best reflect the Western diet. The third enterotype has a wider range of bacteria with *Ruminococcus* as the most dominant genus, which may reflect the consumption of a diet rich in dietary fibre and resistant starches, and is also often found in rural, farming communities.

Stool consistency seems connected with the enterotype and species richness: shorter transit times were found to be linked to increased levels of the fast-growing species in samples with enterotypes 2 and 3, suggesting a strategy to prevent washout of certain species by faster growth, a strategy absent in people with the *Prevotella*-enterotype, there possibly compensated for by a closer adhesion of the microbiota to host tissue, as a way to prevent washout (Vandeputte *et al.*, 2016). For further reading on enterotypes, see also Wu *et al.* (2011) and Cheng and Ning (2019).

Besides the enterotype, the microbiota may also be characterised by so-called 'keystone' species. The concept of keystone taxa is an ecological concept that goes back to the sixties (Scott Mills *et al.*, 1993). It is only more recently, when considering the human microbiota, that the aspect of low abundance was added to the concept (Hajishengallis *et al.*, 2011; Power *et al.*, 1996) as well as the specific aspect of microbial turnover (Faust and Raes, 2012). Banerjee *et al.* (2018) more recently studied more than 200 microbial keystone taxa identified in soil, plant and marine ecosystems and proposed that keystone taxa are:

> taxa which have a major influence on the microbiome composition and function at a particular space or time. These taxa often, but not always, have an over-proportional influence in the community, relative to their abundance.

keystone species   It is now well accepted that some key species are depleted in dysbiotic ecosystems (Tudela *et al.*, 2021). As, by definition, they carry unique functions that are essential for the microbiota balance, they are consistently being studied as crucial actors in the understanding of the microbiome dynamics, whether in health or in disease. Replenishing functions of keystone species through food or drugs could become an important step in the development of microbiome-based prophylaxis or therapy.

All the above observations have only fortified the idea that diversity is crucial in establishing and maintaining a healthy microbiota and that contact with external sources of these bacteria, the 'old friends', can help to create and maintain diversity. When talking about providing food sources of these bacteria, fermented foods come into the picture.

11.2.3 What the science is telling us about fermented foods

The production of fermented foods goes back several thousand years, and was practised in many parts of the world (Ray and Joshi, 2014) as a way to preserve perishable foods, mainly milk and vegetables. While this role of extending the shelf life of fresh products is still important, fermentation underwent a tremendous technological evolution, with the improvement of fermentation conditions or the use of selected starter cultures, which all contribute to making fermentation safer and more reliable (Leroy and De Vuyst, 2004). As a result of that, fermentation is now used not only to extend the shelf life of fresh foods but also to improve texture and organoleptic properties of the food, reduce food waste and improve the health of the consumer by e.g. increasing vitamin content, removing toxins, improving digestibility and nutritional value, or improving the bio-availability of iron and minerals, to name just a few (Sanlier *et al.*, 2019).

health benefits Interestingly, the health benefits are situated on two levels. Most of the benefits just mentioned are the well-known nutritional benefits, linked to the fermented food product itself. A second level, however, remains less well known and is situated in the context of the microbiota health of the host consuming these fermented foods.

As has been illustrated for probiotics, bacteria from fermented foods might interact with the host on at least six levels: direct interaction with the microorganisms of the host interaction, directly or indirectly with the immune system, participation at the cross-talk processes, producing metabolites that can influence the physiology, endocrinology or neurology of the host (Case 11.4).

Case 11.4. The benefits of beneficial microorganisms.

The benefits of beneficial microorganisms in interaction with the host (Chen *et al.*, 2022).

Microorganisms can interact positively with the host in many different ways:
1. Microbiologically (competition with pathogens for places and nutrients; production of antimicrobials).
2. Metabolically (bioconversions).
3. Physiologically (e.g. vitamin production).
4. Immunologically (e.g. reduction of inflammation).
5. Endocrinologically (hormone production).
6. Neurologically (e.g. serotonin production).

This overwhelming wealth of possible health effects, while clearly in favour of the host, renders the study and documentation of the specific health effects of selected microorganisms quite difficult. As is generally the case in nutrition studies, the simultaneous impact of many confounding factors like environment, stress, lifestyle, medication and other dietary parameters, besides host-related parameters like age, (epi)genetics or microbiota composition, may impact the outcome of the study. Each of these factors, as well as combinations thereof, may determine whether a subject will be a 'responder' or a 'non-responder' in a planned clinical study. As an example, a study participant who already has high levels of bifidobacteria in the colon might not respond significantly to a probiotic product containing bifidobacteria.

An important challenge, furthermore, is to isolate the effect of the fermented food consumption from other confounding factors. To illustrate this, it may be interesting to learn that while the Hadza people have a healthy microbiota and are far less prone to the non-communicable diseases observed in the Western world, their life expectancy is considerably lower (Frackowiak *et al.*, 2020). Of course, modern health care systems and their availability in the Western world contribute to an extended life expectancy, but this is by no means easy to prove.

11.3 Lack of a clear regulatory framework for foods containing probiotic microorganisms and the implications for communication on the probiotic product category

The category of probiotic foods is not clearly defined and regulated in the EU. This has implications for communication about this product category, in particular the use of the term 'probiotics'. Below, we briefly discuss the EU regulatory situation for foods with probiotic microorganisms and the effect of this regulatory situation on communication about probiotics.

11.3.1 Regulatory situation of food with probiotic microorganisms in the EU

There is no specific harmonised EU legislation on probiotics. This concerns the composition of foods as well as the labelling, presentation and advertising.

composition of food Most EU Member States do not have national provisions on the addition of probiotic microorganisms to foods. A small number of EU countries have national provisions that must be considered when placing a food product containing probiotic microorganisms on the market. Generally, probiotic microorganisms can be used in foods in all EU countries, provided they are safe, not covered by the Novel Food Regulation and not considered medicinal.

The three underlying general EU legislative frameworks regarding the addition of probiotic microorganisms to foods, applicable to all EU Member States, are: (1) the General Food Law Regulation (GFL; Regulation 178/2002); (2) the Novel Foods Regulation (Regulation (EU) 2015/2283); and (3) the Regulation on the addition of vitamins and minerals and of certain other substances to foods (Regulation (EC) 1925/2006).

General Food Law The General Food Law Regulation (Regulation 178/2002) (GFL) lays down the general principles and requirements of EU food law, establishes the European Food Safety Authority (EFSA) and establishes food safety procedures (European

Commission, 2002). This Regulation applies to all stages of production, processing and distribution of food as a whole, requiring the food business operator who places the product on the market to ensure that the food is safe for the consumer, meaning that it must not cause any harm to health (Article 14(3) and 4 of the GFL) or be unfit for human consumption (Article 14(3) and (5) of the GFL). Please consult the GFL for other basic criteria for determining whether a food is considered unsafe.

Consequently, when a food contains probiotic microorganisms, the food business operator must ensure that the ingredients used, including the probiotics, and the product as a whole are safe for human consumption.

novel foods Regulation (EU) 2015/2283 lays down the rules for the marketing of novel foods within the EU (European Commission, 2015). Foods are considered to be novel foods if they have not been used to any significant extent in the EU for human consumption before 15 May 1997 and fall into at least one of the ten categories referred to in Article 3.2(a) of the Regulation. Only novel foods that are authorised and included on the Union list may be placed on the EU market as such, used in or on foods, in accordance with the conditions of use and labelling requirements specified therein.

Foods consisting of, isolated from or produced from microorganisms, fungi or algae and have not been used in food lawfully sold on the EU market before May 1997 would fall under the scope of the novel food Regulation. However, new strains of organisms that had a known history of use in the European Union before May 1997 are not considered novel foods. Also, new strains of microorganisms that have been assigned QPS status (Qualified Presumption of Safety) by EFSA are generally not considered novel foods. EFSA has published over time a list of those biological agents intentionally added to food and feed that have been granted QPS and keeps this list up to date on a six monthly basis (Koutsoumanis *et al.*, 2022).

When placing foods on the EU market, it should be considered that the fact that a microorganism is listed on the QPS list is not in itself sufficient to exclude it from the scope of the Novel Food Regulation. It must also be confirmed that the microorganism has been used as a food, or in food, before 15 May 1997 and the history of food consumption needs to be documented.

addition of vitamins and minerals Regulation (EC) 1925/2006 harmonises the rules for the addition of vitamins and minerals and certain other substances to foods (European Parliament, 2006), defining requirements, restrictions, purity criteria, conditions and even provisions for labelling, presentation and advertising. For the addition of certain other substances, which could be applicable to foods with added probiotic microorganisms, Chapter III and Annex III of this Regulation should be considered. Chapter III describes the establishment of Annex III, a list of substances banned for use in food (Part A), substances which may only be used under specific conditions of use (Part B) or substances under community scrutiny, meaning substances with the possibility of harmful effects on health but where scientific uncertainty remains (Part C).

When a substance is included in Annex III, Part C, all interested parties, including food business operators, can submit a dossier to EFSA containing the scientific data that demonstrates the safety of the substance. This scientific dossier should be taken into account by the European Commission, who must take a decision within four years on whether the use of a substance should be permitted in general or whether it should be included in Annex III, Part A or B, as appropriate.

Although no probiotic microorganisms are currently listed in Annex III, in specific cases some EU Member States may have a notification requirement or a marketing authorisation procedure to be considered for foods with added probiotic microorganisms.

The general principles in the three legal frameworks described above provide some guidance about the addition of probiotic microorganisms to food, guidance which is different when we discuss the labelling, presentation and advertising of such foods.

food advertising　In principle, probiotic microorganisms in foodstuffs can be advertised in the EU, provided that the information is correct (and based on scientific evidence), understandable and not misleading to consumers.

At least two general EU legislative frameworks are applicable to all EU Member States regarding the labelling, presentation and advertising of probiotic microorganisms in foods: (1) The Food Information to Consumers Regulation (Regulation (EU) No 1169/ 2011); and (2) the Nutrition and Health Claims Regulation (Regulation (EC) No 1924/2006).

food information　Regulation (EU) No 1169/2011 lays down the general principles, requirements and responsibilities for food information. The definition of food information is very broad and includes not only food labels and/or other accompanying materials, but also encompasses verbal communication and social media.

To allow consumers to make well-informed choices the Food Information Regulation obliges food business operators to provide food information that is:

- not misleading;
- accurate;
- clear and easy to understand;
- not attributing or referring to the property of preventing, treating or curing a human disease, as the latter is seen as making a medicinal claim, which is prohibited on foods (Article 7(3) of the Regulation (EU) No 1169/2011).

Therefore, food products with probiotic microorganisms will often mention the name of the microorganism on their labels, as well as the product category to which it belongs. Since not all consumers are familiar with the names of various microorganisms, it could be desirable to use the term 'probiotic' in addition to the name of the microorganism, when the respective strain is supported by specific research showing the health benefit of the microorganism. However, the use of the term 'probiotic' is currently prohibited in communications, including through food labels, to consumers in the EU. The reason for this prohibition lies in the European Commission guideline on the implementation of Regulation No. 1924/2006 on Nutrition and health claims made on foods (see below), which states that because of the FAO/WHO definition (Gilliland *et al.*, 2001; Hill *et al.*, 2014), probiotics inherently carry a health benefit, which should be judged by the EFSA and approved by the European Commission. Although this guideline is not legally binding, most EU member states say they follow it in their enforcement activities.

food claims

The Regulation (EC) No 1924/2006 harmonises the legal requirements which relate to nutrition and health claims. These requirements should ensure that nutrition and health claims made in the EU are clear, accurate and based on scientific evidence. The Regulation applies to nutrition and health claims made in commercial communications, whether in the labelling, presentation or advertising of foods, to be delivered as such to the final consumer.

Nutrition claims (any claim which states, suggests or implies that a food has particular beneficial nutritional properties) and health claims (any claim that states, suggests or implies that a relationship exists between a food category, a food or one of its constituents and health) can only be communicated when they are mentioned in the Annex of the Claims Regulation and the Union list of approved health claims, respectively. Health claims that are not included in this Union list of authorised health claims should first be scientifically assessed by the EFSA and approved by the European Commission.

Rules which relate to nutrition and health claims are theoretically the same in all EU member states. A Commission guidance was published to make sure that there is the same interpretation across the EU. However, more and more fragmented national enforcement practice is being seen, including different national guidance documents on compliance, or diverging actions from member state authorities. This is also the case for the use of the term 'probiotics', as illustrated in Figure 11.3 from IPA Europe, the non-profit organisation that represents the interests of the European Probiotic Industry. For many years IPA Europe has been asking for the development of a more consistent approach in the EU on the use of the term 'probiotic', providing useful guidance for stakeholders and consumers.

The lack of a clear legal framework for the communication on products with probiotic microorganisms and the existence of non-legal guidance documents continues to generate hostility among food business operators. While the Food Information Regulation requires food business operators to provide consumers with accurate and clear information, the Food and Health Claims Regulation prohibits this to a certain extent.

Clearly, there is no specific harmonised EU legislation on probiotics. This concerns the composition of foods as well as the labelling, presentation and advertising. A small number of EU countries have national provisions regarding the composition and the labelling, presentation and advertising that must be considered when placing a food product containing probiotic microorganisms on the market.

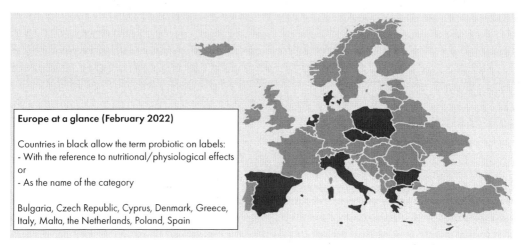

Europe at a glance (February 2022)

Countries in black allow the term probiotic on labels:
- With the reference to nutritional/physiological effects
or
- As the name of the category

Bulgaria, Czech Republic, Cyprus, Denmark, Greece, Italy, Malta, the Netherlands, Poland, Spain

Figure 11.3. Fragmented national enforcement practice on the use of the term 'probiotic' in Europe (https://www.ipaeurope.org/).

11.4 The current challenges and solutions

In contrast to the 'dietary fibres' category, the 'dietary microbes' category is not yet a recognised category. Consequently, in contrast to the dietary fibre category, well known by most consumers and health care professionals as important for a balanced diet, and with well-accepted health benefits, the health benefits of live microorganisms in our diet are not very well known. As a matter of fact, specific dietary recommendations for live microbes or fermented foods are absent in almost all national nutritional guidelines, except in the Indian guidelines with a special mention for pregnant women (Bell *et al.*, 2017). This is rather unexpected, as the link between the reduced intake of dietary microbes, resulting in a reduced diversity and an increase in non-communicable disease, has been raised on several occasions. A study from 2014 suggested that, in our Western diet, a reduced intake of microbiota-accessible carbohydrates (MACs) has led to an altered microbiota, both in terms of composition and functionality, as compared to the microbiota typical of a more traditional lifestyle (Sonnenburg and Sonnenburg, 2014). The authors explain the loss in diversity in a historical context, a hypothesis they called the 'Multiple-Hit hypothesis' (Case 11.5).

Case 11.5. The multiple-hit hypothesis.

The multiple-hit hypothesis explains how the microbiota of Western societies has lost considerable diversity at multiple stages of human evolution.

While diet was suggested as an important factor, the rise of agriculture, industrial food production, with increasingly sterilised foods, as well as increased hygiene and the use of antibiotics have added to the diversity loss.

maintenance of microbiota diversity
These multiple insults to the microbiota prevented the maintenance of microbiota diversity over recent generations and have been linked to increased levels of non-communicable diseases (Sonnenburg and Sonnenburg, 2014). The diversity loss, in combination with the consumption of fewer MACS in our Western diet, has resulted in a reduced production of short-chain fatty acids (SCFAs) in the gut. The latter have the potential to attenuate inflammation, as had some of the (lost?) commensal bacteria.

In a later paper, the same research group (Wastyk *et al.*, 2021) tried to correct this unfavourable situation by setting up a human comparative dietary intervention, where two groups of 18 people consumed a diet rich in plant-based fibre or rich in fermented foods respectively, for a period of 10 weeks. The outcomes measured were at the level of the microbiome and the immune system. Surprisingly, only the group that consumed the fermented foods showed a significant increase in the microbiota diversity and decreased inflammatory markers. The authors concluded that 'fermented foods may be valuable in countering the decreased microbiome diversity and increased inflammation pervasive in the industrialized society' (Wastyk *et al.*, 2021), pleading for greater importance to be attributed to fermented foods and fibre in the daily Western diet.

Of course, an added conclusion was that more extensive dietary interventions are needed to substantiate these statements in a reproducible way.

nutrition studies
However, studies in the nutrition field have historically been shown to be difficult, often yielding opposite results. While it is generally accepted that saturated fat (solid fats in full-fat dairy products like butter, fat cheese and ice cream, eggs, beef fat, coconut and palm oil) increases the risk of heart diseases (American Heart Association (2021), a paper was published in March 2014 (Chowdhury *et al.*, 2014) stating that eating less saturated fat doesn't actually lower the risk of heart disease. This has in the meantime been further documented in a systematic Cochrane review and meta-analysis of 15 randomised controlled trials with 56,675 participants, looking at hard endpoints such as heart attacks or death (Hooper *et al.*, 2020). The study showed that reducing total saturated fat intake decreased the risk of cardiovascular events by 17% but did not affect the risk of dying from heart disease or other causes. The conclusion therefore was that 'people who reduced their saturated fat intake were just as likely to die from heart disease and other causes as those who ate more saturated fat.' The use of poly-unsaturated fats or carbohydrates to replace saturated fats may, however, still have a preventive effect on the development of heart disease and may improve heart health.

This example, again, clearly illustrates that nutritional studies are difficult, as they need to cope with many confounding factors that cannot usually be easily controlled (stress, diet, drug intake, travelling, age, (epi)genetics, ethnicity, geography, etc.). As mentioned before, in the case of the microbiota, all of these factors may actually be relevant. Research on the impact of the microbiota may be complicated further with the observation that no two individuals have the same microbiota composition, although all can be considered 'healthy'. There is currently no real definition of what a 'healthy microbiota' is. Research has estimated that around 40,000 human samples would be needed in order to capture a more or less complete picture of gut microbiota diversity, and more detailed studies would be needed to confirm the associations detected with diseases, using specific experiments necessary to clarify whether these associations were a cause or a consequence of the disease (Falony *et al.*, 2016).

Clearly these results also illustrate that there are many 'functional' combinations of microorganisms that render a host 'healthy'. Factors that have been identified to contribute to health are diversity, the obvious absence of pathogens, and the presence of some keystone bacteria (Tudela *et al.*, 2021). A future challenge therefore will be to collect sufficient microbiome data, with proper meta-data to allow the full inventory of diversity and to identify the risk factors for disease. The start of the 'Million Microbiome of Humans Project' in October 2019 (MGI, 2022) has set the tone for this type of large-scale research.

human genetics

exposome

A more recent microbiome study (Gacesa *et al.*, 2020) tried also to identify, besides lifestyle and diet, the impact of human genetics and the exposome on the microbiome in relation to health and disease. The microbiota of 8,208 Dutch individuals from a three-generational cohort of 2,756 families were profiled and correlated with 241 host and environmental factors, including diet, physical and mental health, drug intake, socio-economic factors as well as childhood and current exposome measurements. Results revealed 2,856 associations between microbiome and health, 7,519 associations between microbiome features and diet, socio-economics, early life and current exposome.

A few, rather unexpected examples are that childhood exposures to smoking, pets and rural environment are associated with the composition of the adult microbiome, indicating that these environmental exposures at a young age may have a lasting effect on the gut microbiome. In contrast, exposures during and shortly after birth (breastfeeding, birth mode and preterm birth) had less impact on the adult microbiome, probably because these exposures happen early in life and are followed by other environmental pressures like diet, environment, cohousing, smoking, etc. Results confirm that the gut microbiome is highly individual and much more research is needed to inventorise and understand the factors that shape the microbiome. Indeed with 241 variables only about 15% of the variation could be explained, indicating that cross-sectional measurements alone were not sufficient. Similarly, as in nutrition research, large-scale and long-term longitudinal studies will be necessary to increase our knowledge on the many links between microbiota and health or disease; but increasing our knowledge will almost certainly facilitate future development of microbiome-directed therapies or become the basis for guidelines that could help to prevent disease in at-risk populations.

communication

A last challenge that may need to be overcome is the apparent sensitivity in the communication around probiotics. While it is common in the nutrition field to see a lot of unverified pseudo-scientific 'knowledge' appear in non-scientific magazines, on social media or in pseudo-scientific blogs, it remains surprising that medically trained staff are also tempted to make (mostly negative) conclusions about probiotics, often based on observations that could somehow be called 'unfair'. Without wanting to give specific examples, it is clear that probiotics are very often 'used' in order to treat 'diseases' that are difficult to treat with traditional medication, or for which traditional drugs have serious, long-term side effects: obesity, inflammatory bowel disease, irritable bowel syndrome, diabetes, even cancer and more recently depression, mood disturbances, autism spectrum disorder, Parkinson's disease and Alzheimer's disease. When results remain below expectation, the conclusion is very quickly: 'probiotics don't work'. Quite often it is forgotten that the probiotics currently on the market, in most cases, were developed as foods (in Europe there are very few registered probiotic drugs or Live Biotherapeutic Products (LBPs)). As foods, probiotic strains should be perfectly safe (no side effects, independent of the dose taken, for any age and health status), but will not be manufactured under strict GMP conditions. The functionality of a probiotic food, by definition, should be evaluated on a healthy population. Since it is indeed a challenge to show a health benefit on an already healthy population, the legislator (EFSA, FDA) will mostly allow the product to be evaluated on an 'at-risk' or 'sub-healthy' population, albeit not on a 'diseased' population. The use of an at-risk or sub-healthy population to investigate probiotic functionality increases the risk of higher expectations, suggesting that the food could also work for more severe disease situations. The use of probiotics on a diseased population, however, requires different safety research (Rouanet *et al.*, 2020) and different supporting research and manufacturing conditions (Cordaillat-Simmons *et al.*, 2020). The risk therefore is high that the consumer gets confused about the 'expected benefits' for probiotic products.

11.5 Looking ahead: what are the expected economic benefits?

Does the above mean that there are no health benefits related to probiotic intake? Lenoir-Wijnkoop and co-workers have developed a model (Indrio *et al.*, 2014) in which the results of one or more clinical trials were extrapolated to the total population of e.g. France, Canada or the USA, and this for e.g. antibiotic-associated diarrhoea (Lenoir-Wijnkoop *et al.*, 2014) or for common respiratory tract infections (Lenoir-Wijnkoop *et al.*, 2015, 2016, 2019). In both cases, the cost savings are in the millions of euros or dollars.

This is perhaps not unexpected, as the estimated average annual total economic burden of e.g. influenza to the healthcare system and society was $11.2 billion ($6.3-$25.3 billion) (Putri *et al.*, 2018).

In a Dutch study a meta-analysis was performed on clinical study results from probiotic interventions on constipation in institutionalised elderly people, showing that the constipation-related expenses of an average-sized nursing home with 100 residents and a constipation prevalence of 42%, amounted to approximately €90,000 per year (Flach *et al.*, 2018). As the meta-analysis suggested that a daily probiotic intervention could lower the constipation prevalence by 28%, a supplementation of all nursing home residents with probiotics had the potential to reduce these expenses by €8,000 to €25,000 annually, not taking into account the workload for the nursing staff.

At the other end of the age spectrum, the prophylactic use of a probiotic strain in the prevention of colic, regurgitation and functional constipation during the first three months of life, significantly reduced the onset of functional gastrointestinal disorders and reduced private and public costs for the management of this condition (Koponen *et al.*, 2012).

fermentation When considering the benefits of fermentation, besides probiotics, we can mention the fact that fermentation is a natural process, nowadays much requested by the consumer, with very little or no impact on the environment. Fermentation allows perishable foods to be preserved for a much longer time than fresh food, while at the same time improving the nutritional and organoleptic quality (taste, texture, smell, colour) of the product. Different variants of fermentation allow diversification and innovation of food production, allowing them to respond effectively to a rapidly expanding market, which was expected to grow by 6% CAGR from 2020 to 2027 to reach USD $875.21 billion by 2027 (Emergen Research, 2020).

11.6 Conclusions

While 'good bugs' can contribute to health, they are yet to be fully recognised, scientifically and legally. While the possible mechanisms to promote health are well known, their application remains the subject of some debate. This holds for the nutritionist as well as for the clinician.

confounding factors The reasons for this are, on the one hand, the many confounding factors (stress, diet, environment, antibiotics or other medications, etc.), which all impact the microbiota, the main playing field of the good bugs. On the other hand, there are also the many non-controllable factors like individual microbiota composition, age, (epi)genetics, ethnics, culture, etc.

These variables necessitate studies with long time-spans, with large populations, and the monitoring of as many parameters and confounding factors as possible. For the food industry these studies are expensive and therefore scarce.

dietary microbes

Still, there are very strong indications that good bugs are essential for maintaining our health, and that foods, through 'dietary microbes', can be a valuable and necessary provider of these good bugs.

More specific regulation for live microorganisms, whether in foods or in drugs, could help to improve communication and therefore raise more awareness, leading to a larger consumption of these beneficial microbes, for which the economic benefits have been studied, covering many topics, in different population groups and in different countries. Crucial to further acceptance will be qualitative research, good regulation and correct communication to the consumer and the health care professional.

References

Abris, G.P., Provido, S.M.P., Hong, S., Yu, S.H., Lee, C.B. and Lee, J.E., 2018. Association between dietary diversity and obesity in the Filipino Women's Diet and Health Study (FiLWHEL): A cross-sectional study. PloS One 13: e0206490. https://doi.org/10.1371/journal.pone.0206490

Aguilar-Toala, J.E., Arioli, S., Behare, P., Belzer, C., Berni Canani, R., Chatel, J.M., D'Auria, E., de Freitas, M.Q., Elinav, E., Esmerino, E.A., Garcia, H.S., da Cruz, A.G., Gonzalez-Cordova, A.F., Guglielmetti, S., de Toledo Guimaraes, J., Hernandez-Mendoza, A., Langella, P., Liceaga, A.M., Magnani, M., Martin, R., Mohamad Lal, M.T., Mora, D., Moradi, M., Morelli, L., Mosca, F., Nazzaro, F., Pimentel, T.C., Ran, C., Ranadheera, C.S., Rescigno, M., Salas, A., Sant'Ana, A.S., Sivieri, K., Sokol, H., Taverniti, V., Vallejo-Cordoba, B., Zelenka, J. and Zhou, Z., 2021. Postbiotics – when simplification fails to clarify. Nature Reviews: Gastroenterology & Hepatology 18: 825-826. https://doi.org/10.1038/s41575-021-00521-6

American Heart Association, 2021. Saturated fat. Available at: https://www.heart.org/en/healthy-living/healthy-eating/eat-smart/fats/saturated-fats.

Araya, M., Morelli, L., Reid, G., Sanders, M.E., Stanton, C., Pineiro, M. and Ben Embarek, P., 2002. Guidelines for the evaluation of probiotics in food. Report of a Joint FAO/WHO Working Group on Drafting Guidelines for the Evaluation of Probiotics in Food, pp. 1-11.

Arumugam, M., Raes, J., Pelletier, E., Le Paslier, D., Yamada, T., Mende, D.R., Fernandes, G.R., Tap, J., Bruls, T., Batto, J.M., Bertalan, M., Borruel, N., Casellas, F., Fernandez, L., Gautier, L., Hansen, T., Hattori, M., Hayashi, T., Kleerebezem, M., Kurokawa, K., Leclerc, M., Levenez, F., Manichanh, C., Nielsen, H.B., Nielsen, T., Pons, N., Poulain, J., Qin, J., Sicheritz-Ponten, T., Tims, S., Torrents, D., Ugarte, E., Zoetendal, E.G. Wang, J., Guarner, F., Pedersen, O., de Vos, W.M., Brunak, S., Dore, J., Meta, H.I.T.C., Antolin, M., Artiguenave, F., Blottiere, H.M., Almeida, M., Brechot, C., Cara, C., Chervaux, C., Cultrone, A., Delorme, C., Denariaz, G., Dervyn, R., Foerstner, K.U., Friss, C., van de Guchte, M., Guedon, E., Haimet, F., Huber, W., van Hylckama-Vlieg, J., Jamet, A., Juste, C., Kaci, G., Knol, J., Lakhdari, O., Layec,

S., Le Roux, K., Maguin, E., Merieux, A., Melo Minardi, R., M'Rini, C., Muller, J., Oozeer, R., Parkhill, J., Renault, P., Rescigno, M., Sanchez, N., Sunagawa, S., Torrejon, A., Turner, K., Vandemeulebrouck, G., Varela, E., Winogradsky, Y., Zeller, G., Weissenbach, J., Ehrlich, S.D. and Bork, P., 2011. Enterotypes of the human gut microbiome. Nature 473: 174-180. https://doi.org/10.1038/nature09944

Bach, J.F., 2002. The effect of infections on susceptibility to autoimmune and allergic diseases. New England Journal of Medicine 347: 911-920.

Bach, J.F., 2020. Revisiting the hygiene hypothesis in the context of autoimmunity. Frontiers in Immunology 11: 615192. https://doi.org/10.3389/fimmu.2020.615192

Bach, J.F. and Chatenoud, L., 2012. The hygiene hypothesis: an explanation for the increased frequency of insulin-dependent diabetes. Cold Spring Harbor Perspectives in Medicine 2: 1-10. https://doi.org/10.1101/cshperspect.a007799

Baedke, J., Fabregas-Tejeda, A. and Nieves Delgado, A., 2020. The holobiont concept before Margulis. Journal of Experimental Zoology Part B: Molecular and Developmental Evolution 334: 149-155. https://doi.org/10.1002/jez.b.22931

Banerjee, S., Schlaeppi, K. and van der Heijden, M.G.A., 2018. Keystone taxa as drivers of microbiome structure and functioning. Nature Reviews: Microbiology 16: 567-576. https://doi.org/10.1038/s41579-018-0024-1

Bell, V., Ferrao, J. and Fernandes, T., 2017. Nutritional guidelines and fermented food frameworks. Foods 6. https://doi.org/10.3390/foods6080065

Berg, G., Rybakova, D., Fischer, D., Cernava, T., Verges, M.C., Charles, T., Chen, X., Cocolin, L., Eversole, K., Corral, G.H., Kazou, M., Kinkel, L., Lange, L., Lima, N., Loy, A., Macklin, J.A., Maguin, E., Mauchline, T., McClure, R., Mitter, B., Ryan, M., Sarand, I., Smidt, H., Schelkle, B., Roume, H., Kiran, G.S., Selvin, J., Souza, R.S.C., van Overbeek, L., Singh, B.K., Wagner, M., Walsh, A., Sessitsch, A. and Schloter, M., 2020. Microbiome definition re-visited: old concepts and new challenges. Microbiome 8: 103. https://doi.org/10.1186/s40168-020-00875-0

Bloom, D.E., Cafiero, E.T., Jané-Llopis, E., Abrahams-Gessel, S., Bloom, L.R., Fathima, S., Feigl, A.B., Gaziano, T., Mowafi, M., Pandya, A., Prettner, K., Rosenberg, L., Seligman, B., Stein, A.Z. and Weinstein, C., 2011. The global economic burden of noncommunicable diseases. Geneva: World Economic Forum. Available at: https://www3.weforum.org/docs/WEF_Harvard_HE_GlobalEconomicBurdenNonCommunicableDiseases_2011.pdf.

Bolte, E.E., Moorshead, D. and Aagaard, K.M., 2022. Maternal and early life exposures and their potential to influence development of the microbiome. Genome Medicine 14: 4. https://doi.org/10.1186/s13073-021-01005-7

Centers for Disease Control and Prevention (CDC), 2013. Global Noncommunicable Diseases, Centers for Disease Control and Prevention. Available at: https://www.cdc.gov/globalhealth/healthprotection/ncd/index.html; https://www.cdc.gov/globalhealth/healthprotection/fetp/training_modules/new-8/overview-of-ncds_ppt_qa-revcom_09112013.pdf

Chen, O., Heyndrickx, M., Meynier, A., Ouwehand, A., Pot, B., Stahl, B., Theis, S., Vaughan, E. and Miani, M., 2022. Dietary probiotics, prebiotics and the gut microbiota in human health. ILSI monograph. ILSI, Brussels, Belgium. https://doi.org/10.5281/zenodo.6394213

Cheng, M. and Ning, K., 2019. Stereotypes about enterotype: the old and new ideas. Genomics, Proteomics & Bioinformatics 17: 4-12. https://doi.org/10.1016/j.gpb.2018.02.004

Chowdhury, R., Warnakula, S., Kunutsor, S., Crowe, F., Ward, H.A., Johnson, L., Franco, O.H., Butterworth, A.S., Forouhi, N.G., Thompson, S.G., Khaw, K.T., Mozaffarian, D., Danesh, J. and Di Angelantonio, E., 2014. Association of dietary, circulating, and supplement fatty acids with coronary risk: a systematic review and meta-analysis. Annals of Internal Medicine 160: 398-406. https://doi.org/10.7326/M13-1788

Cordaillat-Simmons, M., Rouanet, A. and Pot, B., 2020. Live biotherapeutic products: the importance of a defined regulatory framework. Experimental and Molecular Medicine 52: 1397-1406. https://doi.org/10.1038/s12276-020-0437-6

De Filippo, C., Cavalieri, D., Di Paola, M., Ramazzotti, M., Poullet, J.B., Massart, S., Collini, S., Pieraccini, G. and Lionetti, P., 2010. Impact of diet in shaping gut microbiota revealed by a comparative study in children from Europe and rural Africa. Proceedings of the National Academy of Sciences of the USA 107: 14691-14696. https://doi.org/10.1073/pnas.1005963107

Durack, J. and Lynch, S.V., 2019. The gut microbiome: relationships with disease and opportunities for therapy. Journal of Experimental Medicine 216: 20-40. https://doi.org/10.1084/jem.20180448

Emergen Research, 2020. Fermented food and ingredients market by food type (fermented dairy products, fermented beverages), by ingredient type (organic acids, amino acids, industrial enzymes), by distribution channel (online stores, supermarkets), forecasts to 2027. Available at: https://www.emergenresearch.com/industry-report/fermented-food-and-ingredients-market

European Commission, 2002. Regulation (EC) No 178/2002 of the European Parliament and of the Council of 28 January 2002 laying down the general principles and requirements of food law, establishing the European Food Safety Authority and laying down procedures in matters of food safety. Available at: http://data.europa.eu/eli/reg/2002/178/oj.

European Commission, 2006. Regulation (EC) No 1925/2006 of the European Parliament and of the Council of 20 December 2006 on the addition of vitamins and minerals and of certain other substances to foods. Available at: https://eur-lex.europa.eu/legal-content/EN/ALL/?uri=CELEX%3A32006R1925.

European Commission, 2015. Regulation (EU) 2015/2283 of the European Parliament and of the Council on novel foods, amending Regulation (EU) No 1169/2011 of the European Parliament and of the Council and repealing Regulation (EC) No 258/97 of the European Parliament and of the Council and Commission Regulation (EC) No 1852/2001. Official Journal of the European Union L 327, 11.12.2015: 1-22.

European Commission, 2022. Non-communicable-diseases: an overview. DG Public Health and Food Safety, Brussels, Belgium Available at: https://health.ec.europa.eu/non-communicable-diseases/overview_en.

Falony, G., Joossens, M., Vieira-Silva, S., Wang, J., Darzi, Y., Faust, K., Kurilshikov, A., Bonder, M.J., Valles-Colomer, M., Vandeputte, D., Tito, R.Y., Chaffron, S., Rymenans, L., Verspecht, C., De Sutter, L., Lima-Mendez, G., D'Hoe, K., Jonckheere, K., Homola, D., Garcia, R., Tigchelaar, E.F., Eeckhaudt, L., Fu, J., Henckaerts, L., Zhernakova, A., Wijmenga, C. and Raes, J., 2016. Population-level analysis of gut microbiome variation. Science 352: 560-564. https://doi.org/10.1126/science.aad3503

Faure, D., Simon, J.-C. and Heulin, T., 2018. Holobiont a conceptual framework to explore the eco-evolutionary and functional implications of host-microbiota interactions in all ecosystems. New Phytologist 218: 1321-1324.

Faust, K. and Raes, J., 2012. Microbial interactions: from networks to models. Nature Reviews: Microbiology 10: 538-550. https://doi.org/10.1038/nrmicro2832

Flach, J., Koks, M., Van der Waal, M.B., Claassen, E. and Larsen, O.F.A., 2018. Economic potential of probiotic supplementation in institutionalized elderly with chronic constipation. PharmaNutrition 6: 198-206. https://doi.org/10.1016/j.phanu.2018.10.001

Fødevarestyrelsen, 2021. Kosttilskud. Available at: https://www.foedevarestyrelsen.dk/Foedevarer/Kosttilskud/Sider/Kosttilskud.aspx.

Foligne, B., Nutten, S., Grangette, C., Dennin, V., Goudercourt, D., Poiret, S., Dewulf, J., Brassart, D., Mercenier, A. and Pot, B., 2007. Correlation between *in vitro* and *in vivo* immunomodulatory properties of lactic acid bacteria. World Journal of Gastroenterology 13: 236-243.

Frackowiak, T., Groyecka-Bernard, A., Oleszkiewicz, A., Butovskaya, M., Zelazniewicz, A. and Sorokowski, P., 2020. Difference in perception of onset of old age in traditional (Hadza) and modern (Polish) societies. International Journal of Environmental Research and Public Health 17: 7079. https://doi.org/10.3390/ijerph17197079

Gacesa, R., Kurilshikov, A., Vich Vila, A., Sinha, T., Klaassen, M.A.Y., Bolte, L.A., Andreu-Sánchez, S., Chen, L., Collij, V., Hu, S., Dekens, J.A.M., Lenters, V.C., Björk, J.R., Swarte, J.C., Swertz, M.A., Jansen, B.H., Gelderloos-Arends, J., Hofker, M., Vermeulen, R.C.H., Sanna, S., Harmsen, H.J.M., Wijmenga, C., Fu, J., Zhernakova, A. and Weersma, R.K., 2020. Environmental factors shaping the gut microbiome in a Dutch population. Nature 604: 732-139. https://doi.org/10.1101/2020.11.27.401125

Gibson, G.R., Hutkins, R., Sanders, M.E., Prescott, S.L., Reimer, R.A., Salminen, S.J., Scott, K., Stanton, C., Swanson, K.S., Cani, P.D., Verbeke, K. and Reid, G., 2017. Expert consensus document: The International Scientific Association for Probiotics and Prebiotics (ISAPP) consensus statement on the definition and scope of prebiotics. Nature Reviews: Gastroenterology & Hepatology 14: 491-502. https://doi.org/10.1038/nrgastro.2017.75

Gilbert, J.A., Blaser, M.J., Caporaso, J.G., Jansson, J.K., Lynch, S.V. and Knight, R., 2018. Current understanding of the human microbiome. Nature Medicine 24: 392-400. https://doi.org/10.1038/nm.4517

Gilliland, S.E., Morelli, L. and Reid, G., 2001. Report of a Joint FAO/WHO Expert consultation on evaluation of health and nutritional properties of probiotics in food, including powder milk with live lactic acid bacteria, probiotics in food: Health and nutritional properties and guidelines for evaluation, pp. 1-50. WHO, Geneva, Switzerland.

Guarner, F., Bourdet-Sicard, R., Brandtzaeg, P., Gill, H.S., McGuirk, P., van Eden, W., Versalovic, J., Weinstock, J.V. and Rook, G.A., 2006. Mechanisms of disease: the hygiene hypothesis revisited. Nature Reviews: Gastroenterology and Hepatology 3: 275-284. https://doi.org/10.1038/ncpgasthep0471

Haahtela, T., Alenius, H., Lehtimaki, J., Sinkkonen, A., Fyhrquist, N., Hyoty, H., Ruokolainen, L. and Makela, M.J., 2021. Immunological resilience and biodiversity for prevention of allergic diseases and asthma. Allergy 76: 3613-3626. https://doi.org/10.1111/all.14895

Haahtela, T., Holgate, S., Pawankar, R., Akdis, C.A., Benjaponpitak, S., Caraballo, L., Demain, J., Portnoy, J., von Hertzen, L., Change, W.A.O.S.C.o.C. and Biodiversity, 2013. The biodiversity hypothesis and allergic disease: world allergy organization position statement. World Allergy Organization Journal 6: 3. https://doi.org/10.1186/1939-4551-6-3

Hajishengallis, G., Liang, S., Payne, M.A., Hashim, A., Jotwani, R., Eskan, M.A., McIntosh, M.L., Alsam, A., Kirkwood, K.L., Lambris, J.D., Darveau, R.P. and Curtis, M.A., 2011. Low-abundance biofilm species orchestrates inflammatory periodontal disease through the commensal microbiota and complement. Cell Host & Microbe 10: 497-506. https://doi.org/10.1016/j.chom.2011.10.006

Hill, C., Guarner, F., Reid, G., Gibson, G.R., Merenstein, D.J., Pot, B., Morelli, L., Canani, R.B., Flint, H.J., Salminen, S., Calder, P.C. and Sanders, M.E., 2014. Expert consensus document. The International Scientific Association for Probiotics and Prebiotics consensus statement on the scope and appropriate use of the term probiotic. Nature Reviews: Gastroenterology & Hepatology 11: 506-514. https://doi.org/10.1038/nrgastro.2014.66

Holmboe-Ottesen, G. and Wandel, M., 2012. Changes in dietary habits after migration and consequences for health: a focus on South Asians in Europe. Food & Nutrition Research 56. https://doi.org/10.3402/fnr.v56i0.18891

Hooper, L., Martin, N., Jimoh, O.F., Kirk, C., Foster, E. and Abdelhamid, A.S., 2020. Reduction in saturated fat intake for cardiovascular disease. Cochrane Database of Systematic Reviews 8: CD011737. https://doi.org/10.1002/14651858.CD011737.pub3

Indrio, F., Di Mauro, A., Riezzo, G., Civardi, E., Intini, C., Corvaglia, L., Ballardini, E., Bisceglia, M., Cinquetti, M., Brazzoduro, E., Del Vecchio, A., Tafuri, S. and Francavilla, R., 2014. Prophylactic use of a probiotic in the prevention of colic, regurgitation, and functional constipation: a randomized clinical trial. JAMA Pediatrics 168: 228-233. https://doi.org/10.1001/jamapediatrics.2013.4367

IPA Europe, 2021. Denmark's new document on the use of the term probiotic – Non-official translation May 2021. Available at: https://www.ipaeurope.org/wp-content/uploads/2021/05/IPA-EUROPE-Denmark.pdf.

Koopman, J.J., van Bodegom, D., Ziem, J.B. and Westendorp, R.G., 2016. An emerging epidemic of noncommunicable diseases in developing populations due to a triple evolutionary mismatch. American Journal of Tropical Medicine and Hygiene 94: 1189-1192. https://doi.org/10.4269/ajtmh.15-0715

Koponen, A., Sandell, M., Salminen, S. and Lenoir-Wijnkoop, I., 2012. Nutrition economics: towards comprehensive understanding of the benefits of nutrition. Microbial Ecology in Health and Disease 23: 46-50. https://doi.org/10.3402/mehd.v23i0.18585

Koutsoumanis, K., Allende, A., Alvarez-Ordóñez, A., Bolton, D., Bover-Cid, S., Chemaly, M., Davies, R., De Cesare, A., Hilbert, F., Lindqvist, R., Nauta, M., Peixe, L., Ru, G., Simmons, M., Skandamis, P., Suffredini, E., Cocconcelli, P.S., Escámez, P.S.F., Prieto-Maradona, M., Querol, A., Sijtsma, L., Suarez, J.E., Sundh, I., Vlak, J., Barizzone, F., Hempen, M. and Herman, L., 2022. Update of the list of QPS-recommended biological agents intentionally added to food or feed as notified to EFSA 15: suitability of taxonomic units notified to EFSA until September 2021. EFSA Journal 20: e07045. https://doi.org/https://doi.org/10.2903/j.efsa.2022.704

Leeming, E.R., Johnson, A.J., Spector, T.D. and Le Roy, C.I., 2019. Effect of diet on the gut microbiota: rethinking intervention duration. Nutrients 11: 2862. https://doi.org/10.3390/nu11122862

Leeming, E.R., Louca, P., Gibson, R., Menni, C., Spector, T.D. and Le Roy, C.I., 2021. The complexities of the diet-microbiome relationship: advances and perspectives. Genome Medicine 13: 10. https://doi.org/10.1186/s13073-020-00813-7

Lenoir-Wijnkoop, I., Gerlier, L., Bresson, J.L., Le Pen, C. and Berdeaux, G., 2015. Public health and budget impact of probiotics on common respiratory tract infections: a modelling study. PloS One 10: e0122765. https://doi.org/10.1371/journal.pone.0122765

Lenoir-Wijnkoop, I., Gerlier, L., Roy, D. and Reid, G., 2016. The clinical and economic impact of probiotics consumption on respiratory tract infections: projections for Canada. PloS One 11: e0166232. https://doi.org/10.1371/journal.pone.0166232

Lenoir-Wijnkoop, I., Merenstein, D., Korchagina, D., Broholm, C., Sanders, M.E. and Tancredi, D., 2019. Probiotics reduce health care cost and societal impact of flu-like respiratory tract infections in the USA: an economic modeling study. Frontiers in Pharmacology 10: 980. https://doi.org/10.3389/fphar.2019.00980

Lenoir-Wijnkoop, I., Nuijten, M.J., Craig, J. and Butler, C.C., 2014. Nutrition economic evaluation of a probiotic in the prevention of antibiotic-associated diarrhea. Frontiers in Pharmacology 5: 13. https://doi.org/10.3389/fphar.2014.00013

Leroy, F. and De Vuyst, L., 2004. Lactic acid bacteria as functional starter cultures for the food fermentation industry. Trends in Food Science & Technology 15: 67-78.

Ley, R.E., Bäckhed, F., Turnbaugh, P., Lozupone, C.A., Knight, R.D. and Gordon, J.I., 2005. Obesity alters gut microbial ecology. Proceedings of the National Academy of Sciences of the USA 102: 11070-11075. https://doi.org/10.1073/pnas.0504978102

Ley, R.E., Turnbaugh, P.J., Klein, S. and Gordon, J.I., 2006. Microbial ecology: human gut microbes associated with obesity. Nature 444: 1022-1023. https://doi.org/10.1038/4441022a

Lozupone, C.A., Stombaugh, J.I., Gordon, J.I., Jansson, J.K. and Knight, R., 2012. Diversity, stability and resilience of the human gut microbiota. Nature: 220-230. https://doi.org/10.1038/nature11550

Marco, M.L., Sanders, M.E., Ganzle, M., Arrieta, M.C., Cotter, P.D., De Vuyst, L., Hill, C., Holzapfel, W., Lebeer, S., Merenstein, D., Reid, G., Wolfe, B.E. and Hutkins, R., 2021. The International Scientific Association for Probiotics and Prebiotics (ISAPP) consensus statement on fermented foods. Nature Reviews: Gastroenterology & Hepatology 18: 196-208. https://doi.org/10.1038/s41575-020-00390-5

Marlowe, F.W., 2005. Hunter-gatherers and human evolution. Evolution Antropology 14: 54-67.

Marlowe, F.W., 2010. The Hadza – hunter-gatherers of Tanzania. University of California Press Ltd, London, UK.

Marlowe, F.W. and Berbesque, J.C., 2009. Tubers as fallback foods and their impact on Hadza hunter-gatherers. American Journal of Physical Anthropology 140: 751-758. https://doi.org/10.1002/ajpa.21040

MGI, 2022. 'Million microbiome of humans project' will provide important basis for analyzing the relationship between microbiome and human health. Available at: https://en.mgi-tech.com/news/114/.

Okada, H., Kuhn, C., Feillet, H. and Bach, J.F., 2010. The 'hygiene hypothesis' for autoimmune and allergic diseases: an update. Clinical and Experimental Immunology 160: 1-9. https://doi.org/10.1111/j.1365-2249.2010.04139.x

Power, M.E., Tilman, D., Estes, J.A., Menge, B.A., J., B.W., Scott Mills, L., Daily, G., Castilla, J.C., Luhchenco, J. and Paine, R.T., 1996. Challenges in the quest for keystones. Bioscience 40: 609-620.

Putri, W.C.W.S., Muscatello, D.J., Stockwell, M.S. and Newall, A.T., 2018. Economic burden of seasonal influenza in the United States. Vaccine 36: 3960-3966. https://doi.org/https://doi.org/10.1016/j.vaccine.2018.05.057

Ragonnaud, E. and Biragyn, A., 2021. Gut microbiota as the key controllers of 'healthy' aging of elderly people. Immunity & Ageing 18: 2. https://doi.org/10.1186/s12979-020-00213-w

Ray, R.C. and Joshi, V.K., 2014. Fermented foods: past, present and future. In: Ray, R.C. and Montel, D. (eds) Microorganisms and Fermentation of traditional foods. CRC Press, Boca Raton, FL, USA, pp. 1-36. https://doi.org/10.13140/2.1.1849.8241

Roswall, J., Olsson, L.M., Kovatcheva-Datchary, P., Nilsson, S., Tremaroli, V., Simon, M.C., Kiilerich, P., Akrami, R., Kramer, M., Uhlen, M., Gummesson, A., Kristiansen, K., Dahlgren, J. and Backhed, F., 2021. Developmental trajectory of the healthy human gut microbiota during the first 5 years of life. Cell Host & Microbe 29: 765-776. https://doi.org/10.1016/j.chom.2021.02.021

Rouanet, A., Bolca, S., Bru, A., Claes, I., Cvejic, H., Girgis, H., Harper, A., Lavergne, S.N., Mathys, S., Pane, M., Pot, B., Shortt, C., Alkema, W., Bezulowsky, C., Blanquet-Diot, S., Chassard, C., Claus, S.P., Hadida, B., Hemmingsen, C., Jeune, C., Lindman, B., Midzi, G., Mogna, L., Movitz, C., Nasir, N., Oberreither, M., Seegers, J., Sterkman, L., Valo, A., Vieville, F. and Cordaillat-Simmons, M., 2020. Live biotherapeutic products, a road map for safety assessment. Frontiers in Medicine 7: 237. https://doi.org/10.3389/fmed.2020.00237

Salminen, S., Collado, M.C., Endo, A., Hill, C., Lebeer, S., Quigley, E.M.M., Sanders, M.E., Shamir, R., Swann, J.R., Szajewska, H. and Vinderola, G., 2021. The International Scientific Association of Probiotics and Prebiotics (ISAPP) consensus statement on the definition and scope of postbiotics. Nature Reviews: Gastroenterology & Hepatology 18: 649-667. https://doi.org/10.1038/s41575-021-00440-6

Sanlier, N., Gokcen, B.B. and Sezgin, A.C., 2019. Health benefits of fermented foods. Crit Rev Food Sci Nutr 59: 506-527. https://doi.org/10.1080/10408398.2017.1383355

Schnorr, S.L., Candela, M., Rampelli, S., Centanni, M., Consolandi, C., Basaglia, G., Turroni, S., Biagi, E., Peano, C., Severgnini, M., Fiori, J., Gotti, R., De Bellis, G., Luiselli, D., Brigidi, P., Mabulla, A., Marlowe, F., Henry, A.G. and Crittenden, A.N., 2014. Gut microbiome of the Hadza hunter-gatherers. Nature Communications 5: 3654. https://doi.org/10.1038/ncomms4654

Scott Mills, L., Soule, M.E. and Ooak, D.F., 1993. The Keystone-species concept in ecology and conservation. Bioscience 43: 219-224.

Shi, J., Barakat, M., Chen, D. and Chen, L., 2018. Bicellular tight junctions and wound healing. International Journal of Molecular Sciences 19: 3862. https://doi.org/10.3390/ijms19123862

Sonnenburg, E.D. and Sonnenburg, J.L., 2014. Starving our microbial self: the deleterious consequences of a diet deficient in microbiota-accessible carbohydrates. Cell Metabolism 20: 779-786. https://doi.org/10.1016/j.cmet.2014.07.003

Swanson, K.S., Gibson, G.R., Hutkins, R., Reimer, R.A., Reid, G., Verbeke, K., Scott, K.P., Holscher, H.D., Azad, M.B., Delzenne, N.M. and Sanders, M.E., 2020. The International Scientific Association for Probiotics and Prebiotics (ISAPP) consensus statement on the definition and scope of synbiotics. Nature Reviews: Gastroenterology & Hepatology 17: 687-701. https://doi.org/10.1038/s41575-020-0344-2

Thomas, L., 2022. The microbiome of a newborn. News Medical Life Sciences. Available at: https://www.news-medical.net/health/The-microbiome-of-a-newborn.aspx

Tudela, H., Claus, S.P. and Saleh, M., 2021. Next generation microbiome research: identification of keystone species in the metabolic regulation of host-gut microbiota interplay. Frontiers in Cell Development and Biology 9: 719072. https://doi.org/10.3389/fcell.2021.719072

Turnbaugh, P.J., Hamady, M., Yatsunenko, T., Cantarel, B.L., Duncan, A., Ley, R.E., Sogin, M.L., Jones, W.J., Roe, B.A., Affourtit, J.P., Egholm, M., Henrissat, B., Heath, A.C., Knight, R. and Gordon, J.I., 2009. A core gut microbiome in obese and lean twins. Nature 457: 480-484. https://doi.org/10.1038/nature07540

Vandeputte, D., Falony, G., Vieira-Silva, S., Tito, R.Y., Joossens, M. and Raes, J., 2016. Stool consistency is strongly associated with gut microbiota richness and composition, enterotypes and bacterial growth rates. Gut 65: 57-62. https://doi.org/10.1136/gutjnl-2015-309618

Vaskoska, R., 2022. Hostile microbiology. In: Wernaart, B.F.W. and Van der Meulen, B.M.J. (eds) Applied Food Science. Wageningen Academic Publishers, Wageningen, the Netherlands, pp. 247-266.

Vieira-Silva, S., Sabino, J., Valles-Colomer, M., Falony, G., Kathagen, G., Caenepeel, C., Cleynen, I., van der Merwe, S., Vermeire, S. and Raes, J., 2019. Quantitative microbiome profiling disentangles inflammation- and bile duct obstruction-associated microbiota alterations across PSC/IBD diagnoses. Nature Microbiology 4: 1826-1831. https://doi.org/10.1038/s41564-019-0483-9

Vinderola, G., Sanders, M.E. and Salminen, S., 2022. The concept of postbiotics. Foods 11. https://doi.org/10.3390/foods11081077

Warinner, C., Speller, C., Collins, M.J. and Lewis, C.M., Jr., 2015. Ancient human microbiomes. Journal of Human Evolution 79: 125-136. https://doi.org/10.1016/j.jhevol.2014.10.016

Wastyk, H.C., Fragiadakis, G.K., Perelman, D., Dahan, D., Merrill, B.D., Yu, F.B., Topf, M., Gonzalez, C.G., Robinson, J.L., Elias, J.E., Sonnenburg, E.D., Gardner, C.D. and Sonnenburg, J.L., 2021. Gut-microbiota-targeted diets modulate human immune status. Cell 184: 4137-4153. https://doi.org/10.1016/j.cell.2021.06.019

Wilmanski, T., Diener, C., Rappaport, N., Patwardhan, S., Wiedrick, J., Lapidus, J., Earls, J.C., Zimmer, A., Glusman, G., Robinson, M., Yurkovich, J.T., Kado, D.M., Cauley, J.A., Zmuda, J., Lane, N.E., Magis, A.T., Lovejoy, J.C., Hood, L., Gibbons, S.M., Orwoll, E.S. and Price, N., 2021. Gut microbiome pattern reflects healthy aging and predicts survival in humans. Nature Metabolism 3: 274-286. https://doi.org/10.1038/s42255-021-00348-0

Windey, K., De Preter, V. and Verbeke, K., 2012. Relevance of protein fermentation to gut health. Molecular Nutrition & Food Research 56: 184-196. https://doi.org/10.1002/mnfr.201100542

Wu, G.D., Chen, J., Hoffmann, C., Bittinger, K., Chen, Y.Y., Keilbaugh, S.A., Bewtra, M., Knights, D., Walters, W.A., Knight, R., Sinha, R., Gilroy, E., Gupta, K., Baldassano, R., Nessel, L., Li, H., Bushman, F.D. and Lewis, J.D., 2011. Linking long-term dietary patterns with gut microbial enterotypes. Science 334: 105-108. https://doi.org/10.1126/science.1208344

Yatsunenko, T., Rey, F.E., Manary, M.J., Trehan, I., Dominguez-Bello, M.G., Contreras, M., Magris, M., Hidalgo, G., Baldassano, R.N., Anokhin, A.P., Heath, A.C., Warner, B., Reeder, J., Kuczynski, J., Caporaso, J.G., Lozupone, C.A., Lauber, C., Clemente, J.C., Knights, D., Knight, R. and Gordon, J.I., 2012. Human gut microbiome viewed across age and geography. Nature 486: 222-227. https://doi.org/10.1038/nature11053

Zhang, X., Li, L., Butcher, J., Stintzi, A. and Figeys, D., 2019. Advancing functional and translational microbiome research using meta-omics approaches. Microbiome 7: 154. https://doi.org/10.1186/s40168-019-0767-6

12. Hostile microbiology

This is what you need to know about pathogens and spoilage

Rozita Vaskoska

The University of Melbourne, Royal Parade 30, 3010 Parkville, Victoria, Australia; CSIRO, 671 Sneydes Road, Werribee 3030, Victoria, Australia; rozita.spirovskavaskoska@csiro.au

Abstract

Foodborne illnesses significantly affect the public health systems of countries around the world, both through the negative impact on people's health and the large economic burden on their healthcare systems. Foodborne illnesses are caused by pathogenic microorganisms or, in short, pathogens. Pathogens differ in their characteristics, incubation periods, symptoms, and complications. They also differ in the ways they cause an illness (infection or intoxication), the numbers or amounts that need to be ingested to cause an illness (infection or toxic dose), and their preferential conditions for growth. Microbiological spoilage is the occurrence of deterioration of the quality of food through the action of microorganisms. Contemporary issues in food microbiology include emerging pathogens, non-culturable viable state and stress response, food adulteration, alternative food proteins trends, biofilms, antibiotic resistance and – in response – predictive microbiology. Pathogens are detected using microbiological testing methods based on the specific characteristics of the pathogen, while spoilage is indirectly assessed through the total viable counts of the food. Control measures are implemented on a government, business, and consumer level to prevent or inactivate pathogens and spoilage microorganisms in food. Food businesses prevent the emergence and growth of pathogens using chemical, physical, and biological interventions. In understanding the growth and inactivation kinetics of pathogens, process parameters can be set for many essential steps in the food industry. The prevention of spoilage relies on the process of food preservation that includes physical, chemical and biological interventions. While traditional methods are still being used extensively in microbiological testing, whole genome sequencing finds more extensive application in food microbiology, particularly in source attribution of outbreaks of foodborne illnesses. Finally, rapid tests and metabolomics are other important developments in the area of microbiological testing.

Key concepts

- ► Pathogenic microorganisms are microorganisms that can cause an illness.
- ► Foodborne pathogenic microorganisms are pathogenic microorganisms that are transmitted through food.
- ► Gram-positive and Gram-negative microorganisms differ from each other in type of cell membrane.
- ► Infection is harm caused directly by microorganisms.
- ► Intoxication is harm caused indirectly by microorganisms through the toxins they produce.
- ► Spores are a dormant state of microorganisms in which they can survive difficult circumstances and are difficult to detect.
- ► Dose is the number of cfu to which a person is exposed.
- ► Cfu/colony-forming units is the number of viable microorganisms.
- ► WGS, whole genome sequencing is a method to decode the DNA of an organism.
- ► log. is a factor 10 ($3 \log = 10^3 = 1000$).
- ► D-value is the time or dose needed to reduce the presence of an organism by a decimal. The time or dose required to reduce the number of microorganisms by 1 log (i.e. to 1/10 i.e. by 90%) = 1D.
- ► Incubation time is the time between the initial infection and the emergence of symptoms.
- ► Disability Adjusted Life Years (DALYs) is a unit for measuring the damage caused by illness.
- ► Sp. (plural spp.) in the name of an organism stands for species.

12.1 Introduction

Several chapters in this book address elements of biology and microbiology. This chapter focuses on microorganisms that may negatively affect food safety or food quality.

12.1.1 Pathogens

foodborne pathogenic microorganisms

Foodborne pathogenic microorganisms mainly come from the following groups: bacteria, viruses and parasites. There are very rare reports of pathogenicity of yeast, while moulds present more of a toxicological then a biological concern, as they cause illnesses through the chemicals they produce

– the mycotoxins (see also Chapter 13 on toxicology; Bast, 2022). When the conditions of the food are favourable, bacteria can grow in the food. Viruses, however, need living cells for growth, thus they get transmitted through food but do not grow in it. Food is also a vehicle for parasites that end up on food through their complex life cycles. The most commonly found pathogens in food are presented in Table 12.1.

Table 12.1 Common pathogenic microorganisms, their characteristics, associated foods and their incubation times and illness symptoms (adapted from Bintsis, 2017; Food & Drug Administration, 2012; Institute of Food Technologists, 2004; Mohammad et al., 2018).

Bacteria	Characteristic	Common foods	Illness
Bacillus cereus	Gram-positive Facultatively anaerobic Spore-forming Can cause infection and intoxication (toxin cereulide)	• Diarrhoeal: meats, milk, vegetables, fish • Vomiting: rice products, potato, pasta, and cheese • Other foods: salads, sauces, puddings, soups, casseroles, pastries	Incubation: 8-16 h (diarrhoeal), 2-4 h (emetic) Symptoms: diarrhoea, nausea, vomiting, abdominal cramps Complications: systemic infections, infections of the brain (meningitis), fat cells, eye cells, lung, heart, gangrene, and infant death
Campylobacter	Gram-negative Microaerophilic	• Poultry meat, non-pasteurised milk, cheeses made from non-pasteurised milk, water	Incubation (jejuni): 2-5 days Symptoms: fever, headache, joint pain, weakness, malaise, swelling of lymph nodes Complications: sepsis, meningitis, liver (hepatitis), gallbladder (cholecystitis), pancreas (pancreatitis)
Cronobacter sakazakii	Gram-negative	• Infant formula (powdered and rehydrated) • Bread, cereal, rice, fruit and vegetables, legumes, herbs, spices, milk, cheese, meat, and fish	Incubation: few days Symptoms (in infants): poor feeding, irritability, grunting respiration, jaundice, varying body temperature, Complications: seizures, brain abscess, hydrocephalus, and developmental delay
Clostridium botulinum	Gram-positive Anaerobic Spore-forming Toxin-forming	• Canned vegetables, soups, mushrooms, ripe olives, spinach, tuna fish, chicken, chicken liver, liver pate, luncheon meats, ham, sausage, stuffed aubergine, lobster, and smoked and salted fish • Honey (infants)	Incubation: 2-6 h (often 12-36 h) Symptoms: vertigo, blurred or double vision, loss of light reflex, dryness in mouth, weakness Complications: swallowing difficulty, respiratory paralysis, paralysis of extremities

Table 12.1 Continued.

Clostridium perfringens	Gram-positive Anaerobic (but aerotolerant) Spore-forming Toxin-forming	• Meats (beef, poultry, gravies and stews, vegetable products, spices and herbs, raw and processed foods)	Incubation: 2-36 h (on average 6-12 h) Symptoms: abdominal cramps, diarrhoea (putrefactive), occasional vomiting and nausea Complications: necrosis of small intestine, infection of the peritoneum (peritonitis), sepsis
Listeria monocytogenes	Gram-positive Facultative anaerobic	• Raw milk, improperly pasteurised milk, chocolate milk, soft cheese, ice cream, raw vegetables, raw poultry and meats, fermented raw-meat sausages, hot dogs and deli meats, raw and smoked fish, seafood	Incubation: few hours to 3 days Symptoms: fever, headache, arthralgia, weakness, malaise, swelling of lymph nodes Complications: meningitis, abortion, stillbirth
Escherichia coli H7:O157	Gram-negative Facultative anaerobic Toxin-forming (Shiga toxins Stx)	• Raw or undercooked ground beef • Raw milk	Incubation: 3-4 days Symptoms: severe abdominal pain, nausea or vomiting, and watery and bloody diarrhoea Complications: Haemolytic Uremic Syndrome (HUS)
Salmonella typhi, paratyphi and other species	Gram-negative Facultative anaerobic	• Meats, poultry, eggs, milk and dairy products, fish, shrimp, spices, yeast, coconut, sauces, freshly prepared salad dressings made with unpasteurised eggs, cake mixes, cream-filled desserts and toppings that contain raw egg, dried gelatine, peanut butter, cocoa, fruits and vegetables, sprouts, tomatoes, peppers, cantaloupes, chocolate	Incubation (*typhi*): 7-28 days Symptoms (*typhi, paratyphi*): fever, malaise, headache, cough, nausea, vomiting, constipation, abdominal pain, chills, rose spots, bloody stools Symptoms (other *Salmonella*): fever, abdominal cramps, nausea, vomiting, headache Complications: arthritis, enteric fever, septicaemia, death
Shigella	Gram-negative Facultative anaerobic	• Drinking water, lettuce, raw vegetables, salads (potato, tuna, shrimp, macaroni, and chicken), milk and dairy products, bakery products, sandwich fillings, poultry	Incubation: 12-96 hours Symptoms: fever, abdominal cramps and pain, diarrhoea, blood, mucus or pus in stools, vomiting, tenesmus Complications: reactive arthritis, haemolytic uremic syndrome, Reiter's syndrome
Staphylococcus aureus	Gram-positive Facultative anaerobic Toxin forming	• Meat and meat products, poultry and egg products, salads (egg, tuna, chicken, potato, and macaroni), bakery products (cream pastries, cream pies, chocolate éclairs, sandwich fillings), milk and dairy products	Incubation: 1-7 h (average 2-4 h) Symptoms: nausea, vomiting, abdominal pain, weakness, Complications: usually resolves uneventfully

Table 12.1 Continued.

Vibrio parahaemolyticus	Gram-negative Facultative anaerobic	• Raw, improperly cooked or recontaminated fish and shellfish	Incubation: 12-96 h Symptoms: low fever, chills, nausea, vomiting, diarrhoea Complications: septicaemia
Yersinia enterocolitica	Gram-negative Aerobic, facultative anaerobic	• Meat, oysters, fish, crabs, and raw milk	Incubation: 3-7 days Symptoms: fever, diarrhoea and/or vomiting, bloody stools, headache, abdominal pain Complications: arthritis, lymphadenitis, erythema nodosum, meningitis
Viruses			
Norovirus	RNA virus	• Water, salads, fruit, and oysters	Incubation: 12-48 hours Symptoms: explosive vomiting, watery diarrhoea, abdominal cramps, nausea, muscle pain, chills Complications: dehydration
Hepatitis A virus	RNA virus	• Water, shellfish, salads, sandwiches, fruits and fruit juices, milk and milk products, uncooked food, cooked foods, vegetables	Incubation: 30 days Symptoms: fever, headache, anorexia, nausea, vomiting, diarrhoea, dark urine, flu-like symptoms, myalgia, and jaundice Complications: hepatitis, hepatic necrosis, liver failure
Parasites			
Taenia	Tapeworm	• Beef and pork	Incubation: 2 months Symptoms: asymptomatic, abdominal pain, nausea, diarrhoea, malaise, changed appetite Complications: seizures, altered mental status, pressure in the head
Toxoplasma	Obligate intracellular parasite	• Raw or undercooked meats (e.g. pork, lamb, or wild game), fruits or vegetables or water contaminated with cat faeces or cysts	Incubation: 5-23 days Symptoms: muscle pain and sore lymph nodes Complications: miscarriage, stillbirth or damage to baby brain in pregnant women
Trichinella	Roundworm	• Undercooked meat, particularly pork	Incubation: 1-4 weeks Symptoms: diarrhoea, abdominal discomfort, nausea, vomiting Complications: invasion of brain, heart and lungs, death

For bacteria to grow in food they need several intrinsic and extrinsic factors to be fulfilled. Intrinsic factors include the nutrients of the food, the water activity (which is the available water) and the acidity (pH) of the food. The lower the nutrient levels in the food, the drier and more acidic it is, the less likely that a bacterium will grow in it and that other microorganisms will survive in it. Extrinsic factors that affect microbial growth include the temperature and aerobic conditions in which the food is stored. Based on the growth temperature, pathogens, and microorganisms in general, can be divided into: psychrophiles (0-20 °C), mesophiles (10-50 °C), and thermophiles (40-110 °C) (Russell and Fukunaga, 1990). Based on their preference for a gaseous environment also known as oxidation-reduction potential (Eh): microorganisms are divided into aerobic (can only grow in the presence of oxygen), anaerobic (can grow in the absence of oxygen), facultatively anaerobic (can grow in both aerobic and anaerobic conditions) and microaerophilic (can grow at low levels of oxygen) (Jay, 1995). Most pathogens are facultatively anaerobic or aerobic; an example of anaerobic bacteria is *Clostridium* spp., while an example of microaerophilic bacteria is *Campylobacter*.

Gram stain

Another important characteristic of pathogens is how they are stained by a Gram stain, which indicates the type of cell membrane the microorganism has. The Gram staining of major foodborne pathogens is shown in Table 12.1. Gram-positive microorganisms have a thick peptidoglycan layer, while gram-negative bacteria have a thin peptidoglycan layer plus a lipid membrane on the outside. This is very important, as it determines the efficacy of antimicrobial treatments when applied to food. For instance, bacteriocins are more effective in preventing the growth of Gram-positive bacteria (Prudêncio *et al.*, 2015), while high-pressure processing is more effective in inactivating Gram-negative bacteria (Fonberg-Broczek *et al.*, 2005).

infection and intoxication

Pathogenic microorganisms cause two main groups of diseases: infection and intoxication (Mohammad *et al.*, 2018). Some bacteria cause illness in their vegetative form, primarily in the form of infections of the gastrointestinal tract. These bacteria colonise, invade and will damage the intestinal mucosa and cause gastroenteritis (Teunis and Havelaar, 1996). Some microorganisms can produce toxins (*Bacillus cereus, Clostridium* spp.., *E. coli* O157:H7) and thus cause an intoxication. The concern with *Bacillus cereus* and *Clostridium perfringens* or *botulinum* extends to their ability to form spores. Spore-forming microorganisms form spores in unfavourable conditions and can survive in the food until they get the chance to transform into their vegetative form and produce the toxin. The higher thermotolerance of spores compared to the vegetative form of the microorganisms is the reason behind the design of many food processes to specifically inactivate bacterial spores.

dose

For an illness to occur, the microorganisms or toxins need to be present at a certain dose. The dose is the number of colony-forming units, spores, cysts, etc. (Teunis and Havelaar, 1996). The relationship between the dose and the illness it causes, is known as dose response. In the early days of research into foodborne illnesses, the dose response of microorganisms was established by feeding human volunteers (often prisoners) with food contaminated with microorganisms at varying doses and observing the symptoms that would emerge. However, with advances in ethics and science, these types of experiments are no longer conducted, and dose response is established based on epidemiological evidence, such as outbreak and food survey data. The infective dose can vary between a few cells (i.e. 15) for *Salmonella typhi*, to 100s of cells for *Listeria monocytogenes* (Institute of Food Technologists, 2004). Interestingly, some individuals that have an infection may not present with symptoms. This is known as asymptomatic infection.

foodborne illness

The illnesses that pathogens cause can be quite similar, but also quite different. Most foodborne illnesses present as gastroenteritis. Gastroenteritis is defined as diarrhoea and vomiting combined with at least two of the following symptoms: fever, nausea, pain or cramps in the abdomen, and bloody or mucous stool within seven days (Teunis and Havelaar, 1996) (Table 12.1). Incubation time is the first important parameter of a foodborne illness. When presented with foodborne illness symptoms, most consumers expect it to be the food they last ate that made them ill. However, while this might be the case for some pathogens that have a shorter incubation time, foodborne illnesses will sometimes be caused by pathogens that have found their way to the consumer plate days or even a month ago (Table 12.1). That is why outbreak epidemiological investigations are not simple and require many different methodologies to be applied for source identification. While most foodborne illnesses present as gastroenteritis, sometimes bacteria in particular may move to the bloodstream and cause infections in various organs that can be quite severe and even fatal. Most commonly, complications associated with foodborne illnesses will occur in the vulnerable consumer groups, including young children, older adults, pregnant women and immunocompromised people. Severe illnesses caused by specific pathogens might target one demographic, for instance *L. monocytogenes* severely affects pregnant women, while *Cronobacter* spp. severely affects infants.

morbidity and mortality

The impact of foodborne illnesses is measured through their morbidity and mortality. Morbidity is the extent of disease that a pathogen would cause, while mortality is the extent of death that a pathogen would cause. Microorganisms may have a high morbidity but low mortality and *vice versa*, low morbidity and high mortality. For instance, *Campylobacter* and norovirus cause the greatest proportion of foodborne diseases internationally, but typhoidal and non-typhoidal *Salmonella* cause a large proportion of the deaths (World Health Organization, 2015). *L. monocytogenes* causes a very low number of diseases, but its mortality is quite high (Farber and Peterkin, 1991). Internationally, the impact of foodborne illnesses is measured by the Disability Adjusted Life Years (DALYs). DALYs are the 'sum of the years of life lost to due to premature mortality and the years lived with a disability due to prevalent cases of the disease or health condition in a population' (World Health Organization, 2021).

Case 12.1. The 2011 EHEC crisis.

A decade ago (at the time of writing this chapter), an *E. coli* O104: H4 outbreak was raging through Europe, particularly in Germany, followed by France. 3,950 people became ill, 53 died and 800 suffered from haemolytic uremic syndrome (HUS), which is a severe kidney disease that can lead to kidney failure. A foodborne outbreak of this extent had not been recorded in Europe before. The strain that caused the outbreaks was not expected to cause a disease of this severity, and the implications for the food chain were immense. The strain was called: Enterohemorrhagic Escherichia Coli (EHEC).

The outbreak was initially mistakenly related to Spanish cucumbers, while the investigation further revealed that in fact the suspected sources of the outbreak were sprouted seeds from Germany, grown from seeds imported from Egypt. Even when the seeds were found to be the source, conflicting evidence of negative and positive samples emerged, showing the importance of sampling, and possessing the right contaminated batch.

The toll on human health was immense, with a large percentage of the infected consumers suffering severe symptoms. The toll on the food chain was also big, with imports and exports of vegetables greatly affected. From the mistaken attribution to Spanish cucumbers alone, Spanish exporters already lost $200 million a week. A series of emergency measures affecting food import and export were taken in response to the outbreak (Spirovska Vaskoska, 2012).

A few things are noteworthy in this case. The strain of *E. coli*, at the time, was not known to produce shiga toxins, the toxins responsible for the severe disease. For this reason, the strain was not only referred to as EHEC, but also as STEC (shiga toxin producing *E. coli*). The strain must have acquired this ability through contact with a bacteriophage (Beutin and Martin, 2012). The severity of the illnesses, the mortality and the expansion of the outbreak were concerning and required better preparedness for preventing this kind of outbreak. Finally, developments such as whole genome sequencing could be very beneficial in linking geographically isolated cases and conducting a more precise source attribution.

12.1.2 Spoilage

Spoilage is a quality deterioration process that cannot be controlled completely even with modern preservation methods (Gram *et al.*, 2002). The terms spoilage and pathogenic microorganisms are sometimes mixed up, but it is important to understand that spoiled food might not always be unsafe (although when spoilage microorganisms are present in high numbers it is likely that the food can also have a pathogen growing), and that food contaminated with a pathogen might not appear spoiled. Thus, smelling food from the refrigerator is not a reliable way of ensuring that we will not get a food infection or poisoning, as some pathogens can cause disease in very small numbers without visible changes in the food. Bacteria, yeasts and moulds are the main causative agents of microbial spoilage of food, as they can grow in food. The intrinsic and extrinsic factors for growth of spoilage microorganisms are the same as those described for pathogens, earlier in this chapter. *Pseudomonas* and lactic acid bacteria are common spoilage microorganisms. While *Pseudomonas* grows in an aerobic environment, lactic acid bacteria are facultatively anaerobic. Yeasts and moulds are mostly found as spoilage microorganisms in food with either high osmotic pressure (high sugar content), low water activity (dry foods) or high acidity (low pH). While pathogens may cause harm at levels such as 100 cfu/g, food containing spoilage microorganisms will normally only show signs of spoilage at much higher levels. Food spoilage, particularly bacterial spoilage, will manifest with visible and sensory changes when the levels of microorganisms exceed 1,000,000 cfu/g (Panigrahi *et al.*, 2006). The mechanism of the occurrence of spoilage is based on the consumption of nutrients by the microorganisms. Bacteria normally start consuming the simple sugars, like lactose, followed by the amino acids, and when these compounds are exhausted, they affect structural nutrients like proteins, resulting in visible changes in structure or colour and the appearance of slime and odours. However, the type of spoilage depends on the types of enzymes the microorganisms produce. If they produce proteases, proteolytic degradation will occur (for instance, bacteria in meat); if they produce pectinolytic enzymes, they will degrade the sugars, for example pectin in fruits; if they produce lipases, they will degrade fat (Steele, 2004). Therefore, the metabolites, appearance, and sensory characteristics of spoilage will differ between causative agents.

12.2 Contemporary issues

According to the World Health Organization (WHO), every year 600 million people experience a foodborne illness which 420,000 of them do not survive. This amounts to 33 million healthy live years (Disability Adjusted Life Years – DALYs) and US$110 billion in loss of productivity and medical-related costs in low- and middle-income countries (WHO, 2020).

Contemporary issues in the area of microbial pathogens include emerging pathogens, non-culturable viable state and stress response of microorganisms, antibiotic resistance, food crime, the alternative food proteins trends, biofilms, and – in response – predictive microbiology. However, considering the complexity of the field, the list of contemporary issues is not exhaustive.

emerging pathogens Pathogens that have not been previously known are seen as new pathogens; pathogens that have been known but have recently emerged as foodborne are known as emerging; and pathogens that change and become more potent/ virulent are known as evolving or re-emerging (Mor-Mur and Yuste, 2010). However, the label emerging pathogens is also used as an umbrella term to refer to all three of these categories. The *E. coli* strain in Case 12.1 was an example of a re-emerging pathogenic microorganism obtaining greater virulence through gene transfer. Emerging pathogens are often of zoonotic origin, which means they normally live in an animal reservoir, and people become exposed incidentally (Behravesh *et al.*, 2012). While listed as a common pathogen in Table 12.1, *Cronobacter* is one of the most well-known emerging pathogens, as it was related to severe illness in infants as recently as the end of the last century. An example of bacteria normally not considered a pathogen that has also become a concern in food is *Arcobacter*; while examples of bacteria known to infect other species but that have become a foodborne concern are *Mycobacterium avium* and *Helicobacter pillory* (Behling *et al.*, 2010; Mor-Mur and Yuste, 2010).

non-culturable viable state Microorganisms can enter a non-culturable viable state, which is a state of reduced vitality, occurring as a consequence of exposure to stress caused by disinfectants, antibiotics, low nutrients level and increase in temperature (Ferro *et al.*, 2018). These microorganisms will not be detected by traditional plating methods. DNA tests would be the only way to detect these microorganisms (Ferro *et al.*, 2018). If these microorganisms end up in food, they might resuscitate and cause illness, although they would be undetectable with traditional microbiological tests.

stress response In addition, as microorganisms get exposed to a series of harsh environmental conditions, they develop an adaptive behaviour, also known as stress response (Den Besten *et al.*, 2010). As a consequence, they become more robust, and can thus become protected from otherwise lethal preservation methods. This jeopardises food safety and human health. A case in point is antibiotic resistance addressed hereafter.

antimicrobial resistance Similarly to bacterial pathogens that affect humans through other routes, foodborne pathogenic bacteria can acquire antimicrobial resistance. As high as 50% of tested common bacterial pathogens have presented an antimicrobial resistance to at least one antibiotic, and there was an alarming level of pathogens showing multidrug resistance (Mayrhofer *et al.*, 2004). Restricted medicinal use of antibiotics in combination with prohibited use of antibiotics as growth promoters are some of the major control measures (Koluman and Dikici, 2013). Major food pathogens such as shiga toxin producing *E coli* (as in the case), *Salmonella enterica* serovar Typhimurium, *Campylobacter jejuni*, *L. monocytogenes* and *Yersinia enterocolitica* have been recognised as emerging antimicrobial resistance phenotypes (White *et al.*, 2002).

food crime

Crime is an issue in the food sector. It may take the form of intentional adulteration or reduced efforts to ensure food safety. Motives may be ideological, but are usually economic. The latter is often referred to as 'food fraud' (Corini, 2020).

While the deliberate addition of pathogenic microorganisms is not the type of food adulteration that takes place nowadays, chemical substitution of ingredients, use of unapproved facilities, expiry date change, and food without accompanying documentation (Montgomery *et al.*, 2020) could very well expose the food to microbiological concerns.

alternative proteins

Plant-based foods and other alternatives for animal-based food such as lab-grown meat (also referred to as cultured meat, cultivated meat or cellular meat) are often presented as 'cleaner' foods in the context of microbiological contamination. Cellular meat has even been called 'clean meat' (Bryant *et al.*, 2019), which might lead to a misconception that meat from animals is dirty.[47] Risks from animal- and plant-based foods are not higher or lower, they are just different. While animal-based foods are likely to have a greater incidence of zoonotic microorganisms, pathogens that are present in animals often end up in the plant food chain by contamination of the field through faecal or water contamination (Erickson and Doyle, 2012). Cellular-based animal food alternatives could use antibiotics in the process that might lead to antimicrobial resistance, can pose a risks of viruses, as well as prions when the cells are grown in bovine growth serum (Hadi and Brightwell, 2021). Plant-based meats can contain allergens, anti-nutrients, thermally produced carcinogens (Hadi and Brightwell, 2021), natural toxicants, and pesticides. Insect protein (see also Chapter 20; Floto-Stammen, 2022) might also pose microbiological risks and can contain allergens (Hadi and Brightwell, 2021). So, although microbiological risks might be more diverse in animal-based foods, alternative protein sources

---

[47] Muscles consumed as meat, coming from healthy animals, are sterile and only become contaminated when exposed to the environment.

are not exempt from food safety risks. In fact, they may present risks that one was not previously aware of and that for this reason may only be detected after they cause harm.

biofilms

Biofilms are complex communities of microorganisms surrounded by a matrix of extracellular compounds (normally polysaccharides)[48], which are formed on surfaces in food production facilities. They may have the opportunity to form due to improper equipment design and/or inadequate cleaning procedures. They can 'shield' pathogens like *Bacillus cereus*, *E. coli*, *Salmonella enterica*, *Staphylococcus aureus* (Galié *et al.*, 2018) and protect them from the action of disinfectants, allowing for the possibility of their detachment from the equipment and contamination of the food, thus causing risk of spoilage, equipment damage and food safety risk (Yuan *et al.*, 2020).

predictive microbiology

Predictive microbiology or quantitative microbial modelling of food is a discipline that aims to proactively assess the safety and quality of food by collecting data on the responses of bacteria to environmental conditions and summarises this data in models and databases (Pradhan *et al.*, 2019). A great amount of work in this area and its application in microbiological risk assessment has been done by the International Commission on Microbiological Specifications of Food (ICMSF). Predictive microbiology often uses existing data of growth of bacteria to estimate the lag time (time until bacteria start growing), growth rate and highest level of growth (asymptote) under a variety of conditions such as temperature, water activity and pH and uses this data to predict the growth of microorganisms that have not been assessed experimentally, but are exposed to the same or extrapolated environmental conditions (Pradhan *et al.*, 2019). Big data and artificial intelligence tools are expected to further advance the progress in the area of predictive microbiology (Pradhan *et al.*, 2019). Predictive microbiology tools and approach can be very beneficial to food businesses that can take advantage of various freely available software and databases online.

12.3 Methods

Pathogenic microorganisms are detected and quantified using traditional and advanced microbiological testing methods. The traditional method goes as follows. A sample of food is mixed with a dilution liquid (normally a buffer). Depending on the level of the contamination it is then mixed with a liquid to homogenise it. A small liquid sample is distributed in an agar plate that consists of selective media, i.e. a solidified liquid that has nutrients specific to the

---

[48] The best-known example of biofilm is probably dental plaque.

growth of that microorganism, and antibiotics against other microorganisms. The sample is left in the agar for a number of hours (often 48) to grow in an incubator at its optimal temperature, and in its optimal gaseous atmosphere (special pouches are used for creating an anaerobic environment). At the end of the incubation time, the plates are observed and colony-forming units are counted. Normally the count is expressed as a natural logarithm. Thus, when microbiological testing results are obtained the numbers are normally expressed as log cfu/g, which represents the tenth logarithm of the colony-forming units per gram. For example, if the level of contamination of a food is 3 log cfu/g, it means that the food contains (10^3 =) 1000 cfu/g.

Other microbiological tests used for pathogen identification include: gel electrophoresis of bacterial proteins to identify a fingerprint of the microorganism, chemical reactions for identification, immunological methods – enzyme linked immunoassays (ELISA) – and most popular recently, genetically based tests such as polymerase chain reaction (PCR) and whole genome sequencing.

However, microbiological testing is not a guarantee for the safety of the product. Absence of a pathogen in a microbiological test does not guarantee that the food is free of the pathogen. There is always a residual risk resulting from the choice of a sampling plan for that food (Zwietering *et al.*, 2021). Food sampling has inherent limitations: we cannot test everything to ensure safety, as there would be nothing left to eat, and it would be very costly. Microbiological testing nowadays is considered more of a verification and validation procedure rather than an actual control measure. More often, food is not even tested for pathogen presence, but for what we call an indicator microorganism. These include total viable count, *Enterobacteriaceae*, coliform microorganisms, *E. coli*, etc. If these microorganisms are present at high levels, it means that the hygiene of the processing environment is not maintained, thus there is a greater risk that pathogens are also present. The total viable count test is also an indicator of the spoilage occurrence in food.

12.4 Management of foodborne pathogens

Food businesses need to prevent contamination by taking care of multiple aspects. They need to ensure the safety of their raw materials with their suppliers. They have to formulate their foods such that they can control microbial growth, and handle the food in a safe manner through the production process. This includes pathogen inactivation steps, implementing quality assurance programmes covering all the above-mentioned aspects, including establishing critical control points as part of the Hazard Analysis and Critical Control Points system (if a pathogenic microorganism is likely to grow at some stage;

see also Chapter 10; Ferris, 2022). Finally, storage instructions and directions for use may need to be provided to consumers to maintain safety at their end. There are chemical, physical and biological methods to control the growth of pathogenic and spoilage microorganisms. Chemical methods include the use of food additives from synthetic or natural origin or natural preservatives to prevent the growth of microorganisms or inactivate them. Physical methods include processing techniques such as heating (cooking), freezing, high pressure processing, irradiation, etc. Some processing steps are specifically designed to inactivate microorganisms. Examples of heating processes include pasteurisation of milk to inactivate vegetative cells and sterilisation to inactivate spores; listericidal processes that ensure 6D reduction of *L. monocytogenes*; and sterilisation of canned food to inactivate spores of *Clostridium botulinum* (12D reduction). Sometimes, processing steps are designed to obtain some sensory characteristics of a product or are an inevitable step in the manufacturing process of such food products, but simultaneously control microbial growth, e.g. fermentation, which has the ability to decrease pH and thus prevent growth; drying, which decreases water activity and thus prevents growth; and nitrate curing, which at the right level prevents the growth of *Clostridium botulinum*. Biological methods include the use of biological agents to affect microorganisms, e.g. bacteriophages (viruses that can control bacteria) (Garcia *et al.*, 2008).

On a government level, food pathogens are managed through regulations that require microbiological testing (Van der Meulen and Wernaart, 2020), as well as through pathogen reduction programmes, which when implemented properly can be very efficient and can reduce the burden of foodborne diseases quite significantly. The EU, for example, has put a strict policy in place to combat *Salmonella*.

The last links in the chain of measures to control pathogens are in our own kitchens. Practices such as regular hand washing are essential both in domestic and industrial conditions, as the faecal-oral route is the main contamination route for food. In addition, washing vegetables and fruits, properly cooking food at the required temperatures (e.g. foods containing eggs and meat), properly storing food, and preventing cross-contamination (e.g. not cutting meat and preparing salad on the same cutting board) are essential to maintaining food safety.

12.5 Looking ahead

Humans and pathogens will continue to coexist on our planet for a long time to come. Humans will continue to look for ways to limit the negative health impacts on people while pathogens will evolve to overcome the hurdles that humans place in their way. We are thus facing a future of continuous development and adaptation. New scientific approaches and technologies are invaluable. Even new trends in consumer preferences may bring new challenges with regard to pathogens.

whole genome sequencing

One of the greatest developments in relation to food pathogens of recent times is the application of whole genome sequencing (WGS). WGS has moved from being a research method to a regular testing method for some applications. In particular, the use of WGS has become an invaluable method for source attribution and investigation of foodborne outbreaks. It has primarily enabled epidemiologists to identify the same pathogen in the food sample and the clinical sample. In addition, it has allowed for outbreaks at multiple locations and seemingly unrelated cases to be connected in the same outbreak (Ronholm et al., 2016). In addition, WGS can be used for better characterisation of microorganisms, in the step of hazard identification of the process of risk assessment (Jagadeesan et al., 2019; see also Chapter 6; De Boer, 2022). In addition to outbreaks, it can also serve as a hygiene control measure, although its penetration in the area is limited compared to traditional microbiological methods due to the practicality, complexity and associated costs (Jagadeesan et al., 2019). Finally, the information obtained about the pathogen can also serve as an input in predictive food microbiology.

rapid methods

In addition to whole genome sequencing, another important prospect in the area of microbiological testing is the use of rapid methods. Traditional microbiological methods are laborious and time consuming. Therefore, it is essential that faster, sensitive, specific and labour-saving methods are developed (Law et al., 2015). Improvements in analytical sample pre-processing are very important to ensure that these methods are more sensitive than traditional methods (Dwivedi and Jaykus, 2011). PCR, microarray, biosensors, and immunoassays are some of the areas of development in rapid microbiological testing.

metabolomics

Metabolomics is an omics discipline[49] that detects and identifies the metabolites produced by a cell, organism in a biological sample (Pinu, 2016). Metabolomics allows the identification of biomarkers related to food spoilage and presence of pathogens that will allow better detection and control (Pinu, 2015). Metabolomics mainly relies on the mass spectrometry and nuclear magnetic resonance approach (Johanningsmeier et al., 2016). Using metabolomics in

---

[49] i.e. a sub-discipline in biology.

relation to food spoilage is very important, as spoilage should be seen as an effect caused by a 'community' of microorganisms rather than by isolated causative agents.

clean labelling

Consumers are increasingly objecting to the presence of substances in their food that they perceive as unnatural, such as in particular food additives or 'e-numbers'. This has given rise to a trend of clean labelling on food products, which means fewer ingredients and fewer chemicals included in the product and listed on the labels (Asioli *et al.*, 2017). This has largely changed the landscape of food preservation in recent years, whereby natural preservatives are sought as a replacement for chemically synthesised or isolated food additives. Since natural preservatives are often less potent than chemical (compulsory to label) additives, the approach that is taken in response is to apply multiple hurdles that will act synergistically and protect the food from the growth of spoilage and pathogenic microorganisms (Leistner, 2000).

12.6 Conclusions

Foodborne pathogens cause a great burden to human health and the healthcare system. Common pathogenic microorganisms differ in their characteristics and can cause illnesses that range in severity from asymptomatic to severe complications and death. Spoiled food is not always unsafe and unsafe foods are not always recognisable through spoilage. Control of microorganisms is conducted through physical, chemical and biological methods. Contemporary issues and developments in the field have been discussed and their importance for food safety and quality has been highlighted.

References

Asioli, D., Aschemann-Witzel, J., Caputo, V., Vecchio, R., Annunziata, A., Næs, T. and Varela, P., 2017. Making sense of the 'clean label' trends: a review of consumer food choice behavior and discussion of industry implications. Food Research International 99: 58-71. https://doi.org/10.1016/j.foodres.2017.07.022

Bast, A., 2022. Food toxicology. In: Wernaart, B.F.W. and Van der Meulen, B.M.J. (eds) Applied Food Science. Wageningen Academic Publishers, Wageningen, the Netherlands, pp. 267-288.

Behling, R.G., Eifert, J., Erickson, M.C., Gurtler, J.B., Kornacki, J.L., Line, E., Radcliff, R., Ryser, E.T., Stawick, B. and Yan, Z., 2010. Selected pathogens of concern to industrial food processors: infectious, toxigenic, toxico-infectious, selected emerging pathogenic bacteria. In: Kornacki, J.L. (ed.) Principles of microbiological troubleshooting in the industrial food processing environment. Springer, New York, NY, USA, pp. 5-61. https://doi.org/10.1007/978-1-4419-5518-0_2

Behravesh, C.B., Williams, I.T. and Tauxe, R.V., 2012. Emerging foodborne pathogens and problems: expanding prevention efforts before slaughter or harvest, Improving food safety through a one health approach: workshop summary. National Academies Press, Washington, DC, USA.

Beutin, L. and Martin, A., 2012. Outbreak of Shiga toxin-producing Escherichia coli (STEC) O104:H4 infection in Germany causes a paradigm shift with regard to human pathogenicity of STEC strains. Journal of Food Protection 75: 408-418. https://doi.org/10.4315/0362-028X.JFP-11-452

Bintsis, T., 2017. Foodborne pathogens. AIMS Microbiology 3: 529. https://doi.org/10.3934/microbiol.2017.3.529

Bryant, C., Szejda, K., Parekh, N., Deshpande, V. and Tse, B., 2019. A survey of consumer perceptions of plant-based and clean meat in the USA, India, and China. Frontiers in Sustainable Food Systems 3: 11. https://doi.org/10.3389/fsufs.2019.00011

Corini, A.A, 2020. Food fraud and the EU food safety law framework. In: Van der Meulen, B. and Wernaart, B. (eds) EU Food Law Handbook, Wageningen Academic Publishers, Wageningen, the Netherlands, pp. 677-704.

De Boer, A., 2022. Risk analysis for foods. In: Wernaart, B.F.W. and Van der Meulen, B.M.J. (eds) Applied Food Science. Wageningen Academic Publishers, Wageningen, the Netherlands, pp. 99-123.

Den Besten, H.M., Arvind, A., Gaballo, H.M., Moezelaar, R., Zwietering, M.H. and Abee, T., 2010. Short-and long-term biomarkers for bacterial robustness: a framework for quantifying correlations between cellular indicators and adaptive behavior. PloS ONE 5: e13746. https://doi.org/10.1371/journal.pone.0013746

Dwivedi, H.P. and Jaykus, L.-A., 2011. Detection of pathogens in foods: the current state-of-the-art and future directions. Critical Reviews in Microbiology 37: 40-63. https://doi.org/10.3109/1040841X.2010.506430

Erickson, M.C. and Doyle, M.P., 2012. Plant food safety issues: linking production agriculture with One Health, Improving Food Safety Through a One Health Approach: Workshop Summary. National Academies Press, Washington, DC, USA.

Farber, J. and Peterkin, P., 1991. Listeria monocytogenes, a food-borne pathogen. Microbiological Reviews 55: 476-511. https://doi.org/10.1128/mr.55.3.4000000000076-511.1991

Ferris, I.M., 2022. Hazard analysis and critical control points (HACCP). In: Wernaart, B.F.W. and Van der Meulen, B.M.J. (eds) Applied Food Science. Wageningen Academic Publishers, Wageningen, the Netherlands, pp. 187-213.

Ferro, S., Amorico, T. and Deo, P., 2018. Role of food sanitising treatments in inducing the 'viable but nonculturable' state of microorganisms. Food Control 91: 321-329. https://doi.org/10.1016/j.foodcont.2018.04.016

Floto-Stammen, S., 2022. Food marketing. In: Wernaart, B.F.W. and Van der Meulen, B.M.J. (eds) Applied Food Science. Wageningen Academic Publishers, Wageningen, the Netherlands, pp. 453-479.

Fonberg-Broczek, M., Windyga, B., Szczawiński, J., Szczawińska, M., Pietrzak, D. and Prestamo, G., 2005. High pressure processing for food safety. Acta Biochimica Polonica 52: 721-724.

Food & Drug Administration, 2012. Bad bug book. Center for Food Safety and Applied Nutrition (CFSAN) of the Food and Drug Administration (FDA), U.S. Department of Health and Human Services.

Galié, S., García-Gutiérrez, C., Miguélez, E.M., Villar, C.J. and Lombó, F., 2018. Biofilms in the food industry: health aspects and control methods. Frontiers in Microbiology 9: 898. https://doi.org/10.3389/fmicb.2018.00898

Garcia, P., Martinez, B., Obeso, J. and Rodriguez, A., 2008. Bacteriophages and their application in food safety. Letters in Applied Microbiology 47: 479-485. https://doi.org/10.1111/j.1472-765X.2008.02458.x

Gram, L., Ravn, L., Rasch, M., Bruhn, J.B., Christensen, A.B. and Givskov, M., 2002. Food spoilage – interactions between food spoilage bacteria. International Journal of Food Microbiology 78: 79-97. https://doi.org/10.1016/S0168-1605(02)00233-7

Hadi, J. and Brightwell, G., 2021. Safety of alternative proteins: technological, environmental and regulatory aspects of cultured meat, plant-based meat, insect protein and single-cell protein. Foods 10: 1226. https://doi.org/10.3390/foods10061226

Institute of Food Technologists, 2004. Bacteria associated with foodborne diseases. Institute of Food Technologists, Chicago, IL, USA.

Jagadeesan, B., Baert, L., Wiedmann, M. and Orsi, R.H., 2019. Comparative analysis of tools and approaches for source tracking *Listeria monocytogenes* in a food facility using whole-genome sequence data. Frontiers in Microbiology 10: 947. https://doi.org/10.3389/fmicb.2019.00947

Jay, J.M., 1995. Intrinsic and extrinsic parameters of foods that affect microbial growth. In: Jay, J.M. (ed.) Modern food microbiology. Springer US, Boston, MA, USA, pp. 38-66. https://doi.org/10.1007/978-1-4615-7476-7_3

Johanningsmeier, S.D., Harris, G.K. and Klevorn, C.M., 2016. Metabolomic technologies for improving the quality of food: practice and promise. Annual review of food science and technology, 7: 413-438. https://doi.org/10.1146/annurev-food-022814-015721

Koluman, A. and Dikici, A., 2013. Antimicrobial resistance of emerging foodborne pathogens: Status quo and global trends. Critical Reviews in Microbiology 39: 57-69. https://doi.org/10.3109/1040841X.2012.691458

Law, J.W.-F., Ab Mutalib, N.-S., Chan, K.-G. and Lee, L.-H., 2015. Rapid methods for the detection of foodborne bacterial pathogens: principles, applications, advantages and limitations. Frontiers in Microbiology 5: 770. https://doi.org/10.3389/fmicb.2014.00770

Leistner, L., 2000. Basic aspects of food preservation by hurdle technology. International Journal of Food Microbiology 55: 181-186. https://doi.org/10.1016/s0168-1605(00)00161-6

Mayrhofer, S., Paulsen, P., Smulders, F.J.M. and Hilbert, F., 2004. Antimicrobial resistance profile of five major food-borne pathogens isolated from beef, pork and poultry. International Journal of Food Microbiology 97: 23-29. https://doi.org/10.1016/j.ijfoodmicro.2004.04.006

Mohammad, A.-M., Chowdhury, T., Biswas, B. and Absar, N., 2018. Food poisoning and intoxication: A global leading concern for human health. In: Grumezescu, A.M. and Holban, A.M. (eds) Food safety and preservation. Academic Press, Washington, DC, USA, pp. 307-352. https://doi.org/10.1016/B978-0-12-814956-0.00011-1

Montgomery, H., Haughey, S.A. and Elliott, C.T., 2020. Recent food safety and fraud issues within the dairy supply chain (2015-2019). Global Food Security 26: 100447. https://doi.org/10.1016/j.gfs.2020.100447

Mor-Mur, M. and Yuste, J., 2010. Emerging bacterial pathogens in meat and poultry: an overview. Food and Bioprocess Technology 3: 24-35. https://doi.org/10.1007/s11947-009-0189-8

Panigrahi, S., Balasubramanian, S., Gu, H., Logue, C. and Marchello, M., 2006. Design and development of a metal oxide based electronic nose for spoilage classification of beef. Sensors and Actuators B: Chemical 119: 2-14. https://doi.org/10.1016/j.snb.2005.03.120

Pinu, F.R., 2015. Metabolomics – The new frontier in food safety and quality research. Food Research International 72: 80-81. https://doi.org/10.1016/j.foodres.2015.03.028

Pinu, F.R., 2016. Metabolomics: Applications to food safety and quality research. In: Beale, D.J., Kouremenos, K.A. and Palombo, E.A. (eds) Microbial Metabolomics. Springer, Cham, Switzerland, pp. 225-259. https://doi.org/10.1007/978-3-319-46326-1_8

Pradhan, A.K., Mishra, A. and Pang, H., 2019. Predictive microbiology and microbial risk assessment. In: Doyle, M.P., Diez-Gonzalez, F, and Hill, C. (eds) Food microbiology, ASM Press, Washington, DC, USA, pp. 989-1006. https://doi.org/10.1128/9781555819972.ch39

Prudêncio, C.V, Dos Santos, M.T. and Vanetti, M.C.D., 2015. Strategies for the use of bacteriocins in Gram-negative bacteria: relevance in food microbiology. Journal of Food Science and Technology 52: 5408-5417. https://doi.org/10.1007/s13197-014-1666-2

Ronholm, J., Nasheri, N., Petronella, N. and Pagotto, F., 2016. Navigating microbiological food safety in the era of whole-genome sequencing. Clinical Microbiology Reviews 29: 837-857. https://doi.org/10.1128/CMR.00056-16

Russell, N. and Fukunaga, N., 1990. A comparison of thermal adaptation of membrane lipids in psychrophilic and thermophilic bacteria. FEMS Microbiology Letters 75: 171-182. https://doi.org/10.1016/0378-1097(90)90530-4

Spirovska Vaskoska, R., 2012. 'Sprouting' food law. A decade of emergency measures and crisis. Wageningen University, Wageningen, the Netherlands. https://edepot.wur.nl/206868

Steele, R., 2004. Understanding and measuring the shelf-life of food. Woodhead Publishing, Sawston, UK.

Teunis, P. and Havelaar, A., 1996. The dose-response relation in human volunteers for gastro-intestinal pathogens. RIVM, Bilthoven, the Netherlands. https://www.rivm.nl/bibliotheek/rapporten/284550002.pdf

Van der Meulen, B.M.J. and Wernaart, B. (eds), 2020. EU food law handbook. Wageningen Academic Publishers, Wageningen, the Netherlands.

White, D.G., Zhao, S., Simjee, S., Wagner, D.D. and McDermott, P.F., 2002. Antimicrobial resistance of foodborne pathogens. Microbes and Infection 4: 405-412. https://doi.org/10.1016/s1286-4579(02)01554-x

World Health Organization, 2015. WHO estimates of the global burden of foodborne diseases: foodborne disease burden epidemiology reference group 2007-2015. World Health Organization, Geneva, Switzerland.

World Health Organization, 2020. Food safety. World Health Organization, Geneva, Switzerland. Available at: https://www.who.int/news-room/fact-sheets/detail/food-safety#:~:text=An%20estimated%20600%20million%20%E2%80%93%20almost,healthy%20life%20years%20(DALYs).&text=Children%20under%205%20years%20of,125%20000%20deaths%20every%20year.

World Health Organization, 2021. Disability-adjusted life years (DALYs). World Health Organization, Geneva, Switzerland. Available at: https://www.who.int/data/gho/indicator-metadata-registry/imr-details/158.

Yuan, L., Hansen, M.F., Røder, H.L., Wang, N., Burmølle, M. and He, G., 2020. Mixed-species biofilms in the food industry: Current knowledge and novel control strategies. Critical Reviews in Food Science and Nutrition 60: 2277-2293. https://doi.org/10.1080/10408398.2019.1632790

Zwietering, M.H., Garre, A., Wiedmann, M. and Buchanan, R.L., 2021. All food processes have a residual risk, some are small, some very small and some are extremely small. Current Opinion in Food Science 39: 83-92. https://doi.org/10.1016/j.cofs.2020.12.017.

13. Food toxicology

This is what you need to know about the safety aspects of food by studying the properties, effects, and detection of natural and man-made substances in food

Aalt Bast

Maastricht University, Department of Pharmacology and Toxicology, Campus Venlo, P.O. Box 616, 6200 MD Maastricht, the Netherlands; a.bast@maastrichtuniversity.nl

Abstract

Toxicology is an old scientific discipline that tries to describe and understand the effect of compounds on life forms. Toxicology integrates knowledge and techniques on physiology (normal function of the body), chemistry (characteristics and measurement of the putative toxic compounds) and pathology (study of the cause and development of disease). In food toxicology, the health risk of both contaminants and natural toxins in food is investigated. In understanding the health effect of all these chemicals (both natural as well as man-made) it is also important to know how these chemicals are converted into so-called metabolites both in the environment as well as in the body. Frequently, not only the parent compounds but also their metabolites contribute to the eventual health effect of chemicals.

Key concepts

- Toxicology helps us to understand harmful effects of substances on humans, animals and the environment with the aim of minimising the risk of exposure to these compounds.
- *In vitro*, *in vivo* and *in silico* methods are used as models in toxicology.
- Well-known toxicological dose descriptors are Lethal Dose 50 (LD50), No Observed Adverse Effect Level (NOAEL), Lowest Observed Effect Level (LOAEL), Acceptable Daily Intake (ADI), Benchmark Dose (BMD) and Benchmark response (BMR).
- Hazard is the potential of a substance to produce adverse effects (toxicity).

► Risk is the probability that a hazard will be realised under certain conditions of use (exposure).

► Exposure involves route of exposure (dermal, oral, inhalation, injection), amount (dose) and duration.

► The term metabolism is confusing in literature. In physiology or human biology metabolism is the process by which your body converts what you eat or drink into energy. In toxicology, the term metabolism is used to indicate the breakdown of compounds (xenobiotics). This is also described as biotransformation. This process primarily takes place in the liver.

► Xenobiotics are compounds that are foreign to the body; compounds which are found in an organism but which are not normally produced or expected to be present in the body. Examples of xenobiotics include plant constituents, pesticides, drugs, cosmetics, industrial chemicals and environmental pollutants.

► The indication 'bio' or 'natural' does not automatically mean safe just as 'synthetic' or 'chemical' do not necessarily mean unsafe.

► Plants protect themselves by making natural pesticides among other things. Many of these phytotoxins are also toxic for humans.

► Mycotoxins are produced by certain type of moulds (fungi). Many mycotoxins are chemically stable and survive food processing.

► What is the amount of compound that is safe to eat (drink or inhale) lifelong every day?

► The Maximal Residue Level (MRL) is the concentration of a chemical that is allowed to be present as a residue in a product.

► The Threshold of Toxicological Concern (TTC) is a rather pragmatic parameter that assumes that the body can cope with compounds until a certain threshold dose (the TTC) is reached whereupon concern arises.

► Pesticide is an umbrella term to describe plant protection agents and biocides.

► Hormesis is a term used to describe adaptive responses. Low levels of damage or stress can result in an increased power of the organism to protect itself against the inflicted damage.

Case 13.1. The dangerous courgette stew.

A 79-year-old German pensioner and his wife were hospitalised with complaints of nausea and acute stomach pain after eating a stew of courgettes. The courgettes were given to them by their neighbours, who were hobby gardeners. The pensioner died soon after hospitalisation. His wife slowly recovered. She said the stew tasted very bitter and she had eaten only a little.

The editorial staff of the daily newspaper discussed this incoming message. Several conclusions were debated: (1) intentional intoxication by the neighbours; (2) improper use of pesticides by the amateur gardeners; and (3) a natural toxin found in plants of the Cucurbitaceae family such as pumpkins, and gourds such as courgettes.

Option (1) seemed very unlikely because it was a friendly gift from the neighbours. Deliberate intoxication via food was a favourite method of killing someone in the Roman Empire. The job of a pre-taster for the emperor could be regarded as a risky occupation. There are theories that Napoleon suffered from arsenic poisoning. Also in more recent times, there have been political poisonings, with toxins or even nerve agents being hidden in food or drinks. During his campaign for the Presidency in Ukraine in 2004, Viktor Yushchenko was poisoned with dioxin. And in 2020 the Russian opposition leader Alexei Navalny was allegedly poisoned with Novichok.

Pesticides are by definition toxic compounds, they are intended to kill insects, fungi or unwanted weeds. In general, strict regulations on use will limit human exposure to residues of pesticides in fruits and vegetable. Amateur gardeners are not always aware of these regulations and improper use, option (2), may pose a risk. However, these gardeners often use organic techniques rather than pesticides.

The last option (3), i.e. the presence of naturally occurring toxic cucurbitacins, is very probable. Hobby gardeners sometimes use seeds kept from previous harvests and thus continue with seeds from their own crops. This can be dangerous because cross-pollination may infect the cultivated varieties in which the natural toxins have been bred out. The bitter taste of the stew was an indication that the toxins were indeed present. Eating a large portion results in a high dose of the natural toxins and thus a high risk.

The Cucurbitaceae family includes not only courgettes but also pumpkin and cucumber. Hobby gardeners should be aware that spontaneous mutations or cross-pollination with wild growing species can reintroduce the toxic compounds into the next generation grown from saved seed.

Recently, a batch of courgette seeds was recalled by a British supplier because it produced bitter tasting fruits, a rare but dangerous occurrence. Apparently, it's not just amateurs who run this risk.

The editor of the daily newspaper decided that the following headline would be the most suitable: 'Do not eat bitter tasting courgettes, pumpkins or cucumbers'.

13.1 Toxicology models and toxicological dose descriptors

Early in toxicological research, pieces of tissue, isolated cells or subcellular fractions were used to study the mechanism of toxicity of chemicals. The human relevance of these *in vitro* studies is of course limited. The use of animals is thought to provide more relevant data. By studying the adverse effect of compounds *in vivo*, the toxicologist is also able to investigate the role of the breakdown products in the overall toxic response. The liver, as metabolic organ, is important in this respect. Finally, population studies or obtaining exposure data gives further insight into the relevance of the observed *in vitro* toxicity. Current toxicology makes use of co-cultures of various cell types. In recent years, the toxicological read-out parameters have also improved. Image techniques make it possible to see real live intracellular changes in, for example, calcium or pH. Moreover, there has been a tremendous progression in computer-aided prediction of toxicity, with the result that *in vitro* and *in vivo* models have been supplemented with in silico methods. Promising new methods are now emerging. It is becoming increasingly possible to grow small organ systems on a chip, equipped with flow transport of compounds over the chip via so-called microfluidics. The interest in further defining mechanistic adverse biochemical pathways is also attracting quite a lot of attention. All these developments are expected to accumulate in a 'virtual human' on whom to test and predict the toxicity of compounds (Table 13.1).

The relationship between the effect of a chemical compound and the dose (or concentration) at which it takes place is characterised by so-called toxicological dose descriptors. These dose descriptors are used in the assessment of risk or to quantify chemical hazards. Some of the descriptors are criticised because of their limited practical value. A common dose descriptor familiar to many of **LD50 (Lethal Dose** us is the LD50 (Lethal Dose 50%). The dose of a compound at which 50% of **50%)** the exposed animals will die. The problem with using the LD50 is that the

Table 13.1. Toxicology models.

Early toxicology	Current toxicology	Future toxicology
Tissue, cells, subcellular fractions, genes	Organo-typic cell culture (coculture, organ function)	Human on a chip (multi-organ models with micro fluidics)
Animal models	Cell culture + omics or image analysis	Toxicity mechanisms (adverse outcome pathways, 'human toxome')
Human (population based studies, human exposure data)	Integrated test strategies (combined tests)	Systems toxicology (virtual human)
	In silico prediction (structure activity relationships)	

Applied food science

LC50 (Lethal Concentration 50%)	duration and route of exposure will be important. It requires animal testing and a choice of the species that best reflects the human. Moreover, the LD50 does not provide insight into the mode of action of the chemical. The unit of the LD50 is mg/kg bw/d, which stands for mg of substance administered per kg bodyweight per day. Of course, the route of administration is important as well, for example p.o. (per os, which stands for the oral route) or i.v. (the intravenous route). In inhalation toxicology the LC50 (Lethal Concentration 50%) is used, and is derived from concentration-dependent death of an animal by compounds administered via the air. The unit of LC50 is mg/l, the concentration of chemical in the air. Sometimes the unit ppm is used.
The No Observed Adverse Effect Level, the NOAEL **Acceptable Daily Intake (ADI)**	The No Observed Adverse Effect Level, the NOAEL, reflects the highest exposure level (a dose or a concentration) at which there is no sign of an increase in the severity of adverse effect between the exposed species and the appropriate, i.e. unexposed, controls. This toxicological dose descriptor is commonly acquired from repeated dose toxicity studies as a 28-day repeated dose study or a 90-day repeated dose study, respectively indicated as a subchronic or a chronic toxicity study with animals (Vrolijk *et al.*, 2020). The long-term NOAEL, obtained in the most sensitive animal, is used to calculate among others the Acceptable Daily Intake (ADI). The ADI of a compound is the maximum amount of that compound that can be ingested lifelong on a daily basis without a noticeable health risk. To this end a safety or uncertainty factor is applied (usually 100) to take into account the difference in sensitivity between the animal species and human (a factor 10) and to account for inter-individual differences between humans, again a factor 10. The ADI is thus calculated as the long-term NOAEL divided by 100.
Lowest Observed Adverse Effect Level (LOAEL)	If a NOAEL cannot be attained, the Lowest Observed Adverse Effect Level (LOAEL) can be used to calculate threshold safety exposure levels in humans. Of course, more strict safety factors should be applied.
benchmark dose (BMD) **benchmark response (BMR)**	Besides the NOAEL for chemical risk assessment, the benchmark dose (BMD) is also used. The word is not entirely correct because the BMD can also be a concentration instead of a dose. In fact, a BMD is the dose or the concentration that produces a predetermined change in the response rate of an adverse effect. The predetermined response itself is known as the benchmark response (BMR). Usually, the default BMR for an adverse effect is set at 5% or 10%.
	The NOAEL very much depends on the doses selected in the experiment, while the BMD does not. The shape of the dose response curve is not taken into account for the NOAEL, while it is for the BMD. Unfortunately, it is time consuming to derive a BMD whereas the NOAEL is relatively easy to obtain. Finally, the NOAEL is more widely known than the BMD.

Further debate on other toxicological dose descriptors can be found in Section 13.4 in the discussion on pesticides.

13.2 Food toxicology

Food toxicology focuses on substances found in food that, when consumed, may cause harm. Because acquiring safe food is an essential part of human survival, it is easy to conclude that food toxicology is one of the oldest scientific disciplines (Bast and Hanekamp, 2017). Our ancestors depended on the gathering of food, and knowledge about what was safe or unsafe to eat was crucial. Unfortunately, much of this old knowledge has vanished. Today, in the Western world only a few people forage for food in the wild. Modern humans cultivate crop plants and thus reduce the content of endogenous toxins present in the wild varieties. We now rely on the quality and safety of the products we buy. So, the modern consumer automatically expects the food to be 100% safe. In the EU this has led to the application of the so-called precautionary principle in case of scientific uncertainty, also known as the 'better safe than sorry principle'. Uncertainty regarding food safety has become very difficult to accept and the organisation in the EU involved in food safety, the European Food Safety Authority (EFSA), tries to regulate this. Unfortunately, this is not possible and attempts to reach 100% safety are bound to fail. It is just impossible to know all the substances that constitute our food. In a daily diet there are thousands of compounds, most of them not chemically characterised and with unknown bioactivity. Moreover, although we mainly use cultivated crop plants there still are many endogenous (i.e. naturally occurring) potentially toxic compounds in these plants. The classical approach in toxicology to quantify a health risk is to define the hazard of the substance and the exposure to that substance.

Risk = Hazard × Exposure

Only a very limited number of substances in food have been characterised. This impedes the guarantee of total safety. At best, a risk/benefit ratio of food products can be estimated: both the benefit and the risk associated with food products are more than guesstimates but not certainties. Although risk assessment is still regarded as a scientific endeavour, the quality of the science should not be overestimated. As indicated, there is much that is unknown, not only with regard to the individual substances but also with regard to the combination of compounds in the food matrix. This hampers adequate hazard identification and characterisation. For man-made chemicals hazard characterisation is easier. The chemical is generally available in pure form and in large quantities, which facilitates research. For scientific risk assessment, the exposure should be determined as well. Exposure includes knowledge about

risk assessment

route of administration (for food, mostly the oral route), single dose or multi-hit (for food, mostly a single daily exposure), sensitisation, and absorption (uptake via the intestine), distribution over the body or even within certain organs, metabolism and excretion. These processes determine the concentration at the site of toxic action and thereby the risk. The toxicological scientific toolbox is further used in risk assessment to define the probability of the risk, in other words how likely is the adverse event. Moreover, the consequences of the risk are determined, i.e. what is the likely damage.

risk management   Risk management is mainly policy based and control opportunities over the risks and legal considerations govern risk management decisions. Several other economic or social factors will have an influence on the risk management decisions. The relation to other risks will be taken into account in a comparative risk analysis. Issues on potential risk reduction will involve questions about the desired level of reduction. Topics on the strategy of risk reduction will be part of the risk management. Of course, the available budget will play a role: what is the financial commitment, how much money should be spent?

risk perception   Citizens are not good at perceiving risks. Distortions of facts by the (social) media or previous accidents influence the way citizens assess risks. New discoveries are generally distrusted, and therefore the risks involved appear more serious. On the other hand, the familiar leads to an underestimation of risks. Risks, imposed involuntarily or viewed as unnecessary, are generally perceived as big. In addition, the feeling of unfairness in distribution of risks will influence risk perception. Therefore, better education of the public on risks is needed. Interactive exchange of information and opinions concerning risks will improve risk communication (Figure 13.1).

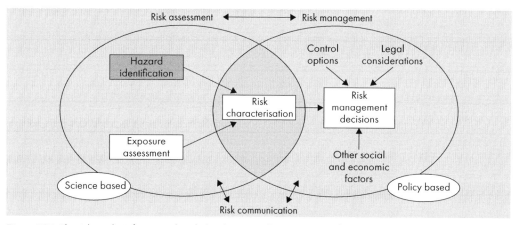

Figure 13.1. The risk analysis framework with the elements risk assessment, risk management and risk communication (adapted from Nauta et al., 2018).

Considering the difference in food risk perception between food specialists and non-specialists, it is not surprising that both have different concerns about food safety. A recent Japanese study compared the changes in risk perception over a period of 15 years in Japan. In 2004 public concern was highest with regard to pesticide residues and chemical contamination, but this has decreased over time. In 2004 awareness about food poisoning by harmful microorganisms was negligible, yet in 2018 this risk received the highest score. For food specialists contaminants do not represent a food safety issue in 2018, but they are still present on the public list. The concern scores for radioactive materials were highest in 2011 because the survey was conducted just after the nuclear accident in Fukushima. The comparison between food specialists and the general public reveals the gap between them. It also shows the challenge that exists for adequate risk communication in the food safety field.

13.3 Natural toxins

Case 13.2. Poison garden.

The Alnwick Garden was created by the Duchess of Northumberland and is a delight, according to the website, consisting of a modern garden combining sculptures, water features and beautiful plants to create a wonderful family attraction. It further reads: 'The Alnwick Garden plays host to the small but deadly Poison Garden – filled exclusively with around 100 toxic, intoxicating, and narcotic plants. The boundaries of the Poison Garden are kept behind black iron gates, only open on guided tours.'

The website information is correct. You are warned by a sign over the fence at the entrance to the garden you want to visit stating 'The poison garden'. Skull and crossbones accompany a sign board with the text 'These plants can kill'.

It is indeed a beautiful garden, situated north of Newcastle in the UK. However, a visit to this site makes you wonder why there are so many toxic plants.

phytotoxins

Plants protect themselves against herbivores with thorns or a bitter taste. Moreover, many plants contain a wide variety of toxins, also known as phytotoxins. These toxins form a natural defence against predators, insects or microorganisms. In essence, plants make their own pesticides. It has been calculated that 99.99% (by weight) of our intake of pesticides through food is of natural origin (Ames *et al.*, 1990). Within the context of the massive exposure to these naturally occurring chemicals, these phytotoxins, it is easy to look at the toxicological significance of exposure to synthetic chemicals from another perspective (Bast, 2002). The ability to and performing of comparison between the toxic load of chemicals in the natural world and of synthetic chemicals such as pesticides, generates a difference in perception about potential hazards in humans. This difference is nicely illustrated in Figure 13.2, which shows the different level of concern about pesticide residues in food between citizens and food specialists.

Examples of natural toxins are (Table 13.2):

cyanogenic
glycosides

Eating insufficiently processed cassava is probably the cause of a paralytic disease that occurs in rural populations in Africa which is known as konzo. The diet in these populations is particularly low in sulphur amino acids. These amino acids are key in the detoxification and subsequent urinary excretion of cyanide. The high levels of cyanogenic glycosides in the bitter cassava, which constitutes a major part of their diet, is the source of cyanide. Low intake of sulphur amino acids with concurrent high consumption of cyanogenic glycosides is a pathogenic combination.

lectins

Lectins are primarily found in plants but also occur in animals or bacteria. Lectins are sometimes referred to as anti-nutrients. A well-known example is phytohaemagglutinin in raw beans which is destroyed by cooking. Other lectins are more resistant to heating but are non-toxic like those in potatoes. The very toxic ricin from beans from the castor oil plant (*Ricinus communis*) is also a lectin. Ricin inhibits protein synthesis. It can take several hours before

Level of concern	2004		2011		2018	
Very concerned	Specialists	Others	Specialists	Others	Specialists	Others
	Contaminants	Pesticide residues	Radioactive materials	Radioactive materials	Food poisoning	Food poisoning
	Food poisoning	Contaminants	Food poisoning	Food poisoning	Health foods	Antimicrobial resistance
	Pesticide residues	Antimicrobial resistance	Contaminants & Health foods[1]	Pesticide residues	Antimicrobial resistance	Contaminants
	Antimicrobial resistance	Food additives		Antimicrobial resistance & Food addit.[2]	Mycotoxin	Mycotoxin
	BSE	Genetically modified food	Pesticide residues		Allergen	Health foods & Radioactive materials[3]
Somewhat concerned	[1] Both at 3rd level of concern.		[2] Both at 4th level of concern.		[3] Both at 5th level of concern (somewhat).	

Figure 13.2. Ranking order of concern level on food safety related hazards conducted in 2004, 2011 and 2018 and comparing food specialists and the general public (others). The five levels of concern vary from very concerned (level 1) to somewhat concerned (level 5). Data are obtained from the publication by Abe et al., 2020.

Table 13.2. Natural toxins in food: source and adverse health effect.

Toxin	Source	Adverse health effects
Cyanogenic glycosides	Cassava, sorghum, stone fruits, almonds.	Cyanide poisoning potentially leads to decreased blood pressure, headache, vomiting, cyanosis, death.
Furocoumarines	Parsnips, celery, citrus plants such as lime, lemon, grapefruit and bergamot.	Gastrointestinal effects (e.g. interaction with biotransformation of drugs), phototoxic (phytophotodermatitis, severe skin reactions under sunlight).
Lectins	Several types of beans, in particular red kidney beans. Boiling destroys lectins.	Vomiting, diarrhoea, nutritional deficiencies and immune reactions.
Mycotoxins	Certain moulds that grow on cereals, dried fruits, nuts, spices.	Severe illness, even death.
Solanines and chaconine	Solanacea plants, potatoes, tomatoes and eggplants.	Nausea, diarrhoea, vomiting and eventually coma and convulsion.
Poisonous mushrooms	A variety of toxins produced by the fungus.	Vary from vomiting, diarrhoea, confusion, hallucinations to death.
Pyrrolizidine alkaloids	Many plants, frequently weeds, that contaminate food crops.	Acute toxicities, such as hepatotoxicity and DNA-damaging effects.
Marine toxins	Produced during blooms of certain algae. Shellfish like mussels, scallops and oysters can contain these toxins.	Diarrhoea, vomiting, tingling, paralysis.

the primarily gastrointestinal toxicity manifests. Early symptoms after inhalation are fever and cough. Lectins bind to some saccharides, which makes these compounds useful as affinity reagents in lab research.

mycotoxins   Mycotoxins are produced by certain type of moulds (fungi). Many mycotoxins are chemically stable and survive food processing. Moulds grow on/in food under warm, damp and humid conditions. Long-term effects of exposure include cancer and immune deficiency.

Aflatoxins are mycotoxins produced by *Aspergillus flavus* and *Aspergillus parasiticus*. Products like corn, peanuts, hazel nuts, pistachio nuts, grain, rice, various spices can be infected. Derived products like peanut butter, peanut flour bread and beer may subsequently also contain aflatoxins. Animals fed aflatoxin-contaminated food pass on aflatoxin and metabolites in for example milk, eggs or meat.

Ochratoxin is a mycotoxin that is known to cause kidney toxicity in animals. The single name ochratoxin indicates a group of structurally related mycotoxins. This is similar for many other names of mycotoxins. Ochratoxin mainly derives from *Aspergillus* and *Penicillium* and can occur in grain, coffee beans, dried fruit, wine and beer.

The major dietary sources of the mycotoxin patulin are apples and apple juice made from infected fruit.

Some other important mycotoxins are produced by *Fusarium* fungi, such as deoxynivalenol (DON), nivalenol (NIV) and zearalenone (ZEN) and can occur on a variety of different cereal crops.

Several cases of exposure to the pneumotoxin 4-ipomeanol have been reported in cattle. This toxin was produced by *Fusarium solani*, a mould growing on rotten sweet potatoes that were fed to cows. It triggered cow deaths in several cases.

solanines and chaconine

Solanines and chaconine, so-called glycoalkaloids, are produced in Solanaceae or nightshades, which include potato, tomato and eggplant (aubergine). High levels are synthesised in potato sprouts as well as in green tomatoes. The phytotoxins are formed as a response to stress, e.g. after bruising or exposure to UV light. These alkaloids have powerful insecticidal and fungicidal properties. In potato breeding, one should always be aware that new cultivars produce toxic levels of alkaloids. A classic example is the Lenape potato cultivar produced in the 1960s. This potato appeared to be very disease resistant. Unfortunately, it also caused nausea and vomiting because of the high alkaloid content, which made it unsuitable for human consumption. It was therefore pulled from the market in 1970.

mushroom poisoning

Mushroom poisoning always receives quite a lot of public attention. However, exposure to toxic mushrooms is probably a greater risk for animals than for humans, which remains underreported. In fact, detailed identification of edible mushrooms can completely prevent human toxicity. Furthermore, the immune-modulating effect of mushroom beta-glucans can potentially be applied both in foods and in pharmaceutical products (Van Steenwijk *et al.*, 2021).

pyrrolizidine alkaloids

Thousands of plants produce pyrrolizidine alkaloids as a defence mechanism against insect damage. A major concern is how to prevent these plants getting into the food chain. Moreover, it is increasingly being realised that some medicinal herbs or more or less exotic teas may contain pyrrolizidine alkaloids. Interestingly, sheep, goats and to some degree cattle tolerate much higher pyrrolizidine alkaloids than humans. This is probably because the rumen microflora offers protection (Wiederfeld and Edgar, 2010).

marine toxins   There are a few known series of marine toxins: for example, okadaic acid and the dinophysistoxins, saxitoxins, brevetoxins, domoic acid, azaspiracids, pinnatoxins, yessotoxins, pectenotoxins and cyclic imines; and most of them are found all over the world. They have no taste or smell and are not eliminated by cooking or freezing. Another toxin found in, for example, mussels is tetrodotoxin (TTX). This neurotoxin is produced by bacteria and algae. Because mussels and oysters filter large amounts of sea water, they concentrate these toxins in their body. In the Netherlands the norm for TTX is 44 microgram TTX/kg in shellfish meat, a very strict safety norm until the precise source of the TTX found in mussels is known.

A notorious example is the puffer fish, which contains TTX. It is regarded as a Japanese delicacy. Only very skilled cooks are allowed to prepare the fish so as to prevent deadly intoxication of the dining customer with TTX.

The most commonly reported foodborne illness caused by marine toxins in the world is ciguatera. The toxin responsible is ciguatoxin, a lipid-soluble polyether ($C_{60}H_{86}O_{19}$). Many natural analogues have been identified to date. They are produced by dinoflagellates. Ciguatera outbreaks typically occur in a circumglobal belt including Hawaii, the South Pacific, Australia, the Caribbean, and the Indo-Pacific. However, transport of contaminated fish has led to cases in North America and Northern Europe as well. No specific treatment of the neurological symptoms of the toxicity has been described. Nevertheless, intravenous administration of mannitol or treatment with anticonvulsants may limit the severity of the poisoning.

In a balanced diet, the levels of natural toxins usually stay below the threshold for human toxicity. Further ways to minimise the health risk from phytotoxins include advising people:

1. not to assume that natural is synonymous with safe;
2. not to eat food that does nor smell or taste fresh or has an unusual taste;
3. to throw away bruised, damaged or discoloured or mouldy food;
4. to only eat wild plants or mushrooms that have been identified as non-poisonous.

Data on natural toxins are obtained from the World Health Organization (WHO, 2018).

For more than 10,000 years, humans have successfully produced increasingly better performing food plants. Most of the currently available crops are the result of conventional plant breeding practices and have led to edible varieties of the wild form, which initially contained undesirable high levels of natural toxins. With the current genomic knowledge, breeders can now further fine-tune the production of crops, optimising the level and moment of biosynthetic generation of natural toxins thus balancing crop disease resistance and human safety (Kaiser *et al.*, 2020).

Interesting situations arise when a zero tolerance for compounds is advised in food products in which the compound also occurs naturally.

One such example is chloramphenicol, an antibiotic with a broad spectrum of activity. It is used in veterinary practice for therapeutic and prophylactic practice. The lack of scientific data hampers the setting of an Acceptable Daily Intake (ADI). Consequently, a maximum residue limit (MRL) could not be established (Hanekamp and Bast, 2015). This then translated into a zero-tolerance approach, which was guided by the achievable detection limit in food. In the meantime, the maximum residue limit is set at 0.0003 mg/kg. With increasing analytical technological capabilities, the detection limit reaches ever-lower levels. Basically, this means that the zero tolerance level shifts to ever lower exposure levels. Increased detection capacities lead to ever 'cleaner' food products. At a certain level, the ecological background level is crossed. This is the case for chloramphenicol, a product of the actinobacteria (prokaryotic organisms, considered an intermediate group of bacteria and fungi) species *Streptomyces venezuelae*. Recently, this species was also detected in a marine environment (Meena *et al.*, 2013). Batches of large quantities of shrimps have been discarded because of detectable chloramphenicol. This is the consequence of a zero-tolerance approach. A detection limit as approval procedure, however, ignores the toxicological relevance. In view of this discussion a so-called Minimum Required Performance Limit (MRPL) was introduced with the intention of harmonising the analytical methodology of measuring a substance.

A more preferable parameter to ensure safety would be the Threshold of Toxicological Concern (TTC), which accommodates new toxicological insights (*vide infra*) (Hanekamp and Bast, 2015).

13.4 Pesticides

Hundreds of pesticide products have been approved for use. There are various ways to classify all these products. One way is via the pest they address or effect they cause as shown in Figure 13.3. It illustrates the huge variety in available pesticide products.

The ideal pesticide (1) is selective, i.e. other organisms should not be affected; (2) is effective in its action on the target; (3) is metabolised in nature, i.e. after the harvest no residues are detected; and (4) remains active after long-term

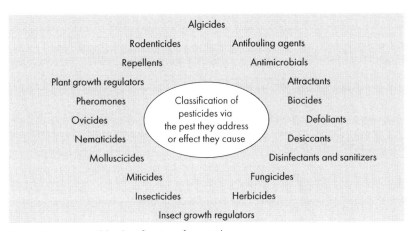

Figure 13.3. A possible classification of pesticides.

usage, i.e. resistance to the pesticide does not occur. There are no pesticides that meet all four criteria. That is where the toxicologist comes in. The risk of using pesticides is determined for other organisms besides the target, which can be the environment (environmental toxicology) and humans (human toxicology). This means that both hazard and exposure have to be quantified in order to know the risk, because it is not only the pesticide molecule that can be toxic but also its breakdown products, which are formed in humans. Preferably the metabolites as well as the parent compound should be investigated. Eventually, these risks and benefits should be weighed up. There are many benefits, like improved yields, quality of appearance, safety, a longer shelf life, less hard manual labour, lower fuel use for weeding, reduced soil disturbance as well as fewer pest epidemics. Also, pests are contained geographically and invasive species are controlled. These primary benefits lead to a variety of impressive secondary benefits like improved food safety and human health, reduced energy use and environmental degradation as outlined by Cooper and Dobson (2007).

benefits

human risk If we limit the risk evaluation to the human risk, then both the manufacturer, the farmer and the consumer are potentially exposed to pesticides. In the case of occupational contact by manufacturers and farmers, adequate safety measurements can be taken to prevent exposure. Consumers might also experience environmental exposure or might be exposed to pesticide residues in crops. In Western counties, the latter hardly occurs anymore. Pesticide use is strictly regulated and proper use is enforced. A critical evaluation of the literature, however, reveals several limitations with regard to characterisation and quantification of health outcome as well as exposure assessment (Bast *et al.*, 2021).

Acceptable Daily Intake

The toxicological boundaries are set by different parameters. For each pesticide, an Acceptable Daily Intake (ADI) has been determined. The ADI is the dose of the pesticide that can be ingested daily for a lifetime without an adverse effect. In most cases, the ADI is derived from animal experiments. In order to extrapolate the data from animal to human a safety factor is applied, and to ensure that ADI also holds for sensitive individuals another safety factor is applied.

Maximal Residue Level

For every pesticide a Maximal Residue Level (MRL) is established. That is the concentration of the pesticide that is allowed to be present as a residue in a product. In this MRL human toxicity data, environmental toxicity, the mean daily consumption as well as political and economic consequences are taken into account.

Acute Reference Dose

The ADI is based on lifelong continuous daily exposure. This means that a single excess dose of the ADI does not immediately lead to a negative health effect. However, the Acute Reference Dose (ARfD) is the dose of the pesticide above which acute intoxication problems can arise. If a person gets a higher dose than the ARfD there is a chance of direct health problems.

When a food safety test detects that the legal standard MRL for a product has been exceeded, checks are made to see if the ADI or ARfD have been exceeded, thus causing harm.

Threshold of Toxicological Concern

It is important to realise that most toxicological knowledge is for individual compounds. Less is known about combinations of compounds. In this case the Threshold of Toxicological Concern (TTC) approach may be helpful.

The TTC is a rather pragmatic method that assumes the body can cope with compounds until a certain threshold dose is reached at which concern arises (Kroes *et al.*, 2004). Based on their chemical structure, even without a full toxicity database, a TTC can be set for compounds with similar structural characteristics. Many years after this suggestion EFSA also cautiously published a guidance document on how to use the TTC in food safety assessment. For pesticides of the organophosphate and carbamate category, a TTC value of 18 µg/person per day is set. This amounts to a TTC value of 0.3 µg/kg body weight per day (EFSA, 2019).

The general public fears man-made chemicals. This so-called chemophobia seems to arise in relation to pesticides sprayed during crop production. Many consumers will go to great lengths to avoid these chemicals, which seems disproportional considering the actual danger posed when these products are employed correctly. For many people, the copious advantages of properly used modern pesticides don't seem to outweigh the fear. Pesticides are indispensable for the sustainable production of high-quality food. The very persistent pesticides of the past, which were cause for much alarm because they accumulated in the food chain, are no longer in use.

Case 13.3. Bio-vitamin C, a toxicological myth.

Why do marketers use the 'bio' prefix so often? Apparently, they prefer bio-vitamin C over vitamin C and bio-magnesium over magnesium. The shelves are loaded with bio-products. Marketers know that consumers love 'bio' because it suggests natural, and natural is thought to be synonymous with safe. Along the same line of thinking, the word chemical is associated with danger. It is a deeply rooted toxicological myth. The vitamin C molecule, also known as ascorbic acid, in an orange is identical to chemically produced vitamin C. Many consumers, however, have the idea that the form found in nature is not only safer but also healthier than the man-made synthetic form. The bio-is-safe myth can even be regarded as one of the major pillars in the defence of organic farming (Bast, 2002).

13.5 Low-dose toxicity is good!

The usual concept in toxicology, also in food toxicology, is that the parameters used to estimate a health risk assume linearity in response. In other words, more toxin leads to more health damage. This notion goes back a long time. Linearity was already suggested by Paracelsus (1493-1541), who is regarded as the grandfather of toxicology. The saying, 'all things are poison and nothing without poison; the dosage alone makes that a thing is no poison' is ascribed to him. He recognised that, depending on the dose, everything becomes toxic if the dose is high enough. In fact, this is linearity *pur sang*. A physician and theologian, his official name was Philippus Aureolus Theophrastus Bombastus von Hohenheim. He travelled in the region of South Germany and Switzerland and used minerals and new remedies which contained mercury, sulphur, iron and copper sulphate. Treating patients with these salts and metals, he probably noticed that increasing the dose of his compounds indeed resulted in toxicity.

In many physiological models an attempt has been made to linearise the findings (Bast and Hanekamp, 2014). We now know that linearity is not a descriptive reality in toxicology and physiology. The reductionist approach in studying the effect of compounds has been very helpful in understanding toxicological processes (Hanekamp *et al.*, 2012). The applied abstraction, however, easily leads to misplaced concreteness. In physiology/toxicology we are not passive recipients of the thousands of compounds that might inflict chaos on our physiology. On the contrary, we adapt. This adaptive response, also known as hormesis, is a moderate overcompensation for a chemical perturbation of the homeostasis of an organism in order to re-establish homeostasis (Leak *et al.*, 2018). Studies on low-dose exposure of what we regard as toxic chemicals show that the dose-response curve is not linear but bell-shaped (Figure 13.4).

Hormesis can be described as an adaptive response to low levels of damage or stress or toxicity, resulting in enhanced robustness of some physiological systems for a finite period of time. Following Paracelsus and our current toxicology models, a compound has graded bad, toxic, properties. More compound means more toxicity. This enables us to derive an ADI and an ARfD. In the more realistic hormesis (adaptive) model, compounds are not only 'bad' or only 'good'. The effect of a compound is determined by the plasticity of the biological organism.

The challenge for toxicology in this hormesis model now is how to translate a risk assessment into risk management actions; after all, bad is not always bad. It will certainly also further complicate communication on risks.

In the hormetic zone an 'enhanced homeostasis' is apparent. This has also been described as increased resilience. In cell models this process has been unravelled (Sthijns *et al.*, 2014). A little bit of toxin even seems to be a good thing.

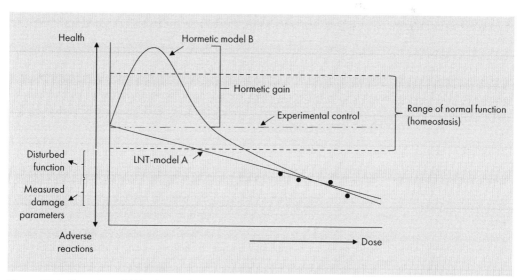

Figure 13.4. Illustration of (A) a linear dose-response curve; and (B) a biphasic or hormetic bell-shaped dose-response curve. Experimental data points are in the range where damage parameters can be measured.

Case 13.4. What doesn't kill you makes you stronger.

In a study published by Mireille Sthijns *et al.* (2014), entitled '... The materialization of the hormesis concept', lung cells were exposed to the toxic compound acrolein, which is a constituent of cigarette smoke. As expected, the lung cells die after exposure to 10 µM of acrolein. If the cells were pre-treated with 3 µM of acrolein 4 hours prior to a second exposure to 10 µM of acrolein, the damaging effect of acrolein was considerably less. The low dose of acrolein induced an adaptive upgrade of the cellular protection capability of the lung cells. Further research unravelled the biochemistry behind this protective mechanism. All kinds of protective factors were synthesised by the lung cells rendering these cells more resilient to the high dose of acrolein. Hormesis is in many ways the physiological equivalent of the philosophical notion that 'what doesn't kill you makes you stronger'.

The WHO definition of health, which reads 'a state of complete physical, mental, and social well-being and not merely the absence of disease or infirmity', describes a static state. Moreover, perhaps not surprisingly, after the World War II, the definition implies absoluteness with the word 'complete' preceding the word well-being. The term 'complete well-being' unintentionally evoked the phenomenon of medicalisation in society. Currently, it is preferable to define health as 'the ability to adapt'. Increased adaptive responses in that way form an integral part of health. Compounds with toxic properties (at a low dose) might increase adaptability and can thus be regarded as health promoting. This holds for industrial chemicals but certainly also for food natural toxins. Examples that illustrate this notion are piling up fast. Flavonoids, which can be found in fruit and vegetables, have been suggested to induce adaptability. In fact, some of the health-promoting activity of flavonoids might be due to their toxic interaction with proteins (Weseler and Bast, 2012, 2017).

13.6 Conclusions

Food is a mixture of (natural) chemicals with all kinds of health effects. We are all raised with the notion that a diversified diet is good for you. This indeed makes sense. By spreading the load of consumed chemicals (natural or man-made) the chances of mistakenly adopting health-impairing consumption habits are reduced. Diet diversity (1) reduces the chance of being exposed to the same food chemicals over and over again. In this way, (2) nutritional adequacy will be better ensured because the chance of missing out on vitamins and minerals that are essential for our health is reduced. Moreover, in the context of this chapter, (3) the chances of being exposed to high concentrations of unwanted man-made or natural chemicals are reduced.

Diversifying the diet will further train the body to deal with all sorts of chemicals. Compounds in fruits and vegetables such as courgette, which are reactive in a small dose, induce protective responses and thereby induce health.

Our toxicological knowledge is focused on man-made chemicals rather than on natural food toxins. This is understandable because man-made chemicals like pesticides can be purchased in large amounts and in pure form. Our toxicological knowledge on the combination of compounds is scarce. This is frequently the topic of public debate and leads to unfounded fear. It should be realised that food consumption itself is the ultimate form of exposure to a combination of compounds, and this should actually be reassuring. The fact that diversification of food intake is beneficial to health, already points to the notion that combinations of compounds are frequently less hazardous to us than single compounds.

new trends
New trends in food processing continuously lead to new challenges for food toxicology. In the transition to a sustainable society, new technologies and materials are being developed with new potential risks. For example, new bio-based plastic food contact materials form new sources of risks (Van der A and Sijm, 2021). Another example that is currently attracting a lot of attention is per- and polyfluoroalkyl substances (PFAS), a group of man-made chemicals that includes PFOA, PFOS, GenX, among others. PFAS is found in food packaged in PFAS-containing materials or grown in PFAS-contaminated soil or water. Certain PFAS can accumulate and stay in the human body for long periods of time and can cause a wide variety of toxicities in animals. In humans, an increase in cholesterol has been related to PFAS exposure.

There is also a trend to use all parts of the plant in order to prevent food waste. In that context, knowledge of plant biochemistry of crop biochemistry is crucial. For example, fruits from the Rosaceae family, such as apples, almonds, apricots, peaches and cherries, are known to produce a natural bitter compound in the seed called amygdalin. High levels can cause cyanide poisoning. Routinely, apple juice processing involves the entire fruit and the seeds may contaminate the juice. Another example is the wild apricot (*Prunus armeniaca*) kernel which, despite being a rich source of protein and oil, is rarely exploited by food industries due to the high amount of anti-nutrients and potentially toxic amygdalin (Tanwar *et al.*, 2019). Apricot cultivars with low amygdalin seed levels could offer new opportunities.

Not only is the fruit and vegetable biochemistry becoming better known and can be used for reducing the risk of toxicities, but also the biochemistry of humans is being better mapped. In the future we will undoubtedly be able to better bridge plant and human biochemistry, thus improving the protection of the individual against certain foodborne toxins.

References

Abe, A., Koyama, K., Uehara, C., Hirakawa, A. and Horiguchi, I., 2020. Changes in the risk perception of food safety between 2004 and 2018. Food Safety 8: 90-96 https://doi.org/10.14252/foodsafetyfscj.D-20-00015

Ames, B.N., Profet, M. and Gold, L.S., 1990. Dietary pesticides (99.99% all natural). Proceedings of the National Academy of Sciences of the Unites States of America 87: 7777-7781. https://doi.org/10.1073/pnas.87.19.7777

Bast, A., 2002. The risk to eat: Natural versus man-made toxins. In: Voss, G. and Ramos, G. (eds) Chemistry of crop protection. Wiley-VCH, Weinheim, Germany, pp. 63-68. https://doi.org/10.1002/3527602038

Bast, A. and Hanekamp, J.C., 2014. 'You can't always get what you want' – Linearity as the golden ratio of toxicology. Dose Response 12(4): 664-672. https://doi.org/10.2203/dose-response.13-032.Hanekamp

Bast, A. and Hanekamp, J.C., 2017. Toxicology. What everyone should know. Academic Press Elsevier, London, UK. http://dx.doi.org/10.1016/B978-0-12-805348-5.00001-6

Bast, A., Semen, K.O. and Drent, M., 2021. Pulmonary toxicity associated with occupational and environmental exposure to pesticides and herbicides. Current Opinion in Pulmonary Medicine 27: 274-283. https://doi.org/10.1097/MCP.0000000000000777

Cooper, J. and Dobson H., 2007. The benefits of pesticides to mankind and the environment. Crop Protection 26: 1337-1348. https://doi.org/10.1016/j.cropro.2007.03.022

European Food Safety Authority (EFSA), 2019. Guidance on the use of the threshold of toxicological concern approach in food safety assessment. EFSA Journal 16: 5708. https://doi.org/10.2903/j.efsa.2019.5708

Hanekamp, J.C. and Bast, A., 2015. Antibiotics exposure and health risks: chloramphenicol. Environmental Toxicology and Pharmacology 39: 213-220. https://doi.org/10.1016/j.etap.2014.11.016

Hanekamp, J.C., Bast, A. and Kwakman, J.H.J.M., 2012. Of reductionism and the pendulum swing: Connecting toxicology and human health. Dose Response 10: 155-176. https://doi.org/10.2203/dose-response.11-018.Hanekamp

Kaiser, D., Douches, D., Dhingra, A., Glenn, K.C., Reed Herzig, P., Stowe, E.C. and Swarup, S., 2020. The role of conventional plant breeding in ensuring safe levels of naturally occurring toxins in food crops. Trends in Food Science and Technology 100: 51-66. https://doi.org/10.1016/j.tifs.2020.03.042

Kroes, B., Renwick, A.G., Cheeseman, M., Kleiner, J., Mangelsdorf, I., Piersma, A., Schilter, B., Schlatter, J., Van Schothorst, F., Vos, J.G. and Würtzen, G., 2004. Structure-based thresholds of toxicological concern (TTC): guidance for application to substances present at low levels in the diet. Food and Chemical Toxicology 42: 65-83. https://doi.org/10.1016/j.fct.2003.08.006

Leak, R.K., Calabrese, E.J., Kozumbo, W.J., Gidday, J.M., Johnson, T.E., Mitchell, J.R., Ozaki, C.K., Wetzker, R., Bast, A., Belz, R.G., Bøtker, H.E., Koch, S., Mattson, M.P., Simon, R.P., Jirtle, R.L. and Andersen, M.E., 2018. Enhancing and extending biological performance and resilience. Dose Response 16: 1559325818784501. https://doi.org/10.1177/1559325818784501

Meena, B., Rajan, L.A., Vinithkumar, N.V. and Kirubagaran, R., 2013. Novel marine actinobacteria from emerald Andaman & Nicobar Islands: a prospective source for industrial and pharmaceutical by products. BMC Microbiology 13: 145. https://doi.org/10.1186/1471-2180-13-145

Nauta, M.J., Andersen, R., Pilegaard, K., Pires, S.M., Ravn-Haren, G., Tetens, I. and Poulsen, M., 2018. Meeting the challenges in the development of risk-benefit assessment of foods. Trends in Food Science and Technology 76: 90-100. https://doi.org/10.1016/j.tifs.2018.04.004

Sthijns, M.M.J.P.E., Randall, M.J., Bast, A. and Haenen, G.R.M.M., 2014. Adaptation to acrolein through upregulating the protection by glutathione in human bronchial epithelial cells: The materialization of the hormesis concept. Biochemical Biophysical Research Communications 446: 1029-1034. https://doi.org/10.1016/j.bbrc.2014.03.081

Tanwar, B., Modgil, R. and Goyal, A., 2019. Effect of detoxification on biological quality of wild apricot (*Prunus armeniaca* L.) kernel. Journal of the Science of Food and Agriculture 99: 517-528. https://doi.org/10.1002/jsfa.9209

Van der A, J.G. and Sijm, D.T.H., 2021. Risk governance in the transition towards sustainability, the case of bio-based plastic food packaging materials. Journal of Risk Research 24: 1639-1651. https://doi.org/10.1080/13669877.2021.1894473

Van Steenwijk, H.P., Bast, A. and De Boer, A., 2021. Immunomodulating effects of fungal beta-glucans: From traditional use to medicine. Nutrients 13: 1333. https://doi.org/10.3390/nu13041333

Vrolijk, M., Deluyker, H., Bast, A. and De Boer, A., 2020. Analysis and reflection on the role of the 90-day oral toxicity study in European chemical risk assessment. Regulatory Toxicology and Pharmacology 117: 104786. https://doi.org/10.1016/j.yrtph.2020.104786

Weseler, A.R. and Bast, A., 2012. Pleiotropic-acting nutrients require integrative investigational approaches: the example of flavonoids. Journal of Agricultural and Food Chemistry 60: 8941-8946. https://doi.org/10.1021/jf3000373

Weseler, A.R. and Bast, A., 2017. Masquelier's grape seed extract: from basic flavonoid research to a well-characterized food supplement with health benefits. Nutrition Journal 16: 5. https://doi.org/10.1186/s12937-016-0218-1

Wiedenfeld, H. and Edgar, J., 2011. Toxicity of pyrrolizidine alkaloids to humans and ruminants. Phytochemistry Reviews 10: 137-151. https://doi.org/10.1007/s11101-010-9174-0

World Health Organisation (WHO), 2018. Natural toxins in food. Available at: https://www.who.int/news-room/fact-sheets/detail/natural-toxins-in-food

14. Food allergies and food allergen control

This is what you need to know about food allergens in a nutshell – the basic concepts to better understand why food allergen declaration and management is important

Harris A. Steinman[*] and Comaine van Zijl

Food and Allergy Consulting and Testing Services, Office 11, The Woodmill, Vredenburg Road, Stellenbosch, South Africa; harris@factssa.com

Abstract

Food allergies are a recognised global public health concern. As a result, many regulators across the world have published legislation, standards and/or guidance for their regions on how the presence of allergens in pre-packed foods must be communicated to consumers. The exact prevalence of food allergies is unknown, but it is estimated to affect between 2 and 10% of the population. On occasion in the past, allergic reactions have claimed the lives of sensitive consumers, and it is known that very small amounts of an offending allergen can cause an adverse reaction. For many years, in many countries, the number-one reason for food recalls has been the presence of undeclared allergens. This is an indication that more work must be done by the food industry to come to grips with what a robust allergen management plan entails. It is important to understand all aspects of food allergies and allergens – not only to comply with food safety management systems and the law, but also to provide safe food to the consumer.

Key concepts

- ► Food intolerance: An adverse reaction not mediated by the immune system. Such reactions are usually dose-dependent, reproducible, and can be divided into three main types depending on their causality: enzymatic, pharmacological and undefined.

Bart Wernaart and Bernd van der Meulen (eds)
Applied food science
DOI: 10.3920/978-90-8686-933-6_14, © Harris A. Steinman and Comaine van Zijl 2022

- Food allergy: A specific immune response that occurs reproducibly on exposure to a given food and can be IgE mediated (type 1 hypersensitivity), non-IgE mediated (cell-mediated reactions) or a combination of both.
- Food antigen: Any substance which is recognised by the immune system is an antigen.
- Food-induced anaphylaxis: An acute, potentially life-threatening allergic reaction.
- Cross-reactivity: Some foods have similar proteins to those of common allergens, and thus will also induce an allergic reaction in a sensitive individual.
- Cleaning in Place (CIP): A method of cleaning equipment requiring minimal dismantling and with minimal operator involvement. CIP is used for the cleaning of pipework or vessels (tanks), by passing cleaning fluids through the pipework or spraying inside the vessel.

Case 14.1. A kiss of death.

In 2021, a young Canadian women died after kissing her new boyfriend. What happened, and how does it relate to food allergies? The young woman had asthma, and known peanut and shellfish allergies. While she was in the bathroom, and unknown to her, her new boyfriend ate a peanut butter sandwich. When they kissed shortly after, there was enough peanut butter protein left in his mouth to trigger a severe allergic reaction that led to her death.

14.1 The medical perspective

Food allergies are a recognised global public health concern that may be life-threatening and significantly impact the quality of life of sufferers and their caregivers. Food allergy symptoms can vary from mild to severe, and in extreme cases, food allergy can lead to anaphylaxis (a life-threatening allergic reaction). There is currently no cure for food allergies.

14.1.1 Overview of adverse foods reactions

Approximately 20-30% of individuals believe that symptoms experienced after ingesting a foodstuff are as a result of that foodstuff, and likely to be an allergy. In fact, very few people are truly allergic when subjected to the suspected food in a controlled oral challenge test (Shek and Lee, 2006). Therefore, even though food may cause an adverse reaction, the reactions may not be due to a food allergy.

Foodstuffs are complex mixtures of a host of components, and multiple components in a single foodstuff can potentially trigger different adverse reactions via different mechanisms. This complexity is magnified by meals (composite foods) that contain multiple ingredients, therefore it may be unclear what triggered a specific reaction. In Table 14.1 the mechanisms of key adverse food reactions are summarised.

Table 14.1. Mechanisms of key food reactions.

Mechanism	Example
Toxic	A reaction to a toxic compound in a foodstuff, e.g. amatoxins in poisonous mushrooms.
Psychosomatic	A reaction manifests when an individual believes a specific food will cause a reaction.
Intolerance	When an individual has a partial to complete lack of the enzyme necessary to metabolise a natural food component, e.g. sucrose, milk lactose, etc.
Pharmacological	When a reaction is caused by a pharmacologically active component, e.g. caffeine, tyramine in cheese, etc.
Immune-mediated (non-allergy)	An abnormal immune response to a food protein (not mediated via the allergy pathways), e.g. coeliac disease triggered by gluten.
Immune-mediated (allergy)	An abnormal immune response to a food protein (mediated via the allergy pathways), e.g. allergy to wheat, milk, peanut, etc.

14.1.2 Food intolerance

The most prominent example of enzymatic food intolerance is lactose intolerance. An adverse reaction occurs following the ingestion of a lactose-containing food, e.g. milk, and as a consequence of a lack of the enzyme lactase, which is required to break down lactose. The 'undigested' lactose ferments in the bowel, resulting in gas formation and osmotic effects, causing symptoms.

The pharmacological form of food intolerance is a reaction to a pharmacologically active component in food, e.g. histamine. Histamine occurs naturally in many foods, e.g. cheese, wines and tomatoes. When certain species of fish are not stored correctly, microbes convert the amino acid histidine into histamine. The resulting high level of histamine causes histamine toxicity, called 'scombroid food poisoning' in this instance.

Undefined food intolerance occurs when someone has adverse reactions to certain products, e.g. food additives or sulphur dioxide. In many cases, the underlying mechanism is unknown.

It is important to note that food intolerances are, unlike food allergies, not mediated by the immune system, but are however commonly confused with food allergies.

14.1.3 Food allergies

The immune system is a complex network of cells, chemicals, proteins and organs that defends the body primarily against infection. Abnormalities of the immune system can lead to allergic diseases, immunodeficiencies and autoimmune disorders.

Essentially, a food allergy is a hypersensitivity response to a food protein (referred to as a food antigen) that are intrinsically harmless, e.g. peanut protein, as discussed in Case 14.1. In healthy individuals, there is normally a state of unresponsiveness to these food antigens, but in allergic individuals, a lack of immunologic tolerance may lead to food allergies (Yu *et al.*, 2016). To date, the definitive cause for this situation is unknown.

food antigen

Food antigens can elicit symptoms of food allergy via two broad immune-mediated mechanisms: antibody-mediated or cell-mediated. The former is directed mainly via an immunoglobulin E (IgE) mechanism. This type of reaction can be precipitated after ingestion of minuscule amounts of the protein to which the individual is sensitised, and symptoms commonly have a rapid onset. Approximately 70% of food-allergic reactions are mediated via IgE antibodies.

IgE mechanism

The IgE mechanism has been well studied. When the body is exposed to a novel food protein that it perceives to be problematic, the B cells present in the blood are stimulated to produce IgE directed at the specific antigen (protein) the body considers problematic, and to stimulate mast-cell expansion. IgE binds to receptors present on the surface of the mast cells. This process is known as 'sensitisation'. When an individual is exposed to the same food protein again, the IgE binds with the specific food protein resulting in degranulation of the mast cell with the release of mediators such as histamine, which elicit the symptoms of IgE-mediated food-allergic reactions (Iweala and Burks, 2016; Sicherer and Sampson, 2010).

eliciting dose

The amount of allergen (the eliciting dose or reactive dose) which provokes a reaction in a food challenge does not predict the severity of the reaction or the likelihood of anaphylaxis (Wainstein *et al.*, 2010). For some individuals, the eliciting dose ('threshold dose') resulting in anaphylaxis can be minuscule. In studies, the ED10 (the eliciting dose predicted to provoke a reaction in 10% of individuals with a specific food allergy) (FARRP, 2021) for celeriac was 1.6 mg protein, compared to 2.8 mg of peanut protein and 8.5 mg of hazelnut protein (Ballmer-Weber *et al.*, 2015). Of the main allergenic foods, mustard (0.05 mg) has one of the lowest eliciting doses (Taylor *et al.*, 2014).

cell-mediated immunity

Although IgE-mediated food-allergic reactions are potentially the most serious, occurring very rapidly and being potentially life-threatening, a non-IgE mediated (cell-mediated immunity) food allergy may markedly affect the quality of life of allergic individuals and is therefore also relevant. Symptoms of IgE-mediated food allergy may manifest within two hours of exposure, whereas non-IgE-mediated reactions manifest between one and 24 hours after exposure to a food and require larger doses of the offending allergen. The latter has proven to be challenging to diagnose and many individuals experience food allergy via both mechanisms (Sicherer and Sampson, 2014).

Although blood and skin-prick tests are useful in diagnosing IgE-mediated food allergy, they are not highly diagnostic. The gold-standard diagnostic technique is the double-blind placebo-controlled oral food challenge (Anagnostou *et al.*, 2015). However, considering the significant resources required to conduct this test, it is not widely or routinely implemented. Figure 14.1 schematically summarises the key types of adverse food reactions.

14.1.4 Food allergy prevalence

Food allergy affects children and adults but is more prevalent in the former. The exact prevalence of food allergies in a population is difficult to determine and differs from one geographical area to another. For example, European and US

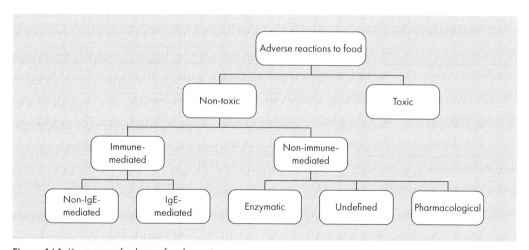

Figure 14.1. Key types of adverse food reactions.

studies have determined that respectively between 6 and 8% of children suffer from food allergies (Gupta *et al.*, 2011; Nwaru *et al.*, 2014), while in Japan this figure may be as high as 12.6% (Ikura *et al.*, 1999).

There is evidence to suggest that food allergy prevalence has increased in the last two to three decades and that allergy to specific foods, e.g. peanut, egg, etc., can vary tenfold between different population groups (Prescott and Allen., 2011; Renz *et al.*, 2019).

14.1.5 Symptoms and quality of life

IgE-mediated allergic reactions can manifest in different ways, from skin reactions (hives) to respiratory (asthma) or gastrointestinal reactions (nausea, vomiting, diarrhoea) to anaphylaxis. Anaphylactic reactions may result in death and are a consequence of a major reaction resulting in shock and affecting all organs of the body. Similar allergic reactions occur with non-IgE-mediated reactions, with the exception of anaphylaxis. Examples of the types of symptoms that food-allergic patients experience is presented in Figure 14.2. It is important to note that some reactions may only manifest when an allergen is ingested and a co-factor – such as exercise, medication, or alcohol – occurs concurrently (Sicherer and Sampson, 2010). Severe reactions may also be elicited from seemingly unconnected foods, as a result of allergen cross-reactivity.

The onset of an IgE-mediated reaction can occur within minutes and is often followed by a secondary, delayed response (also known as a late-phase response), which may only begin four to six hours after exposure to the allergen, and may continue for several days. Uncommonly, some allergic reactions – e.g. to meat – may display first onset only four to six hours after ingestion.

Because food allergies are far more prevalent in children than in adults, their impact goes further than the physical symptoms themselves. They may impact socialisation and schooling in children and the economic activity of parents and other adults. Impacted quality of life and psychological effects are common.

High levels of anxiety and constant vigilance are hallmarks of food-allergic individuals and their caregivers. Highly allergic individuals are required to carry injectable adrenalin; and risk-taking behaviour – for example, eating food that may contain an allergen – is well described, especially in teenagers.

In occupational settings, workers exposed to contact with or dust from foods may develop contact dermatitis (eczema) or occupational asthma, making it difficult for them to continue their employment.

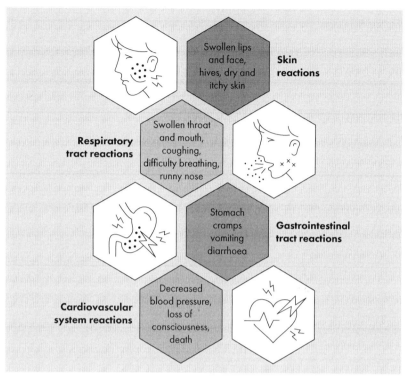

Figure 14.2. Food-allergic reactions.

14.1.6 Treatment and prognosis

Currently, as there is no cure, strict avoidance of the offending allergens is truly the only way food-allergic individuals can protect themselves. Current research focuses on various methods to induce de-sensitisation and allow individuals to have some measure of tolerance to the offending allergen.

food labels

Careful attention to food labels is essential, together with avoidance of food that has had significant cross-contact with allergenic foods (shared processing lines, cutting boards, slicers, mixers) (Sicherer and Sampson, 2010). 'Hidden' allergens in foods may add complexity (Skypala, 2019; Steinman, 1996); these include celery, spices, lupin, pea, natural food colourings, and preservatives that are not readily identifiable on food labels. Hidden allergen also include allergenic material that is present in a foodstuff due to cross-contamination that occurred along the food chain.

Allergy to cow's milk, egg, wheat and soya typically resolves during childhood. Around 50-60% of children with milk and egg allergy demonstrate tolerance by school age. In contrast, allergy to peanut, tree nuts, fish, shellfish and sesame tends to persist lifelong, and rarely resolves (Sicherer and Sampson, 2014).

14.2 Food allergens

14.2.1 Common food allergens

Because all foods contain protein, all foods can potentially induce an allergic reaction; but only around 200 are currently associated with documented reactions. As stated before, the prevalence of food allergies may differ from one geographic region to another (Devdas *et al.*, 2018). Influences include dietary practices, flora, genetics, etc. For example, the prevalence of peanut allergy is high in Western countries but much lower in China. Co-existent eczema (atopic dermatitis), asthma, increased hygiene and reduced exposure to infections may increase the risk of food allergy (Sicherer and Sampson, 2010).

Although an individual could have an allergic reaction to any food that contains protein, 90% of food-allergic individuals react to one or more of a narrow and common list of foods often referred to as the 'big eight' (Sicherer and Sampson, 2010). These are recognised as major allergens by Codex through its General Standard for the Labelling of Pre-packed Foods. The standard gives general guidance on how to declare the presence of these foods if they are deliberately added to pre-packed foods.

These are the 'big eight':

1. cereals containing gluten, i.e. wheat, rye, barley, oats, spelt or their hybridised strains and products made from them;
2. crustacea and products made from them;
3. eggs and egg products;
4. fish and fish products;
5. peanuts and products containing them;
6. soybeans and products containing them;
7. milk and milk products; and
8. tree nuts and nut products.

Although Codex currently recognises these as the major allergens, individual countries are encouraged to identify their own priority allergens that must be regulated. For example, in a number of countries, lupin, sesame seed, mustard, celery and buckwheat would be relevant foods, because a high prevalence of allergic reactions to them has been documented. In India, chickpea (*Cicer arietinum*) is a common allergen (Devdas *et al.*, 2018).

Considering that labelling and advertising are the most important means of communication between consumers and food manufacturers, many regulatory authorities have enacted regulations specifically to alert allergic individuals and thereby protect themselves. The Food Allergy Research and Resource Programme (FARRP) maintains a comprehensive regulatory chart listing the foodstuffs recognised per region.

14.2.2 Novel food allergens

As food consumption habits change and novel food proteins are introduced into a community, allergic reactions to these new allergens may become prevalent. The fungus *Fusarium venenatum* has become the basis for a popular meat substitute and this mycoprotein has been linked to a large number of reported allergic reactions, many involving anaphylaxis (Jacobson and DePorter, 2018; Skypala, 2019). The use of lupin flour in bakeries, particularly for gluten-free products, led to an increased number of peanut-allergic patients reporting reactions to lupin, and also to fenugreek, a legume commonly used in spice mixes (Namork *et al.*, 2011). Other novel food allergens on the horizon include caterpillars, locusts, grasshoppers, etc. (Skypala, 2019).

14.2.3 Cross-reactivity

Individuals are not allergic to a food per se, but to certain proteins within that food and every food contains a range of different proteins. These typically belong to protein families, according to the function that they typically display in the food. For example, many foods contain lipid transfer proteins (LTP, in the defence protein family), albumins (storage protein family), etc. As an example, wheat contains more than 50 types of protein, of which some belong to the prolamin, LTP, profilin and albumin families, among others (Hassan and Venkatesh, 2015; Monaci *et al.*, 2020).

Proteins belonging to a protein family occurring in a number of foods may resemble each other closely but are never an exact match; they differ by a number of amino acids, sometimes only a few, sometimes many. Therefore, the LTP in peach may closely resemble the LTP in wheat, but will not be identical.

The higher the similarity of the protein, the higher the risk of allergic cross-reactivity: an individual allergic to the LTP protein in peach has a high risk of also being co-sensitised to the LTP protein in wheat, as they have close similarity and the body's antibody IgE is unable to differentiate between them (Cox *et al.*, 2021; De Gier and Verhoeckx, 2018; Leoni *et al.*, 2019; McKenna *et al.*, 2016; Ruethers *et al.*, 2018). The consequence is that a peach-allergic individual may experience an allergic reaction to wheat. An individual with hay fever from birch pollen may experience an allergic reaction to eating an apple. Similar factors contribute to the cross-reactivity between peanuts and tree nuts, and between those allergic to latex with cross-reactivity to mango, avocado and kiwi (Cox *et al.*, 2020). Reactions to fenugreek and lupin are usually linked to cross-reactivity with a primary allergy to peanuts (Cabanillas *et al.*, 2018; Che *et al.*, 2017; Faeste *et al.*, 2010).

14.2.4 Protein modification

Proteins (allergens) may be heat-labile or stable and proteins may become more allergenic or less allergenic upon heating. For example, when roasted, peanut becomes more allergenic; although specific individual proteins present in peanut lose their allergenicity, e.g. profilin. Apple contains heat-stable and heat-labile proteins, therefore, apple-allergic individuals sensitised only to the heat-labile proteins can safely eat cooked apple, whereas those allergic to heat-stable proteins cannot. Furthermore, an allergen may not be distributed throughout the whole food but be limited to the stalk (e.g. the LTP in celery) or skin (e.g. the LTP in apple) (Gadermaier *et al.*, 2011).

Other processing methods may also influence the allergenicity of proteins. For example, fermentation or hydrolysis may reduce the allergenicity of a protein, depending on the extent of the process; e.g. wheat protein in beer may still retain allergenicity after the production process, but the wheat in soy sauce may not.

Heat-labile proteins are often broken down rapidly by the body's digestive enzymes. Hence, for example, individuals allergic to heat-labile proteins in apple may display oral symptoms (oral allergy syndrome) from eating apple, but would not experience symptoms if the mouth could be bypassed.

It is essential to recognise that allergens follow a very different paradigm to that of microbiological, chemical and other hazards, and therefore their management requires a different model.

14.3 What about gluten and wheat?

The differences between coeliac disease, gluten or wheat intolerance and wheat allergy are not always clear and may be confusing to the food industry and consumers (Caio *et al.*, 2019; Rubin and Crowe, 2020). Often these terms are used interchangeably. The confusion is further exacerbated by the consolidation of gluten-containing cereals into one group by many regulators. This creates the impression that gluten-containing cereals trigger a single adverse food reaction, and that gluten is the agent of concern.

Clinically, the conditions mentioned are distinctly different. Wheat allergies can result in life-threatening reactions. Coeliac disease symptoms are severe, but not acutely fatal, and gluten or wheat intolerance commonly leads to mild or moderate discomfort.

coeliac disease
Coeliac disease is a lifelong autoimmune systemic disorder triggered by gluten found in barley, rye and wheat, and is not immediately life-threatening; however, symptoms may slowly worsen and be exacerbated over days. Long-term effects include short stature, anaemia and ataxia. Prolonged exposure can lead to an increased risk of developing various cancers.

wheat allergy
Wheat allergy is an IgE-mediated food allergy to wheat proteins. Minute amounts can potentially trigger life-threatening anaphylaxes, although less severe reactions are more common. *Gluten intolerance* or *non-coeliac gluten sensitivity* (NGCS) is a condition in which symptoms are experienced by people who do not have coeliac disease and are not allergic to wheat. It is a non-immune-mediated response, and its cause is not well understood. The inability of the body to break down FODMAPS (Fermentable Oligo-, Di-, Mono-saccharides And Polyols) in wheat, rye or barley, and their subsequent fermentation, has been suggested as a cause.

14.4 Allergen management in a food processing facility

There are broadly four reasons why the food industry needs to manage allergens: (1) to ensure regulatory compliance (Gendel, 2012), (2) to drive food safety standard compliance, (3) to protect consumers from harm, and (4) to protect an organisation's brand reputation.

Food manufacturers are responsible for the safety of the products they produce (Codex Alimentarius, 2020). This includes effective allergen management and clear labelling of food allergens.

14.4.1 Food safety standards requirements

Allergen control and management must be included in a food organisation's prerequisite programmes (PRPs) and is a crucial part of its good manufacturing practices (GMPs) programme. GMP is the basis for a robust HACCP (Hazard

Analysis Critical Control Points) programme and both are a requirement to ensure compliance with a Global Food Safety Initiative (GFSI) benchmarked standard, e.g. Brand Reputation Compliance Global Standards (BRCGS) and Food Safety System Certification (FSSC) 22000. Both BRCGS and FSSC 22000, for example, have defined requirements in terms of allergen management.[50]

14.4.2 Allergen Control Plan

Several guidelines are available as possible approaches to allergen control. It is important to note that, based on a company's specific needs, many variations on these recommendations could achieve acceptable results (Taylor and Hefle, 2005). The focus of any Allergen Control Plan (ACP) should be to prevent the presence of undeclared allergens in a final product. An ACP is an organisation's formal documented plan, describing all the controls in place to manage allergens in its scope of operation.

14.4.3 Initial allergen risk assessment

Food allergens are a hazard, and they must be controlled and managed through astute risk assessment and management to prevent allergen-related food safety incidents (BRC, 2014; Taylor and Hefle, 2005).

The first step towards effective allergen control is the identification of all potential allergen sources and possible cross-contact points (Codex Alimentarius, 1985). The next step is to evaluate existing control measures and determine whether they will reduce the likelihood of allergen contamination to an acceptable level.

A comprehensive risk assessment of an organisation's operations should be conducted to develop an ACP (Codex Alimentarius, 2020). The risk assessment should begin with raw materials, equipment and production, intermittent storage, handling, packing, etc. The assessment should continue through every step in the manufacturing process, from receipt of raw materials through to the packaging, labelling and dispatch of the final product.

Critical points where allergens are introduced must be identified, contamination control established, and implemented and monitored to ensure unintentional cross-contact is avoided.

---

[50] On HACCP see also Chapter 10 (Ferris, 2022) in this book.

14.4.4 Supply chain

Allergen cross-contact can occur at every stage of the supply chain; and even if you follow GMPs, cross-contamination may be unavoidable in practice, i.e. cross-contact due to agricultural co-mingling.

When looking at the supply chain, it is important to identify and scrutinise all its links as far as possible (BRC, 2014). Robust traceability and supply chain management systems allow manufacturers to track the movement of ingredients up and down the chain, providing valuable information on potential points of cross-contact, and the risk of accidental or deliberate adulteration.

It is important to understand the allergenic status of all materials (Taylor and Hefle, 2005). Raw material specifications should be thoroughly assessed, paying attention to identify ingredients derived from allergens, and in particular, hidden allergens that are listed using technical names and allergens present due to cross-contact. Suppliers must be able to supply comprehensive, accurate and reliable information on the allergenic status of the products supplied (Komitopoulou, 2014).

Ingredient and supplier information is generally acquired by reviewing the sources listed in column one of Table 14.2, and suppliers should be able to answer the questions listed in column two.

Table 14.2. Raw material information sources and key information required.

Sources	Information required
• Up-to-date product specifications • Product Information Forms (PIFs) • Supplier Information Forms (SIFs) • Benchtop audit forms • Documented correspondence • Other relevant trade documents	• Does the product contain allergenic ingredients? • Which ingredients are derived from common allergens? • What is the nature of the ingredient and allergen? Are they particulate or readily dispersible? • Which allergens may be present due to cross-contact? • What is the nature of the cross-contact? • What is the probable concentration of the cross-contact allergen? • Which allergen controls are in place to mitigate cross-contact?

14.4.5 Product and ingredient matrices

Once the information in the section above has been received, it's a good idea to summarise it in matrices. These matrices are visual tools that make it easy to see at a glance the allergenic status of raw materials, intermittent and finished products. The tables in Figure 14.3 below are examples of raw material and finished product matrices.

14.4.6 Material receiving and storage

There are several elements to consider when receiving materials from suppliers. Warehouse staff must know what to look out for, how to receive materials, where to store them and when it is appropriate to reject or isolate materials. These details must be addressed in a detailed receiving procedure and accompanying record.

Consignments must be accompanied by appropriate documentation, e.g. waybills, delivery notes, invoices, cleaning tickets, Certificate of Analyses, specifications, etc. and attention must be paid to the loading sequence. Typically, allergen-containing ingredients will be loaded first and off-loaded last.

Identification of allergens is necessary to assist the warehouse, picking and material dispensing staff in identifying the correct procedures to follow. Coloured labels can be used to flag allergen-containing ingredients (Taylor and Hefle, 2005). Different colours can be allocated to the different allergens handled on-site, e.g. blue = milk, yellow = wheat, etc. Or a single colour can be allocated to allergens, e.g. red labels are used to identify ingredients that contain allergens. If labels are used, they must be indelible and not easy to remove. It is important that the identification of products containing allergens is practical for the manufacturing facility and that staff are adequately trained in these procedures.

Ingredient name	Milk	Egg	Soy	Peanuts	Tree nuts	Wheat	Gluten	Fish	Crustacean	Molluscs
Cheese sauce	X	X				X	X			
Lemon spice			X	X						

Ingredient name	Milk	Egg	Soy	Peanuts	Tree nuts	Wheat	Gluten	Fish	Crustacean	Molluscs
Parmesan cheese	X	X								
Rye flour							X			

Figure 14.3. Raw material and finished product matrices.

Effective allergen control relies on the segregation of allergenic foods from other products, from the time of receiving until their introduction onto the production line and beyond (BRC, 2014; Taylor and Hefle, 2005).

On receipt, raw material labels should be reviewed to determine the material's allergenic status, and consignments must be checked for signs of possible cross-contamination between allergenic and non-allergenic raw materials.

Control of allergens during storage can be challenging if there are space constraints. Typically, allergen-containing materials are stored in designated or cordoned-off areas. Good allergen storage practices require: (a) allergen- and non-allergen-containing materials to be separated; (b) different allergens to be separated; (c) allergen-containing materials to be stored on bottom shelves and first on pallets; and (d) like allergens to be stored above each other. How the facility meets these requirements will depend on available resources.

It is important to note that the likelihood of allergen contamination increases significantly when product spills are not cleaned up promptly.

14.4.7 Production controls

In terms of the production facility itself, it is useful to adhere to three basic principles to reduce allergen cross-contamination:

1. segregation of operations and processing, in time and/or space;
2. avoiding cross-over of products;
3. controlling movement of people and products: it is good practice to create an allergen cross-contact map based on the layout of the facility; track the movement of allergens in the plant to identify potential and actual points of cross-contamination, e.g. allergen addition points, shared equipment, etc.

Where possible, the production of allergen-containing and non-allergen-containing products (or products containing different allergens) must be segregated (Codex Alimentarius, 2020). Segregation or separation does not always refer to space, but can also be a separation in time, with allergen cleans in between. Physical separation is not a practical solution for all facilities; most rely on efficient cleaning.

Some allergens are airborne, so cross-contact can occur if the ventilation in the facility is insufficient or ineffective. Air vents and fans must be cleaned regularly to minimise the opportunity for allergen cross-contamination, especially if raw materials may become airborne, e.g. flour dust.

When a product is issued from the warehouse to the production floor, it is important to ensure that traceability in terms of allergen identification is not lost, and that cross-contact is avoided or limited.

Colour-coding of scoops and utensils is often used as an allergen management strategy, but such systems can be overcomplicated and difficult to follow. It is best to use fewer colours and manage them efficiently, than use every colour available but manage them poorly. Colour coding may be more suitable when a facility handles a limited number of allergens, and when it fits in with existing colour coding systems (e.g. HACCP area colour coding). It may be better to use metal scoops that are dedicated to certain ingredients. Where scoops and utensils are shared, they must be washed after they have been used to handle allergenic ingredients.

product scheduling Intelligent product scheduling significantly reduces the risk of allergen cross-contact (Codex Alimentarius, 2020). Production schedules aimed at optimal output and reduced risk start with products containing no allergens or the least number of them, adding new allergens as the schedule progresses (BRC, 2014). Subsequent products must contain all the allergens that appear in the preceding product, to prevent allergen carry-over.

Each phase of the allergen production schedule must terminate in a validated allergen-clean step to enable the total removal of allergenic residues from food contact surfaces (Taylor and Hefle, 2015).

Rework may be overlooked when considering allergen management. Like-into-like strategies are considered best practice in terms of allergen rework control. This refers to reworking only like-products into each other, or reworking products with like – or the same – allergen or allergens into each other. Maintaining traceability of rework materials plays a key role in ensuring that the correct materials are reworked into each other (Codex Alimentarius, 2020; Taylor *et al.*, 2006).

14.4.8 Printed packaging material control

Product labels are the only communication between manufacturers and the consumer. Accurate and comprehensive labelling not only helps to protect consumers, but also helps to protect a company from costly recalls, regulatory scrutiny, and potential liability.

Many international allergen-related recalls are due to packaging errors (see also Case 14.2). Often the wrong packaging material is used, such that the ingredient information displayed on the packaging material does not match the contents.

Case 14.2. Allergens and recall.

In Australia and New Zealand, between 2016 and 2020, the biggest percentage of recalls was due to the presence of undeclared allergens. 43% of the recalls were triggered by a customer complaint, 20% due to routine testing by the government and 19% due to routine testing done by the food business. The leading root causes for the recalls were packing errors, supplier verification audits and accidental cross-contact (FSANZ, 2020).

When developing packaging material control protocols and records, it is helpful to keep the following in mind:

- ensure that product labels reflect current formulations;
- review labels when changes are made to ingredient formulations;
- review incoming printed packaging material before receipt;
- implement proper inventory control procedures for packaging materials;
- monitor, document and verify that the correct labels are loaded onto packaging lines at all changeovers;
- train line personnel to ensure product labels are switched during product changeovers;
- manage obsolete packaging appropriately;
- discard older, out-of-date packaging; and
- it is also advisable to contact the packaging material supplier to ensure they no longer have obsolete printing plates or graphics.

14.4.9 Product development

Allergen control begins with new product concepts (Taylor and Hefle, 2005), and the introduction of new ingredients and new product labels.

Allergen information generated during product development and reformulation must be communicated to the food safety team and all other relevant parties involved in the manufacturing of products at a plant, e.g. procurement, production, warehousing and cleaning teams, as well as the marketing team.

It is helpful to consider the following during the development and reformulation of products:

► Aim to add allergens to new food products only when necessary.
► Risk assessments on ingredients and suppliers must be completed.
► Question raw material suppliers on the functionality and necessity of allergens in their formulations.
► Create a process to review allergens in new products with the manufacturing facility before trials and production. Do not forget to factor in enough time for cleaning in between factory trials.
► Understand which allergens are already being controlled in the facility and assess what the implications of introducing a new allergen will be.
► Establish a protocol that defines how formulas are developed, controlled, and changed.

14.4.10 People management

The effectiveness of an ACP ultimately comes down to employees. Employees who understand when, why and how to control allergens in a plant can be an organisation's greatest asset. Several tactics may be implemented to prevent personnel from becoming a source of allergen cross-contact.

For example, different coloured aprons or personal protective clothing (overalls, dust-coats, mop caps, etc.) can be used in different areas of the facility. However, no intervention can be truly successful without allergen awareness training and control-specific tutoring. This holds for all food safety-related aspects, as well as for building a positive food safety culture in a facility.

Regular training and education in allergen management are advised for all staff (including temporary staff and contractors) involved in handling ingredients, equipment, utensils, packaging and products (Codex Alimentarius, 2020; Komitopoulou, 2014). Allergen awareness should be communicated to employees as part of the GMP and HACCP training programmes. Training should be appropriate for each employee's education level and job-specific tasks.

Visitors and contractors should be briefed on allergen policies and what is expected of them to mitigate contamination. It is considered common practice to ask visitors and contractors to sign a register referring to applicable allergen controls to confirm their understanding.

Often, the concept of food allergies is not common knowledge among floor staff. Take the time to explain what food allergies are, which foods trigger allergic reactions, and what happens to allergic individuals. Use pictures and examples. Engage staff members by asking whether they know anyone who has food allergies. Ask them to explain to the rest of the group what happens to that person when they eat and drink something that they are allergic to, and how having an allergy impacts their life.

14.5 Allergen cleaning

If you cannot see it, you cannot clean it. In other words, if you cannot dismantle and inspect it, how can you be sure it is really clean. In many facilities, hygienic design may be a challenge in terms of the cleanability of the equipment and, at times, of the facility itself.

Cleaning is one of the most crucial aspects of an effective allergen control programme (FSA, 2006). It aims to remove allergen soils from equipment and shared lines and should be effective and efficient (Bagshaw, 2009). In 2009 Bagshaw (p. 114) stated 'Cleaning practices that are satisfactory for hygiene purposes may not be sufficient to remove allergens from surfaces and equipment.'

Unlike microorganisms, allergenic food soils are not readily affected by heat and chemicals (Bagshaw, 2009). The allergen risk assessment will inform what need to be cleaned, how often it needs to be cleaned (Bagshaw, 2009).

Cleaning aims to remove contamination, but if it is not applied and managed, it can result in cross-contact, i.e. overspray and aerosols from high-pressure hosing, etc.

Any allergen residue that is not adequately removed from processing lines – whether due to leftover ingredients in hard-to-reach corners, a half-hearted cleaning effort by employees, or a looming product deadline – can inadvertently contaminate the products that follow.

The safety of your products ultimately comes down to process design, protocol documentation and thorough validation of cleaning procedures. Sanitation starts with ensuring that GMPs (good manufacturing practices) and SSOPs (Sanitation Standard Operating Procedures) are in place. Cleaning must be well controlled, with a documented sanitation programme, clearly written cleaning work instructions and corrective actions.

14.5.1 Allergen cleaning programme

The great variety of available cleaning protocols reflects the challenges involved with cleaning different surfaces and soils. Overall, the nature of the allergenic protein, the food matrix, and the type of processing equipment will dictate an appropriate allergy-cleaning protocol and the efficacy of the protocol. Allergenic foods and ingredients may take various forms (e.g. solids, liquids, pastes, particulates or powders); they can be suspended in water, or fats; and they may be present in foods at low or high concentrations.

Equipment surfaces can be stainless steel, metal, plastic, cloth or textured. The finish and smoothness of the food contact surface and the surface condition (e.g. pitted, cracked, or scratched) may also vary (Schmidt, 2018).

It is important to follow a rational approach when developing an allergen-cleaning protocol and programme.

risk assessment Similarly, to developing an allergen-control plan, the first step in developing an allergen-cleaning programme is to complete a risk assessment. The aim is to identify what needs to be cleaned, and the cleaning frequency required, and is done by looking at the manufacturing process and environment and identifying cross-contact points.

The next step is to set realistic standards for cleaning. The ideal is to remove all allergenic residues. But this is not always practically achievable; thus, at a minimum, cleaning should be aimed at reducing the risk of allergen contamination as much as possible. Cleaning standards are generally based on regulatory, customer or company requirements.

prerequisites When developing a cleaning programme, it is important to consider (a) what type of detergent will work best; (b) can and should hot water be used; (c) does the selected cleaning chemical require contact time with surfaces; and (d) how long will it take to complete a thorough clean? Cleaning consultants may be used to assist in this process.

Once all this detail is determined in theory, the cleaning methodology must be validated, to prove that: (a) the cleaning method will achieve the goal and (b) cleaning meets the standards set.

If validation is successful, the method can be documented into SOPs and work instructions. These documents must be detailed and easy to understand, and describe the cleaning method, the authority and responsibilities of cleaning staff, the cleaning frequency, the verification method and desired results.

cleaning verification Last, but not least, cleaning verification checks the efficacy of cleaning on an ongoing basis. Verification should include a visual inspection of the cleanliness of equipment and the production environment (Coutts and Fielder, 2009). It can be supplemented by rapid onsite testing (Taylor and Hefle, 2005) of surfaces, rinse water and product (if applicable). Internal auditing should also be included in a verification programme. The last step is to review the cleaning schedule and programme at least once annually, or if there are any changes in products or processing.

food soils Food soils, product residues that remaining on food contact surfaces, generally consist of water-soluble and non-water-soluble components. The nature of these components greatly influences the ease with which they can be removed. Although the precise composition of the food soil will depend on the type of food being processed, note that the nature of the food soil is not necessarily the same as that of the food itself. Some components of the food may be more easily removed by cleaning, while others bind more tightly to surfaces.

Food soils vary in composition and can be present as complex films containing a mixture of several components. The nature of the food soil must be understood before an optimum cleaning method can be chosen. In general, food proteins (including allergenic proteins) are by far the most difficult food soils to remove.

The cleaning of proteins is even more of a challenge when they have been heat-denatured and have adhered to food contact surfaces (Itoh *et al.*, 1995; Jackson *et al.*, 2005; Jeurnink and Brinkman, 1994). Proteinous food soils, although not readily removed with water alone, can be removed when alkali detergents are used.

It is known that the removal of protein residues from food equipment presents challenges for cleaning, particularly when processing methods have altered the protein structure to bring about denaturation and aggregation. The ability of allergenic proteins to become tightly bound to surfaces through several attractive forces poses concerns from a cleaning and allergen cross-contamination viewpoint.

attracting forces Attracting forces play an instrumental role in the accumulation of residual proteinaceous (perhaps allergenic) material on equipment surfaces. Forces may arise between the material and the surface (forces of adhesion) and between the components of the material themselves (forces of cohesion) and can give rise to the formation of multilayer attachments to surfaces that become difficult to remove by cleaning. Counter forces (energy) are required to break the bonds between the components of the material and between the material and the surface. Thus, for effective cleaning, enough energy must be employed.

Three types of cleaning energies can be used for cleaning (Bagshaw, 2009), namely:

1. Mechanical energy: shear force caused by swabbing, scrubbing, fluid flow or pressure (e.g. water jet).
2. Thermal energy: e.g. elevated temperature.
3. Chemical energy: chemical action of detergents.

An optimal combination of the three types of energies must be established to enhance the efficacy of cleaning. For example, when it is not possible to use the most suitable detergent increased amounts of mechanical energy can be applied (Bagshaw, 2009.). Bagshaw, (2009. p. 118) concluded that 'the greater the attachment of the soils to the surface, the greater the combined energies required for cleaning'.

14.5.2 Wet cleaning

The most powerful tool for the removal of allergenic food soils is the utilisation of water (Jackson *et al.*, 2008). Water can introduce all three cleaning energies and be used to flush away residues (Bagshaw, 2009). However, water cannot be used for cleaning in all cases. The decision regarding whether to use water is often a function of the water activity of the food being processed at that point.

For example, water is generally used for cleaning at processing points for wet mixes but is used less frequently for cleaning at points where the food being produced has low water activity (Jackson *et al.*, 2008).

Where possible, facilities that manufacture high-water-activity foods should be designed to accommodate water, with equipment that can be disassembled and electronics wired to either withstand or be protected from moisture. The floors and walls should be designed with smooth surfaces to prevent the adsorption of allergenic ingredients and to allow for easy and effective cleaning. Floor drains should be available for drainage of water after wet cleaning.

With wet cleaning, four interrelated factors affect the efficacy of the overall cleaning process:
  ► choice of detergent and concentration;
  ► cleaning time;
  ► cleaning temperature; and
  ► the mechanical force used to apply and agitate the cleaning fluid (Jackson *et al.*, 2008).

detergents

Detergents remove food soils from surfaces by combining emulsification properties with one or more types of a chemical reaction. Current evidence does not suggest that the detergents themselves, or the chemical reaction that they bring about, have any direct effect on protein allergenicity. Effective cleaning is thus more reliant on the ability of the detergent to remove the allergenic soils, rather than on their ability to alter the allergenicity of the residues. Although there are many detergents available, given the solubility of proteins in alkaline solutions one would expect that the removal of proteins from surfaces is best accomplished using alkali-based detergents. Chlorinated alkali detergents are considered excellent choices for the removal of proteins from food contact surfaces (Coutts and Fielder, 2009), particularly when these have been heated. Alkaline detergents with peptising and wetting agents increase the wettability and suspendability of protein soils, and caustic detergents with peroxide boosters or chlorine are recommended to hydrolyse proteins, thereby enhancing cleaning.

contact time

Contact time with a detergent is necessary to maximise its performance. Generally, the longer the contact time with a detergent, the less physical energy will be needed to remove the matter from the surface. It is important to note the total cleaning time and contact time in documentation, procedures and the training of staff so that it is carried out properly every time. A validation study must be done to determine the optimum cleaning time.

temperature

The temperature at which is cleaning is carried out plays an important role in its efficiency. Using elevated temperatures for wet cleaning has several advantages, including:

▸ Deceases strength of bonds between soil and surface;
▸ Increases solubility of soluble materials;
▸ Decreases viscosity and increases turbulent action;
▸ Increases chemical reaction rates; and
▸ Essential for effective emulsification of fats where present.

The temperature range used for the removal of food soils is normally between 32 and 85 °C. An increase in temperature of 10 °C is reported to result in doubled cleaning efficiency. However, too high temperatures can decrease the efficiency of the clean, and a fine balance must be achieved to bring about optimal removal of both protein and fat components. For food soils comprised of fat and protein, cleaning should ideally be carried out in a range between the melting temperature of fat and the temperature at which protein denaturation occurs. Below the optimum temperature, fat will not clean effectively; but above the optimum temperature, proteins bind more tightly to surfaces and decrease cleaning efficiency. The optimum temperature for cleaning is approximately 60 °C.

mechanical energy   Generally, the greater the mechanical energy applied, the greater the effectiveness of cleaning. The optimum energy must be determined through a validation study. For effective allergen cleaning, the factors must be identified that affect the efficacy of removal of allergenic foods from food contact surfaces. The efficacy of the cleaning protocols differs depending on the type of food soil, the food contact surface, the temperature of the cleaning solution, and the concentration of the detergent in the cleaning solution. Food processors should evaluate the efficacy of cleaning protocols for each type of food soil, food contact surface, piece of equipment, and processing line, especially when processing allergenic foods.

14.5.3 Dry cleaning

Dry goods manufacturing environments are often not designed to accommodate water, or may even need to be free of water to facilitate manufacturing. For example, the presence of water could affect the quality and consistency of products such as bread, pastry, biscuits, cereals, chocolate, etc.

The introduction of water into equipment and environments not designed to accommodate it may promote uncontrolled microbial growth, development of harbourage sites for bacterial pathogens, and premature equipment failure.

Dry cleaning refers to cleaning in which no liquid detergents or disinfectants are used. Without the use of water for cleaning, manufacturers of dry goods face a challenge in removing allergenic food residues from processing equipment (Holah and Hall, 2004).

The management of allergens in dry goods plants requires food manufacturers to rethink traditional equipment design, to increase equipment accessibility and the ability to clean. Unlike wet cleaning methods that combine all three cleaning energies (chemical, thermal and mechanical) to remove soils, dry cleaning is primarily a mechanical process.

Dry cleaning relies on the soil being physically removed using a limited number of methods, such as vacuuming, sweeping, scraping, wiping with cloths, brushes, compressed air, dry ice, material flushes, product purging, etc. (Jackson *et al.*, 2008).

These procedures are often focused on the removal of the main soil deposits and product layers by e.g. a vacuum cleaner, followed by brushing and/or scraping the surfaces; or, often, only the latter. Dust formation should be avoided as much as possible during the execution of these methods (Jackson *et al.*, 2008).

The nature of the aforementioned physical methods (e.g. brushes and vacuums) means that 100% removal is not always possible. Following brushing or vacuuming, consideration should be given to wiping with disposable impregnated wet wipes or detergent/disinfectant-dampened cloths, to increase overall soil removal.

Tools used for cleaning can become a major route of allergen cross-contamination. Tools should be allocated to defined areas of use, e.g. floor use only or food-contact surface only. Specific, appropriate tools should be assigned to the cleaning of surfaces that have allergenic material on them. It is vitally important that the system is clearly defined and managed.

When brushes or scrapers are used for cleaning, care should be taken to prevent dust generation and recontamination of neighbouring surfaces. Brushes and other cleaning devices used to clean allergen-containing lines or equipment can be colour-coded or dedicated to preventing use on non-allergen lines or equipment, or both.

Additionally, the tools used for the cleaning of allergens should be fabricated using non-porous materials, to prevent the development and spread of contamination. If multiple allergens are produced in a facility, it would be considered best practice to dedicate cleaning tools to each specific allergen being handled.

Cloths and scourers used for allergen cleaning should be disposed of after use. In some cases, disposable (single-use) cloths or paper wipes saturated with water or alcohol are used to clean food contact surfaces, in areas where water is not compatible with the manufacturing environment. These moistened wipes offer the advantages of localising water and minimising dust generation.

negative-pressure systems

Negative-pressure systems such as vacuums are commonly used for dry cleaning in the food industry. Central vacuum systems are frequently used for cleaning large areas in food plants. High-efficiency particulate air (HEPA) filtration vacuum systems have been developed to remove and contain dust and debris during the dry cleaning of food-manufacturing equipment such as food contact surfaces, ovens and floors.

Smaller mobile vacuuming units may be used in smaller operations or to clean localised areas. However, many food plants still clean large areas with the smaller mobile vacuum systems that exhaust air locally, due to allergen-contamination concerns associated with central systems.

Vacuum cleaners and accompanying attachments (e.g. nozzles and hoses) employed for the removal of allergenic material will become contaminated. If possible, specific hoses and nozzles should be dedicated to the cleaning of allergenic material. If this is not possible, thorough cleaning of these attachments will be required to prevent allergen cross-contamination. Regular cleaning and replacement of vacuum filters are also critical, to ensure that no particulate materials are blown out of the unit.

Compressed air is commonly used in the food industry to dislodge food residue from inaccessible areas of equipment and environments. However, it should be used with discretion, and only when no other cleaning options exist (Jackson *et al.*, 2008). The formation of aerosols and airborne dust associated with such cleaning can cause the redistribution of allergens in the plant, and increase the risk of occupational respiratory problems.

Non-allergenic foods or other inert dry materials (e.g. salt, flour and starch) are sometimes used to 'clean' equipment in dry food manufacturing environments, by purging (pushing through) the allergenic foods from surfaces and equipment (Taylor and Hefle, 2005). Flushing with bulk ingredients is normally carried out for the cleaning of pipework and equipment which is not easily accessible for cleaning by other methods. The flushing step should be followed by thorough cleaning of all accessible equipment parts. Careful consideration of the flushing material is important, e.g. the use of an allergenic food for flushing such as wheat flour would only be appropriate when used to clean a line that processes wheat-containing products.

14.6 Allergen control validation and verification

First, it is important to understand the difference between validation and verification and that they are two distinctly different activities.

14.6.1 Cleaning validation

This refers to the process of assuring that a defined cleaning procedure can effectively and reproducibly remove allergenic foods from a specific food processing line or equipment (Taylor and Hefle, 2005). Validation studies ask the question: will the cleaning method remove allergens effectively? It is normally carried out before the commercial manufacture of a product, and at any time when changes are made to the manufacturing or cleaning process (e.g. reformulation of products, modifications to process or equipment, scheduling times or sequences, or cleaning protocols).

14.6.2 Cleaning verification

Refers to the ongoing process of demonstrating that the previously validated cleaning protocols have been properly performed and remain effective.

Validation and verification of cleaning procedures is an essential component of any effective allergen control programme. Validation of allergen-cleaning efficacy provides feedback on the effectiveness of the cleaning protocol, and very importantly, pinpoints areas of insufficient cleaning. Confidence in the allergen-cleaning protocol increases when validation studies are repeated several times.

Both should include visual inspection of equipment surfaces (Coutts and Fielder, 2009) and are best supplemented with appropriate analysis of e.g. finished product and in-process materials, final rinse water and diagnostic swab samples, or a combination of these items. Verification should also include the monitoring of critical cleaning parameters such as time, temperature, and cleaning-solution concentrations. Figure 14.4 illustrate the key differences between allergen cleaning validation and verification.

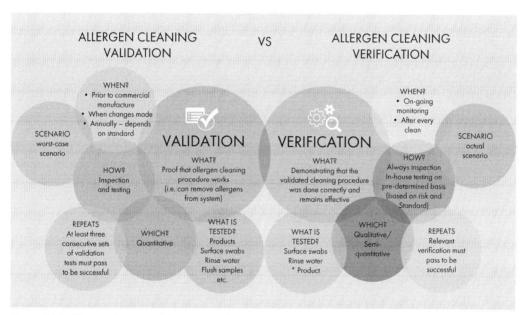

Figure 14.4. Allergen cleaning validation vs verification.

14.6.3 Physical audits

Physical audits or sometimes called visual inspection, is a rapid and simple way to get an indication of whether food-contact surfaces, walls, floors, equipment and employees (hands and coats) are a potential source of allergen cross-contact. It is particularly useful for identifying surfaces that have been ineffectively cleaned, and for targeting areas for sampling for analytical testing.

Areas in the food plant that should be assessed visually include:
- ► Surfaces and equipment, for the presence of settled allergenic dust originating from flours or pre-mixes.
- ► Surfaces and equipment, for allergenic food particles such as prawns, sesame seeds and nuts.
- ► Surfaces and equipment, for films that dry out on surfaces after manual cleaning, indicating inadequate wiping/scrubbing.
- ► Entrapment or 'notoriously' difficult-to-clean areas that may accumulate allergenic soils, e.g. corners in trays and bins, coarse or rough welds, or joints and O-rings.
- ► Conveyor belts, for tears, scratches or defects that could harbour food residue.
- ► Structural areas and fixtures above processing lines, e.g. extractor fans, for material build-up that could fall into the product.
- ► Floors and walls of the plant, for the presence of allergenic residue.

While visual inspection is an essential first-step approach for establishing the effectiveness of a cleaning procedure, it is unlikely that it can be relied upon as the only way to detect the presence of allergen residues that could cross-contaminate the final product. A white layer of milk residue may be easy to see on a stainless-steel surface, but much more difficult to observe on a white plastic surface. The ability to see egg residue on equipment will depend on how the light is reflected off the colour of the surface on which the film is viewed.

The presence of visible residues increases the likelihood that allergenic proteins are present, and tends to suggest a failure in cleaning. Ideally, even in cases where surfaces appear to be visibly clean, analytical methods may need to be used to confirm the absence of allergenic food residues.

14.6.4 Validation strategy

Before completing a validation study, it is highly recommended to compile a validation plan or strategy defining the objective, method, target allergens, sampling plan, testing methods and desired outcome (Coutts and Fielder, 2009).

Validation studies may be based on worst-case scenarios. The rationale is that if you prove that the cleaning method can remove allergenic residues in a worst-case scenario, it will be able to remove allergen residues in all other scenarios. Worst-case scenarios may include, but not be limited to:

▶ Allergenic food soil that is hardest to remove e.g. heated films.
▶ Cleaning after producing product/s that contain the highest number of allergens.
▶ Cleaning after producing product/s containing allergens in the highest concentration.

A validation study cannot be done without completing a visual inspection of equipment and processing lines both pre-and post-cleaning. Ideally, this step should include dismantling equipment. If visual food soils remain on the line after cleaning, the probability that the study will fail is high. In the course of a visual inspection, it is recommended that key inspection or examination points are identified that must be checked and/or swabbed during verification.

Ideally, the validation study should include the validation of the efficacy of the intended verification method. For example, if milk lateral flow devices are going to be used to verify cleaning the validation study will include a section where lateral flow device results are compared with Enzyme-Linked Immunosorbent Assay (ELISA) or liquid-chromatography mass-spectrometry (LC-MS/MS).

Cleaning should be re-validated when there is a significant change in the process, product, structure of the facility, or system; or when new allergens are introduced. However, food facilities typically revalidate once annually.

Verification is completed in actual scenarios, typically after allergen cleaning. It should always include a visual inspection and is best supplemented with testing, e.g. testing rinse water or swabbing identified sampling points. When doing rapid tests, positive controls must be included, at least periodically, and results must be recorded in a preformulated record.

sampling strategy An appropriate sampling strategy is of paramount importance for obtaining a clear depiction of the real state of the situation – i.e. whether the cleaning that has been done is effective. Experience has shown that errors in the results obtained using analytical methods are often related to the way the original sample was taken. As the term 'sample' suggests, it can never reflect all the surfaces that the food encounters during production. Given this restriction, a sampling strategy must be developed that best reflects the effectiveness of the clean.

Process samples taken from equipment to measure cleanliness may include, but are not limited to:

- rinse water samples;
- flush or purge samples;
- direct surface samples (e.g. swabbing);
- final or intermediate product samples;
- settling or air plates; and
- a synthetic process sample.

The collection of a rinse water sample is not always simple or reproducible. It may involve sampling from a fast-moving stream or hot water. Rinse water samples are often taken in closed plant cleaning systems, involving the collection of the last rinse water from a Cleaning in Place (CIP) system (Wilson, 2006). It is also important to note that any fault in the CIP (e.g. failure of a section of pipe to be routed during the clean) may give a clean rinse water sample (because the last rinse would not have flowed through the unrouted section) but will lead to an unsuccessful clean. Such issues must be considered during the interpretation of test results. Once rinse water samples have been collected they should be chilled immediately, as allergens may be unstable in water. Potentially, the presence of carryover cleaning fluids in rinse or wash waters could also affect the outcome of test results, limiting the validity of such samples. Alternative sampling points may be required to confirm the effectiveness of cleaning.

In dry manufacturing environments, flushing or purging a system with dry, inert material (e.g. flour, diatomaceous earth, salt or starch) may be employed as a means of cleaning. Monitoring of the flushing materials after known volumes have passed through the system can indicate the effectiveness of allergen removal. Measuring the decline in allergen levels by sampling, as flushing progresses, may assist the food manufacturer in estimating the extent of flushing required to achieve acceptable allergen levels.

A direct method for sampling the cleanliness of cleaned equipment is sometimes recommended, based on the argument that it is the residues on the cleaned equipment that will affect the products, not what is in the rinse water. Direct surface sampling should target areas where a significant build-up of food residue is expected to occur, and areas that are particularly difficult to clean, e.g. seams, valves, O-ring seals, sampling ports, and porous and irregularly-shaped surfaces. Care should be taken to sample equipment where heating of products occurs, as burned-on allergenic foods can be difficult to clean. Sampling may necessitate some dismantling of equipment for access. Direct surface sampling can also be used to measure the potential for aerial allergens to cross-contaminate surfaces, provided that sufficient time is allowed for the settlement of particles. Alternatively, allergen settling plates at 'worst case' points, points where settling is likely to occur, can be sampled over a defined period. The hands and protective clothing of employees may also be directly sampled as potential sources of allergen contamination.

While the presence of allergenic residues in the rinse water and swabs indicates that the allergen cleaning protocol or its execution requires revision, it does not necessarily indicate the presence of the allergenic protein in the finished product.

The transfer of allergenic protein from equipment surfaces to foods is a complex process, depending on factors such as:

▶ the adhesion properties of the protein on the surface;
▶ the composition of the food contact surface;
▶ the concentration of the allergenic protein;
▶ the properties of the allergenic protein (e.g. physical form and solubility in the food processed subsequently);
▶ processing temperature; and
▶ the abrasiveness of the food processed subsequently.

The first products taken from a processing line (or part of the processing line) after cleaning are a crucial sample for most cleaning validation studies. Any residues from the previous product remaining after cleaning are expected to be concentrated in the first product to exit the line following cleaning (Coutts and Fielder, 2009). In the case of continuous lines, it is recommended that at least the first three units that are manufactured following cleaning should be sampled for testing. Samples taken from the middle and end of the run will indicate if there is a further release of residues. The final product represents the recovery of residual allergen and its dilution in the next product. In general, the presence of allergenic residues in a finished product indicates a failure in the design or execution of the allergen-cleaning protocol. In such a case, it may be necessary to amend and revalidate the cleaning method.

Each of the sampling methods discussed will provide some indication that the cleaning procedure is reducing the levels of residual allergens in the processing system. Cleaning validation relies on a comparison of the results from different sampling methods to accurately show a decrease in the level of allergen residues.

14.7 Allergen testing methods

There are various reasons why one would need to test samples for the presence of allergens. One of the reasons is to prove that a 'free-from' claim can be made, for example, 'gluten-free' (FDA, 2006). Other reasons include checking for accidental contamination, and as a part of a due diligence testing schedule (Crevel, 2006). However, the most frequent reason for testing is for the validation and verification of the efficacy of allergen cleaning.

It is important to know what the objective of testing is before embarking on the process of analysis. Here are just some examples of relevant questions to ask yourself before testing or submitting samples for analysis:

- ► What information do I need/What do I want to know?
- ► Do I need qualitative or quantitative results?
- ► Which method best suits my needs?
- ► What does the testing method target or test for?
- ► What will the test achieve?
- ► What is the desired outcome of the testing?
- ► How will testing assist in maintaining a rigid food safety system?
- ► Does the test need to be accredited?
- ► Does it make sense to test?
- ► Will testing provide the answer to my question?

Laboratory-based testing is generally used for validation studies. These types of tests take longer to complete, require more technical skills, are expensive, and require specialised resources.

Broadly speaking there are two types of allergen detection methods used: methods that target the protein portion of allergenic foods, referred to as direct methods; and methods that target a compound or molecule associated with an allergenic food, referred to as indirect methods.

14.7.1 Enzyme-Linked Immunosorbent Assay (ELISA)

ELISA is an example of a direct detection method. Ideally, analytical methods used to detect and measure allergenic residues after cleaning should be specific to a particular allergen.

ELISA offers an accurate, specific, sensitive and relatively rapid means of detecting and quantifying allergens in finished products, rinse waters and on swabs taken from contact surfaces.

While ELISA kits offer various advantages for allergen detection, they also have certain limitations. These limitations include method cross-reactivity, as well as the interference food matrices, have on the results of testing (EFSA, 2014).

Since detection by ELISA is achieved through the binding of target proteins and antibodies, any changes in the binding properties (immunoreactivity) of the target proteins will influence test results (Coutts and Fielder, 2009). Thermal processing, hydrolysis, and exposure to oxidising chemicals (hypochlorite) can affect the solubility and immunoreactivity of proteins.

As a result, ELISA may be unable to detect allergenic proteins that have undergone thermal or hydrolysis treatments, or that have been exposed to cleaning or sanitising agents. The composition of the food matrix (oil versus water-based) can also affect protein solubility and therefore recovery during extraction, and consequently the detection of the target proteins by ELISA. These can be considerable limitations when using specific allergen testing methods to validate cleaning.

To obtain accurate results with ELISA methods, the analyst must understand the limitations of these tests and apply them to the samples being tested, e.g. if a protein is denatured, the assay may not work. Additionally, an understanding of each specific allergen ELISA is required, including factors such as:

- ► What the antibodies were raised against.
- ► What the kit was standardised against.
- ► Cross-reactivity of antibodies.
- ► Standards and controls used.

14.7.2 Targeted proteomics via liquid-chromatography mass-spectrometry (LC-MS/MS)

Instead of testing for allergenic proteins present in a food matrix, LC-MS/MS directly targets the peptide markers of known allergenic proteins (FoodDrinkEurope, 2013). There are several advantages to the use of LC-MS/MS as opposed to other methods, so testing laboratories are increasingly moving towards its implementation. It can detect multiple allergens in a food product through one single test. This means that for many food products, LC-MS/MS can screen for the presence of and quantify a multitude of allergens at one time. LC-MS/MS is more specific and sensitive than ELISA, and is not affected by cross-reactivity or the challenges brought on by changes in protein structure due to processing.

14.7.3 Polymerase chain reaction (PCR)

Unlike LC-MS/MS and ELISA, which targets specific proteins, the polymerase chain reaction (PCR) targets specific DNA molecules of an allergenic food (EFSA, 2014). The detection of this DNA is considered to be indicative of the presence of the allergen. In some cases, where reliable ELISA methods have not been developed for certain allergens, PCR is probably the best alternative for the detection of allergenic residues (Coutts and Fielder, 2009; EFSA, 2014). Results are typically qualitative and not quantitative.

The availability of such methods also provides a valuable tool for the validation of allergen-specific assays. PCR protocols have been published that allow the detection of many of the major allergenic foods. DNA associated with allergenic protein is specific and sensitive and is not compromised by cross-reactivity.

14.7.4 Rapid onsite testing kits

Rapid onsite test methods are normally recommended for onsite verification (Salter *et al.*, 2005) because they are cost-effective, little technical skill is needed for their operation, and they are mobile. There are a few different kinds of rapid kit that have gained popularity in recent years; they differ according to the desired outcome of the result, the number of allergens that are manufactured, and the financial capability of the facility. Three systems, in particular, are well represented in the food industry. This is summarised in Figure 14.5.

Allergen-specific lateral-flow devices (LFDs) are qualitative or semi-quantitative immunochromatographic tests, based on ELISA (Baumert and Tran, 2015). They are easy to use, compared to their full ELISA counterparts; and they generate results in very little time, without expensive equipment (EFSA, 2014). A set of coloured lines is produced, indicating whether the target allergen was present in the sample or not. The intensity of the line in the test zone is directly proportional to the concentration of the allergen. Semi-quantitative results can be produced by exposing the device to varying amounts of the sample extract, or by using a colour-intensity reader.

Total protein indicators detects protein on surfaces. The rationale is that since most allergens are proteins, a sample found to be protein-free could also be expected to be allergen-free. The system is easy to use, sensitive, and often the most cost-effective option (Stephan *et al.*, 2004), but it is not suitable for all

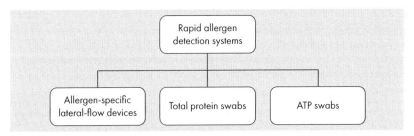

Figure 14.5. Available rapid onsite testing kit systems.

environments – for example, it is unsuitable for the meat and poultry industry. It should be noted though that if a positive test for protein is obtained, specific allergen testing might be required to determine whether the measured proteins are indeed allergens. For validation and verification of cleaning procedures, the sensitivity of general protein tests must be comparable to that of specific allergen tests.

Adenosine triphosphate (ATP) systems do not measure food-allergenic residues directly, but detect the presence of ATP on surfaces and in the rinse water (Coutts and Fielder, 2009; Salter *et al.*, 2005). It is generally used for hygiene monitoring (Al-Taher *et al.*, 2007). Because it detects ATP in microorganisms and food soils, it is not specific to allergens; and the correlation between allergenic residue and ATP results is not well studied.

14.8 Precautionary allergen labelling (PAL)

PAL, in most countries, is applied voluntarily. It serves as a warning to consumers that an allergen, although not part of the composition of the product, may be present due to cross-contact (Gendel, 2013).

Studies show that a high frequency of precautionary labelling may result in an increased likelihood that allergic consumers will ignore these statements. Therefore, to protect consumers, PAL should only be used when due diligence concludes that sporadic cross-contamination of a product cannot be avoided.

This should be based on a thorough process assessment and implementation of an effective ACP. PAL should not be used as a substitute for good manufacturing practices and an allergen control plan (FoodDrinkEurope, 2013; Taylor and Hefle, 2005).

PAL provides little help to food allergic consumers if it is used promiscuously. It can limit allergic consumers' diets and nutritional intake and can lead to increased risk-taking by allergic consumers.

An allergen risk assessment tool, such as VITAL®,[51] developed by the Australia/ New Zealand Allergen Bureau, is a valuable tool in this regard. VITAL® (Voluntary Incidental Trace Allergen Labelling) is a scientifically based tool centred on the most current clinical oral challenge threshold data. This data is reviewed by a scientific expert panel and is based on setting threshold levels that are expected not to elicit a reaction in 99% of allergic consumers.

---

[51] Allergen Bureau, http://allergenbureau.net/ - 03.06.2021

14.9 Looking ahead

Various factors may have a significant impact on the future of allergen control in the food industry. A range of novel protein sources, e.g. insects, seaweed, whose allergenicity has not been well studied and that may prove to be significantly allergenic, are expected to become mainstream. Novel additives and ingredients derived from allergenic sources may also be a consideration.

With the food industry focus on sustainable resources, packaging material made from or containing components derived from allergenic sources may become standard in future. Climate change has been shown to influence plants, and a range of species extends into novel territories, associated cross-reactivity between these plants and food proteins may result in an increased prevalence of specific food allergies. Ragweed is of particular concern and has cross-reactivity with melon and banana.

Ongoing research into reference doses and threshold values will have relevance as well as a potential impact on PAL. International consensus on various allergen control strategies may result in a more uniform approach to allergen management and declaration.

Other noteworthy trends are:
- ► Quantitative risk assessment, using deterministic and probabilistic approaches, to guide PAL and allergen management (Remington, 2021);
- ► Crop engineering to reduce its allergenicity;
- ► Mandatory use of precautionary allergen labelling;
- ► Harmonisation of allergen labelling standards;
- ► Harmonisation of allergen management principles and practices;
- ► Novel food allergens.

Acknowledgements

The content of this chapter has been formulated over 15 years and combines internal FACTS documents and external sources. Over time, the FACTS organisation has adopted this information into its organisational knowledge base, and the origins of source materials may have been lost or diluted. The information presented in the chapter is based in part on practical experience gained at FACTS, but is also firmly established in the information shared by the authors who contributed to Coutts, J. and Fielder R. (eds), 2009, Management of Food Allergens (Wiley-Blackwell, Hoboken, NJ, USA), together with all other works referenced in the chapter, whether credited or uncredited. For the reasons given above, there may be excerpts from other works that have not been credited, as we were not able to establish the sources accurately.

References

Al-Taher, F., Jackson, L. and Salter, R.S., 2007. Detection of dried egg residues on a stainless steel surface using ELISA and sensitive ATP assays. In: Annual Meeting of the Institute of Food Technologists. Chicago, IL, USA, Abstract 193-12.

Anagnostou, K., Stiefel, G., Brough, H., du Toit, G., Lack, L. and Fox, A.T., 2015. Active management of food allergy: an emerging concept. Archives of Disease in Childhood 100: 386-390. https://doi.org/10.1136/archdischild-2014-306278

Bagshaw, S. 2009. Choice for cleaning and cross-contact. In: Coutts, J. and Fielder R. (eds) Management of food allergens. Wiley-Blackwell, Hoboken, NJ, USA, p. 114-137. https://doi.org/10.1002/mnfr.201090008

Ballmer-Weber, B.K., Fernandez-Rivas, M., Beyer, K., Defernez, M., Sperrin, M., Mackie, A.R., Salt, L.J., Hourihane, J.O'B., Asero, R., Belohlavkova, S., Kowalski, M., de Blay, F., Papadopoulos, N.G., Clausen, M., Knulst, A.C., Roberts, G., Popov, T., Sprikkelman, A.B., Dubakiene, R., Vieths, S., Van Ree, R., Crevel, R. and Mills, C., 2015. How much is too much? Threshold dose distributions for 5 food allergens. Journal of Allergy and Clinical Immunology 135: 964-971. https://doi.org/10.1016/j.jaci.2014.10.047

Baumert, J.L. and Tran, D.H., 2015. Lateral flow devices for detecting allergens in food. In: Flanagan, S. (ed.). Handbook of food allergen detection and control. Woodlands Publishing series in food science, technology and nutrition, volume 264. Woodlands Publishing, Cambridge, UK, pp. 219-228. https://doi.org/10.1016/c2013-0-16428-8

British Retail Consortium (BRC), 2014. BRC best practice guideline: allergen management in food manufacturing sites. Version 1. British Retail Consortium, London, UK.

Cabanillas, B., Jappe, U. and Novak, N., 2018. Allergy to peanut, soybean, and other legumes: recent advances in allergen characterization, stability to processing and IgE cross-reactivity. Molecular Nutrition and Food Research 62: 1700446. https://doi.org/10.1002/mnfr.201700446

Caio, G., Volta, U., Sapone, A., Leffler, D.A., De Giorgio, R., Catassi, C. and Fasano, A., 2019. Celiac disease: a comprehensive current review. BMC Medicine 17(1): 142. https://doi.org/10.1186/s12916-019-1380-z

Che, C.T., Douglas, L. and Liem, J., 2017. Case reports of peanut-fenugreek and cashew-sumac cross-reactivity. Journal of Allergy and Clinical Immunology: in Practice 5: 510-511. https://doi.org/10.1016/j.jaip.2016.12.024

Codex Alimentarius, 1985. Codex Alimentarius General standard for the labelling of prepackaged food (CODEX STAN 1-1985). Available at http://www.fao.org/fao-who-codexalimentarius/sh-proxy/en/?lnk=1&url=https%253A%252F%252Fworkspace.fao.org%252Fsites%252Fcodex%252FStandards%252FCXS%2B1-1985%252FCXS_001e.pdf

Codex Alimentarius, 2020. Codex Alimentarius Code of practise on food allergen management for food business operators (CXC 80-2020). Available at http://www.fao.org/fao-who-codexalimentarius/sh-proxy/en/?lnk=1&url=https%253A%252F%252Fworkspace.fao.org%252Fsites%252Fcodex%252FStandards%252FCXC%2B80-2020%252FCXC_080e.pdf

Coutts, J. and Fielder R., 2009. Management of food allergens. Wiley-Blackwell, Hoboken, NJ, USA. https://doi.org/10.1002/mnfr.201090008

Cox, A.L., Eigenmann, P.A. and Sicherer, S.H., 2021. Clinical relevance of cross-reactivity in food allergy. Journal of Allergy and Clinical Immunology: in Practice 9(1): 82-99. https://doi.org/10.1016/j.jaip.2020.09.030

Crevel, R., 2006. Common issues in detecting allergenic residues on equipment and in processed food. In: Koppelman, S.J. and Hefle, S.L. (eds). Detecting allergens in food. Woodhead Publishing, Cambridge, UK, pp. 315-329. https://doi.org/10.1533/9781845690557.4.315

De Gier, S. and Verhoeckx, K., 2018. Insect (food) allergy and allergens. Molecular Immunology 100: 82-106. https://doi.org/10.1016/j.molimm.2018.03.015

Devdas, J.M., Mckie, C., Fox, A.T. and Ratageri, V.H., 2018. Food allergy in children: an overview. Indian Journal of Pediatrics 85: 369-374. https://doi.org/10.1007/s12098-017-2535-6

European Food Safety Authority (EFSA), 2014. Scientific opinion on the evaluation of allergenic foods and food ingredients for labelling purposes. EFSA Journal 12(11): 3894. https://doi.org/10.2903/j.efsa.2014.3894

Faeste, C.K., Christians, U., Egaas, E. and Jonscher, K.R., 2010. Characterization of potential allergens in fenugreek (*Trigonella foenum-graecum*) using patient sera and MS-based proteomic analysis. Journal of Proteomics 73: 1321-1333. https://doi.org/10.1016/j.jprot.2010.02.011

Ferris, I.M., 2022. Hazard analysis and critical control points (HACCP). In: Wernaart, B.F.W. and Van der Meulen, B.M.J. (eds) Applied Food Science. Wageningen Academic Publishers, Wageningen, the Netherlands, pp. 187-213.

Food Drink Europe, 2013. Guidance on food allergen management for food manufacturers. Food Drink Europe, Brussels, Belgium.

Food Standards Agency, 2006. Guidance on allergen management and consumer information. Food Standards Agency publications, London, UK.

Food Standards Australia New Zealand, n.d. Undeclared allergen food recall statistics (1 January 2016 – 31 December 2020). Available at: https://www.foodstandards.gov.au/industry/foodrecalls/recallstats/Pages/allergen-stats.aspx

Gadermaier, G., Hauser, M., Egger, M., Ferrara, R., Briza, P., Santos, K.S., Zennaro, D., Girbl, T., Zuidmeer-Jongejan, L., Mari, A. and Ferreira, F., 2011. Sensitization prevalence, antibody cross-reactivity and immunogenic peptide profile of Api g 2, the non-specific lipid transfer protein 1 of celery. PLoS ONE 6: e24150. https://doi.org/10.1371/journal.pone.0024150

Gendel, S.M., 2012. Comparison of international food allergen labelling regulations. Regulatory Toxicology and Pharmacology 63(2): 279-285. https://doi.org/10.1016/j.yrtph.2012.04.007

Gendel, S.M., 2013. The regulatory challenge of food allergens. Journal of Agricultural and Food Chemistry 61(24): 5634-5637. https://doi.org/10.1021/jf302539a

Gupta, R.S., Springston, E.E., Warrier, M.R., Smith, B., Kumar, R., Pongracic, J. and Holl, J.L., 2011. The prevalence, severity, and distribution of childhood food allergy in the United States. Pediatrics 128: 9-17. https://doi.org/10.1542/peds.2011-0204

Hassan, A.K. and Venkatesh, Y.P., 2015. An overview of fruit allergy and the causative allergens. European Annals of Allergy and Clinical Immunology 47(6): 180-187. http://www.eurannallergyimm.com/cont/journals-articles/391/volume-overview-fruit-allergy-causative-allergens-1041allasp1.pdf

Holah, J.T. and Hall, K., 2004. Cleaning issues in dry production environment. R&D Report 192. Campden & Chorleywood Food Research Association, Gloucestershire, UK.

Institute of Agriculture and Natural Resources Food Allergy Research and Resource Program (FARRP), 2021. Thresholds for allergenic foods. Available at: https://farrp.unl.edu/thresholds-for-allergenic-foods

Itoh, H., Nagata, A., Toyomasu, T., Sakiyama, T., Nagai, T., Saeki, T. and Nakanishi, K., 1995. Adsorption of β-lactoglobulin onto the surface of stainless steel particles. Bioscience, Biotechnology, and Biochemistry 59: 1648-1651. https://doi.org/10.1271/bbb.59.1648

Iweala, O.I. and Burks, A.W., 2016. Food allergy: our evolving understanding of its pathogenesis, prevention, and treatment. Current Allergy and Asthma Reports 16: 37. https://doi.org/10.1007/s11882-016-0616-7

Jackson, L.S., Al-Taher, F.M., Moorman, M., de Vries, J.W., Tippett, R., Swanson, K.M.J., Fu, T-J., Salter, R., Dunaif, G., Estes, S., Albillos, S. and Gendel, S.M., 2008. Cleaning and other control and validation strategies to prevent allergen cross-contact in food-processing operations. Journal of Food Protection 71: 445-458. https://doi.org/10.4315/0362-028X-71.2.445

Jackson, L.S., Schlesser, L.J., Al-Taher, F.M., Gendel, S.M. and Moorman, M., 2005. Effects of cleaning on removal of milk protein from a stainless-steel surface. In: Book of Abstracts: 54I-1. Annual Meeting of the Institute of Food Technologists, Chicago, USA.

Jacobson, M.F. and DePorter, J., 2018. Self-reported adverse reactions associated with mycoprotein (Quorn-brand) containing foods. Annals of Allergy, Asthma and Immunology 120: 626-630. https://doi.org/10.1016/j.anai.2018.03.020

Jeurnink, T.J.M. and Brinkman, D.W., 1994. The cleaning of heat exchangers and evaporators after processing milk or whey. International Dairy Journal 4: 347. https://doi.org/10.1016/0958-6946(94)90031-0

Komitopoulou, E, 2014. SGS White paper: Allergen management as a part of a safe global food supply chain. Available at: https://www.sgs.com/en/white-paper-library/allergen-management-an-integral-part-of-a-safe-food-supply-chain

Leoni, C., Volpicella, M., Dileo, M., Gattulli, B.A.R. and Ceci, L.R., 2019. Chitinases as Food Allergens. Molecules 24(11): 2087. https://doi.org/10.3390/molecules24112087

McKenna, O.E., Asam, C., Araujo, G.R., Roulias, A., Goulart, L.R. and Ferreira, F., 2016. How relevant is panallergen sensitization in the development of allergies? Pediatric Allergy and Immunology 27(6): 560-568. https://doi.org/10.1111/pai.12589

fMonaci, L., Pilolli, R., De Angelis, E., Crespo, J.F., Novak, N. and Cabanillas, B., 2020. Food allergens: classification, molecular properties, characterization, and detection in food sources. Advances in Food and Nutrition Research 93: 113-146. https://doi.org/10.1016/bs.afnr.2020.03.001

Namork, E., Fæste, C.K., Stensby, B.A., Egaas, E. and Løvik, M., 2011. Severe allergic reactions to food in Norway: a ten year survey of cases reported to the food allergy register. International Journal of Environmental Research and Public Health 8: 3144-3155. https://doi.org/10.3390/ijerph8083144

Nwaru, B.I., Hickstein, L., Panesar, S.S., Roberts, G., Muraro, A. and Sheikh, A., 2014. Prevalence of common food allergies in Europe: a systematic review and meta-analysis. Allergy 69: 992-1007. https://doi.org/10.1111/all.12423

Prescott, S. and Allen, K.J., 2011. Food allergy: riding the second wave of allergy epidemic. Pediatric Allergy and Immunology 22: 155-160. https://doi.org/10.1111/j.1399-3038.2011.01145.x

Remington, B., 2021. Food allergy management symposium. ILSI Europe Expert Group. Webinar on Allergen Quantitative Risk Assessment Update.

Renz, H., Allen, K.J., Sicherer, S.H., Sampson, H.A., Lack, G., Beyer, K. and Oettgen, H.C., 2018. Food allergy. Nature Reviews Disease Primers 4: 17098. https://doi.org/10.1038/nrdp.2017.98

Rubin, J.E. and Crowe, S.E., 2020. Celiac disease. Annals of Internal Medicine 172(1): ITC1-ITC16. https://doi.org/10.7326/AITC202001070

Ruethers, T., Taki, A.C., Johnston, E.B., Nugraha, R., Le T.T.K., Kalic, T., McLean, T.R., Kamath, S.D. and Lopata, A.L., 2018. Seafood allergy: a comprehensive review of fish and shellfish allergens. Molecular Immunology 100: 28-57. https://doi.org/10.1016/j.molimm.2018.04.008

Salter, R.S., Hefle, S., Jackson, L.S. and Swanson, K.M.J., 2005. Use of a sensitive adenosine triphosphate method to quickly verify wet cleaning effectiveness at removing food soil and allergens from food contact surfaces. Abstract for the Annual Meeting of the International Association for Food Protection. International Association for Food Protection, Des Moines, IA, USA, pp. P2-06.

Schmidt, R.H., 2018. Basic elements of equipment cleaning and sanitizing in food processing and handling operations. University of Florida Cooperative Extension Service, Institute of Food and Agricultural Sciences, Gainesville, FL, USA. Available at: https://edis.ifas.ufl.edu/publication/FS077

Shek, L.P. and Lee, B.W., 2006. Food allergy in Asia. Current Opinion in Allergy and Clinical Immunology, 6:197–201. https://doi.org/10.1097/01.all.0000225160.52650.17

Sicherer, S.H. and Sampson, H.A., 2010. Food allergy. Journal of Allergy and Clinical Immunology 125: S116-S125. https://doi.org/10.1016/j.jaci.2009.08.028

Sicherer, S.H. and Sampson, H.A., 2014. Food allergy: epidemiology, pathogenesis, diagnosis and treatment. Journal of Allergy and Clinical Immunology 133: 291-307. https://doi.org/10.1016/j.jaci.2013.11.020

Skypala, I.J., 2019. Food-induced anaphylaxis: role of hidden allergens and cofactors. Frontiers in Immunology 10: 673. https://doi.org/10.3389/fimmu.2019.00673

Steinman, H.A., 1996. 'Hidden' allergens in foods. Journal of Allergy and Clinical Immunology 98(2): 241-250. https://doi.org/10.1016/S0091-6749(96)70146-X

Stephan, O., Weisz, N., Vieths, S., Weiser, T., Rabe, B. and Vatterott, W., 2004. Protein quantification, sandwich ELISA, and real-time PCR used to monitor industrial cleaning procedures for contamination with peanut and celery allergens. Journal of AOAC International 87: 1448-1457. https://doi.org/10.1093/jaoac/87.6.1448

Taylor, S.L. and Hefle, S.L., 2005. Allergen control. Food Technology 59: 40-43, 75. Available at: https://www.ift.org/news-and-publications/food-technology-magazine/issues/2005/february/features/allergen-control

Taylor, S.L. and Lehrer, S.B., 1996. Principles and characteristics of food allergens. Critical Reviews in Food Science and Nutrition 36: S91-S118. https://doi.org/10.1080/10408399609527761

Taylor, S.L., Baumert, J.L., Kruizinga, A.G., Remington, B.C., Crevel, R.W., Brooke-Taylor, S. and Allen, K.J., 2014. Establishment of reference doses for residues of allergenic foods: report of the VITAL expert panel. Food and Chemical Toxicology 63: 9-17. https://doi.org/10.1016/j.fct.2013.10.032

Taylor, S.L., Hefle, S.L., Farnum, K., Rizk, S.W., Yeung, J., Barnett, M.E., Busta, F., Shank, F.R., Newsome, R., Davis, S. and Bryant, C.M., 2006. Analysis and evaluation of food manufacturing practices used to address allergen concerns. Comprehensive Reviews in Food Science and Food Safety 5: 138-157. https://doi.org/10.1111/j.1541-4337.2006.00012.x

The Threshold Working Group of the Food and Drug Administration, 2006. Approaches to establish thresholds for major food allergens and for gluten in food. FDA, Silver Spring, MD, USA. Available at: https://www.fda.gov/food/food-labeling-nutrition/approaches-establish-thresholds-major-food-allergens-and-gluten-food

Wainstein, B.K., Studdert, J., Ziegler, M. and Ziegler, J.B., 2010. Prediction of anaphylaxis during peanut food challenge: usefulness of the peanut skin prick test (SPT) and specific IgE level. Pediatric Allergy and Immunology 21: 603-611. https://doi.org/10.1111/j.1399-3038.2010.01063.x

Wilson, D.I., 2005. Challenges in cleaning: recent developments future prospects. Heat Transfer Engineering 26: 51-59. https://doi.org/10.1080/01457630590890175

Yu, W., Freeland, D.M.H. and Nadeau, K.C., 2016. Food allergy: immune mechanisms, diagnosis and immunotherapy. Nature Reviews Immunology 16: 751-765. https://doi.org/10.1038/nri.2016.111

15. Radiation and radioactivity in the food sector

This is what you need to know about radiation and radioactivity in the food sector

Anna Aladjadjiyan[†]

National Biomass Association (BGBIOM), Antim Parvi 22, 4000 Plovdiv, Bulgaria

Abstract

Discharging large amounts of radioisotopes into the environment affects food by either falling onto its surface (like fruits and vegetables) or into animal feed as deposits from the air or as contaminated rainwater/snow. Radioactivity in water can also accumulate in rivers and seas, depositing in fish and seafood. Radioactive material from the environment (whether natural or artificial) can be incorporated into food as it is taken up by plants, seafood or ingested by animals. The aim of this chapter is to offer the reader an introduction to the most important applications of radiation in food technologies. To this end, we will explore contemporary issues in food radioactivity, discuss the basics of its use, introduce positive and negative outcomes, and conclude with some expectations for the near future. In this chapter, pathways for radioactive food contamination are presented. The control of radioactivity is related to its measurement, therefore units and methods for measuring radioactivity are presented. Then, naturally radioactive foods and the reason for their radioactivity are discussed. Developed natural ways to keep homeostasis are mentioned. Harmful radioisotopes are listed, and the application of food irradiation in food processing is considered as well. Finally, regulatory documents for the application of food irradiation are mentioned.

[†] Sadly, Prof. Aladjadjiyan passed away prior to the publication of this work, shortly after submitting this chapter.

Bart Wernaart and Bernd van der Meulen (eds)
Applied food science
DOI: 10.3920/978-90-8686-933-6_15, © Anna Aladjadjiyan 2022

Key concepts

- ▶ Radiation is the emission or transmission of energy.
- ▶ Radioactivity is the property of some unstable atoms due to the large number of nuclei particles.
- ▶ Radioactive disasters are sources of radioactive contamination.
- ▶ Radioactive contamination means the unintended or undesirable deposition of, or presence of radioactive substances on surfaces or within solids, liquids or gases.
- ▶ Radioactivity can be explained through atomic structure, the important feature of which is the period of half-life.

15.1 Radioactivity and food

Radioactivity is a phenomenon that naturally exists in the environment around us. We are immersed in radioactivity daily.

Our modern technologies for satisfying energy needs for humanity include the development of nuclear plants. However, nuclear plants are sources of radioactivity. Atomic bombs, used at the end of the World War II, as well as nuclear weapons tests nowadays have also contributed to radioactive contamination in the environment.

Further development of both military and peaceful use of nuclear energy hides the risks of rising radioactive contamination in nature which will inevitably affect food.

Case 15.1. Fukushima, bananas and irradiation for food control.

In 2011 there was a strong earthquake in Japan. It was level 9 on the Richter scale and caused an accident at Fukushima's nuclear power plant. The earthquake generated a tsunami 14 metres high that arrived shortly afterwards and flooded the lower parts of nuclear reactors 1-4. This disaster led to radioactive contamination.

Many people (154,000) were evacuated from the affected area. In the following days, radiation released to the atmosphere forced the government to declare an ever-larger evacuation zone around the nuclear plant, culminating in an evacuation zone with a 20 km radius as communicated by Fackler and Wald (1 May 2011).

The Fukushima radioactive disaster was not a unique event. Thirty-five years earlier, the accident in Chernobyl in 1986 was even more severe. The nuclear disaster at the Chernobyl nuclear power plant near the city of Pripyat in northern Ukraine is often described as the worst nuclear accident in history. However, rarely is this sensational depiction clarified in more detail (Klug, 2021).

The accident in Fukushima was classified as Level 7 on The International Nuclear and Radiological Event Scale (INES), after initially being classified as Level 5 (Figure 15.1). It has joined Chernobyl as the only other accident to receive this high a classification, as informed by DiSavino (2011).

As you see in Figure 15.1, Level 7 denotes a 'major accident', which means 'a major release of radioactive material with widespread health and environmental effects requiring implementation of planned and extended countermeasures'. Both the Chernobyl 1986 and Fukushima 2011 (Case 15.1) disasters have been categorised as such. But INES is intended for use in non-military applications and only relates to the safety aspects of an event.

If the term 'nuclear disaster' is not only used to describe events, or accidents, in nuclear reactors, but also radioactive emissions caused by humans, then there are many occasions where human-caused nuclear contamination has been greater than that of the Chernobyl disaster, explained Kate Brown, 2021, in an interview with Deutsche Welle (Klug, 2021).

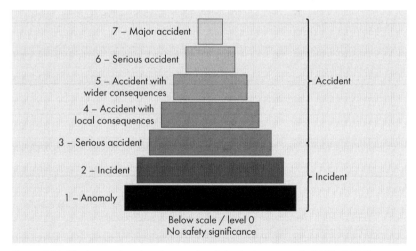

Figure 15.1. The international nuclear and radiological event scale (IAEA, 2008).

foods with natural radioactivity

We regularly come into contact with radiation when we eat. All foodstuffs are slightly radioactive due to their content of carbon-14 and potassium-40 – radioisotopes that are present wherever regular carbon and potassium can be found.

Some foods like Brazil nuts, bananas, potatoes, etc. are naturally radioactive. This is because they contain some radioactive elements such as potassium and carbon. This natural radioactivity is not dangerous, because the content of radioactive elements in natural food is low.

Bananas are rich in potassium (chemical symbol K), and a very small fraction of that naturally occurring potassium is in fact radioactive, about one-hundredth of one per cent (actually 120 parts per million). The radioactive variety of potassium is potassium-40 (K-40, 0.012% of total potassium).

Here's the general information about K-40 and bananas (Edwards, 2021):

> The human body maintains relatively tight homeostatic control over potassium levels. This means that the consumption of foods containing large amounts of potassium will not increase the body's potassium content. As such, eating foods like bananas does not increase your annual radiation dose. If someone ingested potassium that had been enriched in K-40, that would be another story.

15.2 Radiation and radioactivity – basics

As already mentioned, we are immersed in an ocean of radiation throughout our lives. It is an essential feature of nature. To understand the mechanism of radiation and radioactivity, we need to consider the structure of matter. Democritus from ancient Ellada suggested that matter is composed of atoms. Now we know that most often atoms compose molecules and molecules build substances.

We already know that at the heart of an atom there is the nucleus, which is a hundred thousand times smaller than the atom itself. The nucleus itself is composed of particles – protons and neutrons, called nucleons. If Z is the number of protons, and N the number of neutrons, their sum is A ($A=N+Z$). Protons and neutrons have approximately the same mass, but the electric charge of protons is positive while the neutrons are electroneutral. As a result, Z represents the electric charge of the nucleus. This corresponds to the atomic number of the element. Z varies from 1 to 92. The number represented by A corresponds to the mass of the nucleus. The heaviest natural element is Uranium-238, which contains 92 protons for 146 neutrons. The nucleus, therefore, has a combined mass of 238 nucleons, and a charge of +92. Elements with heavier mass (A>238) are artificially synthesised. Negatively charged electrons are distributed around the nucleus. The negative charge of electrons is exactly equal to the positive charge of protons. For a neutral atom (non-ionised) the number of electrons must be equal to the number of protons.

radiation The emission or transmission of energy is defined as radiation. Radiation may be observed in two forms: waves or particles. Both are spread through space or through a medium.

The emission of waves comprises electromagnetic radiation. Electromagnetic radiation includes radio waves, microwaves, infrared, visible light, ultraviolet, X-rays, and gamma radiation. The emission of particles includes alpha and beta particles. Sometimes gamma radiation is also called gamma particles.

radioactivity Radioactivity is a phenomenon which is produced deep within the atom, in its nucleus. This phenomenon is very difficult to observe. Certain unstable atom nuclei are the source of radiation, designated by the first three letters of the Greek alphabet: alpha (α), beta (β) and gamma (γ). This radiation is composed of energetic particles emitted by nuclei. The nature of α, β and γ radiation was revealed by the English physicist Ernest Rutherford at the beginning of the 20th century. He found that α-particles are helium (He) nuclei, β-particles are high-velocity electrons and γ-particles are electromagnetic radiation with wavelengths less than 10^{-11} m.

Radioactivity is the spontaneous act of emitting radiation (γ) or particles (α or β) from the nuclei. Emission is released by an atomic nucleus that, for some reason, is unstable. To shift to a more stable configuration the nucleus needs to give away some energy. In other words, radioactivity represents breaking-up (decay) or rearrangement of an atom's nucleus. Decay occurs naturally and spontaneously to unstable nuclei. A discrepancy between the number of protons and neutrons is usually the reason for this instability. A parent nucleus can release one α particle and become a product nucleus with mass A-4 and an electric charge of Z-2. This is α-decay. If a parent nucleus releases an electron, the product nucleus will have the same mass A, but an electric charge of Z+1. If a parent nucleus releases a positron (electron with a positive charge) the product nucleus will have the same mass A, but an electric charge of Z-1. Both these transformations are called β-decay.

radionuclide The nucleus having this property is called a radionuclide. An element can have some different isotopes with the same number of protons and electrons, but with a different number of neutrons. Some isotopes can be unstable. Radioactive isotopes of an element are called radioisotopes. For example, Carbon-12

(6 neutrons + 6 protons = 12) is stable and never undergoes radioactive decay, while Carbon-14 (6 neutrons + 8 protons= 14), is unstable and undergoes radioactive decay with a half-life of about 5,730 years, meaning that half of its material will be gone after 5,730 years.

Radioactivity, its character, and intensity are a matter of primary importance because of its impact on the life of our planet. Radiation can impact matter in three main ways:

1. Non-ionising radiation: this comprises the low-energy (long wavelength) parts of the electromagnetic spectrum. Here, all the light you can see, radio waves (also known as microwaves – as in the oven) and infrared waves (heat radiation) are included. Ultraviolet rays fall into the high energy end of this category.
2. Ionising radiation: this is the radiation that can remove an electron from its orbit and convert the neutral atom into an ion. Ionising radiation comprises the highest energy (shortest wavelength) part of the electromagnetic spectrum including ultraviolet, Roentgen (X) and gamma waves. Alfa and beta particles, as well as neutrons can also ionise molecules and atoms after collision.
3. Neutrons: are free neutron particles that can collide with other atoms and ionise them.

natural radioactivity

artificial
(anthropogenic)
radioactivity

Two types of radionuclides can be found in the environment – those of natural origin and those of anthropogenic origin. The sources of the natural radioactivity include atmospheric radioactivity, cosmic radiation and radioactive gases released from the ground surface. The artificial (anthropogenic) radioactivity is caused by radioactive environmental contaminants that are a consequence of the use of nuclear power either for peaceful or military purposes, nuclear electric stations and satellites. Accidents in nuclear power plants (the most well known are the Chernobyl and Fukushima disasters, but there have been others such as the Three Mile Island disaster in Pennsylvania USA in 1979) as well as in operating nuclear stations release many radioisotopes in spite of the high-level protection of these systems. These accidents and activities are the main source of artificial radioactivity (World Health Organization (WHO), 2011. Nuclear accidents and radioactive contamination of foods).

When large amounts of radioisotopes are discharged into the environment, they can affect foods. One pathway is by falling directly onto food or animal feed. After the Chernobyl disaster this phenomenon was registered in a number of neighbouring countries, including Bulgaria (Gorbanov, 2000), Cypress (Kritidis *et al.*, 2000), Egypt (El-Naggar, 2000), Germany (Miller, *et al.*, 2000), Greece (Vosniakos, *et al.*, 2000), Italy (Papucci, *et al.*, 2000), Israel (Ne'erman, *et al.*, 2000), Romania (Bologa *et al.*, 2000), Turkey (Akgoz *et al.*, 2000), Syria (Othman, 2000), and Ukraine (Egorov *et al.*, 2000), and led to harmful effects on human health. During the Chernobyl accident (in 1986), large areas of half-natural ecosystems were damaged by radioisotope deposition. Pastures, fruits and vegetable fields and fertile lands were polluted. Later, radioactive contamination spread to animals and directly or indirectly into food products. All this has led to long-term adverse health conditions (Vosniakos, 2012; Zehringer, 2016).

Radioisotopes can be directly inhaled from the atmosphere and ingested with contaminated fruits and vegetables. The other pathway of radioactive pollution is indirectly through contaminated rainwater or snow. Radioactivity in water can accumulate in rivers and seas, contaminating fish and seafood. Once in the environment, radioactive material can also become incorporated in food as it is taken up by plants, seafood or ingested by animals. Recently Japan has announced it will release more than 1 m tonnes of contaminated water from the wrecked Fukushima nuclear power plant into the sea, a decision that has angered neighbouring countries, including China, and local fishermen, as reported by McCurry (2021).

The pathways of food contamination with radioactive pollutants are presented in Figure 15.2. It shows that the radionuclides that are released in the atmosphere are deposited on plants, land and surface waters. From plants, radionuclides pass either to animals and further on to animal products – milk, butter and meat, or directly to humans. From land, radionuclides can pass to surface and underground water, further contaminating drinking water and fish, both finally affecting humans.

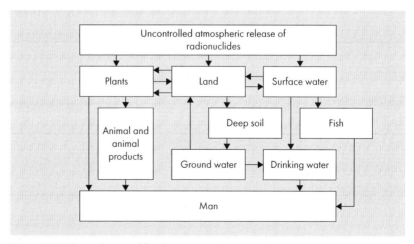

Figure 15.2. The pathways of food contamination.

half-life   An important property of radionuclides is their half-life. The half-life of radionuclides is the time interval necessary for the decay of half of their primary quantity. Half-lives of different nuclides can vary from parts of a second to millions of years. Radionuclides with a longer half-life are more dangerous because of their longer presence in the environment.

Many different kinds of radionuclides can be discharged following a major nuclear emergency. Some of them are very short lived and others do not readily transfer into food. Radionuclides generated in nuclear installations and that could be significant for the food chain include: radioactive hydrogen or tritium (H-3), carbon (C-14), technetium (Tc-99), sulphur (S-35), cobalt (Co-60) strontium (Sr-89 and Sr-90), ruthenium (Ru-103 and Ru-106), iodine (I-131 and I-129), uranium (U-235) plutonium (Pu-238, Pu-239 and Pu-240), caesium (Cs-134 and Cs-137), cerium (Ce-103), iridium (Ir-192), and americium (Am-241). Detailed information about the radioactivity in food is published by the World Health Organization (WHO) 2011 (Nuclear accidents and radioactive contamination of foods).

Of immediate concern is iodine-131 because it is distributed over a wide area, found in water and on crops and is rapidly transferred from contaminated feed into milk. However, iodine-131 has a relatively short half-life and will decay within a few weeks. In contrast, radioactive caesium which can also be detected early on, is longer lived (Cs-134 has a half-life of about 2 years and Cs-137 has a half-life of about 30 years) and can remain in the environment for a long time. Radioactive caesium is also rapidly transferred from feed to milk. Uptake of caesium into food is also of long-term concern. The influence of these radioactive elements on human health was investigated after the Chernobyl accident and it was discovered that iodine-131 is harmful to the endocrine system (causes thyroid cancer) and caesium affects bone strength (Vosniakos, 2012; Zehringer, 2016).

The most important radionuclides causing typical contamination of different kinds of food, according to Vosniakos (2001) are presented in Table 15.1.

Table 15.1. Most important radionuclides having biggest influence on food contamination (Vosniakos, 2001).

Air	^{131}I, ^{134}Cs, ^{137}Cs
Water	^{3}H, ^{89}Sr, ^{90}Sr, ^{131}I, ^{134}Cs, ^{137}Cs
Milk	^{89}Sr, ^{90}Sr, ^{131}I, ^{134}Cs, ^{137}Cs
Meat	^{134}Cs, ^{137}Cs
Other food products	^{89}Sr, ^{90}Sr, ^{134}Cs, ^{137}Cs
Vegetation	^{89}Sr, ^{90}Sr, ^{95}Zr, ^{95}Nb, ^{103}Ru, ^{106}Ru, ^{131}I, ^{34}Cs, ^{137}Cs, ^{141}Cs, ^{144}Cs
Soil	^{90}Sr, ^{134}Cs, ^{137}Cs, ^{238}Pu, $^{239+240}Pu$, ^{241}Am, ^{242}Cm

15.3 Units and methods for measurement of radioactivity

15.3.1 Units

To evaluate the impact of radioactivity on human health it is necessary to measure its quantity.

Radioactivity cannot be detected by any of our senses because it is invisible, has no smell, and makes no sound. However, because it affects the atoms that it passes through, we can easily monitor it using a variety of methods (for more details, see Aladjadjiyan, 2006).

Injury to the human organism by radioactive contamination is due to the absorption of high-energy irradiation by body organs and tissues. For quantitative assessment of the injury, it is necessary to measure the absorbed irradiation.

Physical measurements of irradiation include quantities of the 'absorbed dose' and the 'exposure'. Exposure refers to the radiation incident upon the object, while the absorbed dose refers to the actual interaction of that radiation within the object.

absorbed dose   The absorbed dose can be measured in terms of:
- the energy delivered per volume;
- the energy delivered per mass;
- ion-pairs created per volume; or
- ion-pairs created per mass.

The system (*SI* – Système International) unit of absorbed dose is Gray (*Gy*). The dose *D* is equal to 1 *Gy* when the total energy *E* (*in Joules J*) is absorbed by the object with mass *m* (*kg*).

Previously, *rad*, was used as a unit for the absorbed dose and can be translated into SI units by the formula:

1 *rad* = 10^{-2} *Gy*.

exposure (X)   For externally originating radiation, it is usual to measure the exposure (*X*), i.e. the radiation incident on the sample, in terms of the ionisation produced in air when that radiation passes through it. The exposure is measured by the total electric charge *Q* (measured in *C* – *Coulombs*) of ions produced by X- or gamma-radiation in a given volume of air, divided by its mass m.

Previously Röntgen (R) was used as a unit for X-radiation incident on the sample.

$$1\ R = 2.58 \times 10^{-4} C/kg.$$

15.3.2 Methods for measuring radioactivity

A detailed presentation of radioactivity detection methods and instruments is available in Henriksen (2015).

There are two types of instruments for measuring radioactivity: counting equipment (used to determine the quantity and the quality of radiation) and dosimeters (used to determine radiation dose). Both types of equipment require that 'the radiations result in observable changes in a compound' (whether gas, liquid or solid).

Measuring equipment consists of two parts that are usually connected. The first part consists of a *sensitive volume,* involving a compound that experiences changes when exposed to radiation. The other component is *a device that converts these changes into measurable signals.*

The qualities of radiation that we want to measure are:
- ▸ Activity: the activity or intensity of a source is measured in Becquerel. It can be given as the total activity in *Bq* or given as specific activity in *Bq/kg* for a solid object, *Bq/l* for liquid or *Bq/m3* for gas. When considering pollution of an area, Bq/m^2 is used. Some countries still use Ci/km^2 (Curie per square kilometre).
- ▸ Type of radiation: it is important to distinguish between α- or β-particles, X- or γ-rays and neutrons, because their impact is different as well as the measures of protection needed.
- ▸ The energy: it is usually measured in the unit electron volt (*eV*). The energy of the particles or photons is important for determining the damage.
- ▸ Dose: which is the absorbed radiation energy measured in Gray (*Gy*).

The correct measurement of the dose is important in radiobiological experiments and within radiation therapy.

There are a number of different types of radiation measuring instruments due to the different physical events that can be utilised to assess the impact. They are largely used in laboratories for food control as well as in hospitals using radioactive elements for investigating and treating patients.

film

1. Film: most people have seen an X-ray picture. The picture is the result of radiation hitting a photographic film. The more radiation exposure, the more the blackening of the film. In radiation diagnostics, film has been the detection method. Furthermore, film-badges have been used by people working with radiation in hospitals or in research, keeping track of how much radiation exposure the workers have received.

TLD

2. TLD: these initials stand for *Thermo Luminescence Dosimetry*. A crystal such as LiF (lithium fluoride) containing Mn (manganese) as an impurity is used for detecting radiation. The impurity forms traps in the crystalline lattice where, following irradiation, electrons are held. When the crystal is warmed, the trapped electrons are released, and light is emitted. The amount of light emitted is related to the dose of radiation received by the detector.

ionisation

3. Ionisation: radiation results in the formation of positive and negative ions in a gas as well as in all other materials. Ionisation can be used both for activity *(Bq)* measurements as well as for dose measurement. With knowledge of the energy needed to form a pair of ions, the dose can be obtained. The famous Geiger-Mueller tube, commonly called a Geiger counter, is designed to measure the electrical response produced by the newly formed ions.

scintillation

4. Scintillation: a number of compounds have the property of emitting light when exposed to radiation. The intensity of the emitted light depends on the radiation exposure and the light intensity can be easily measured.

semiconductors

5. Semiconductors: radiation produces an electric current in semiconductors that can be measured.

free radicals

6. Free radicals: radiation produces a class of chemical species known as free radicals. Free radicals, by definition, contain an unpaired electron and, although they are very reactive, they can be trapped in some solid materials. The number of trapped free radicals is a measure of the radiation dose.

redox products

7. Redox products: radiation either reduces (by electron addition) or oxidises (by electron abstraction) the absorbing molecules. Although these changes are initially in the form of unstable free radicals, chemical reactions occur which ultimately result in stable reduction and oxidation products.

15.3.3 Radioactivity and our bodies

Radioactivity can damage our body's DNA. Low doses of radiation can be repaired but higher doses can change our body's cells. When this happens, there may be an increased likelihood of developing cancer. That is why different instruments measuring radioactivity are a necessary part of medical and food control laboratories.

15.4 Natural radioactivity in food and drinking water

As we have already mentioned, radioactivity exists naturally in the atmosphere, soil, seas and rivers. Inevitably some of this gets into the food we eat. Natural radioactivity can be transferred into food in different ways, such as:

- ▶ into crops from radioactive rocks and minerals present in the soil;
- ▶ drinking water can pick up radioactivity from the earth;
- ▶ fish and shellfish can uptake radioactivity from the water or sea floor.

As we see, the pathways for distribution of natural radioactivity are very similar to those presented in Figure 15.2.

Vegetables and fruit contain different quantities of potassium. Grains, dairy products, lentils, meat and fish also have potassium.

Spring water contains radioactivity too. Granite rocks are lightly radioactive owing to the presence of thorium, uranium and several of their radioactive descendants. Spring water passing through these rocks dissolves radioactive elements and becomes radioactive.

During its evolution, the human body has developed mechanisms to adapt to the content of natural radioactive elements. This ability does not work for artificial radioactive elements like caesium, or for any of the radioactive pollutants given off by a nuclear power plant. Most of these materials do not exist in nature at all – and those that do exist in nature are not subject to the same homeostatic mechanism that the body uses to control potassium levels.

We can conclude that any foodstuffs or beverages containing radioactive caesium or other man-made radioactive pollutants will cause an additional annual dose of ionising radiation to the person exposed and hence will be harmful.

But, as explained before, that is not the case with bananas, or with other natural foods rich in potassium. Table 15.2 presents a partial list of such foods rich in potassium.

Artificial radioactivity can also get into food when radioactive materials are discharged into the environment after civil or military nuclear operations (see Figure 15.2). Artificial radioactivity then passes through the food chain in the same way as natural radioactivity.

Table 15.2. Naturally radioactive foods with potassium content in food.[1]

Food	Serving size	Potassium (mg)
Tomato paste	60 ml (¡ cup)	658
Potato, baked	1 (12 cm × 6 cm)	610
Pinto or kidney beans	175 ml (3/4 cup)	566 to 591
Lentils	175 ml (3/4 cup)	579
Avocado	1/2 whole	487
Squash, baked	125 ml (1/2 cup)	473
Banana	1 medium	422
Papaya	1/2 medium	392
Milk, 2%	250 ml (1 cup)	387
Chickpeas	175 ml (3/4 cup)	378
Yogurt	175 g (3/4 cup)	362
Fish	75 g (2 S oz)	313

[1] Source: https://www.unlockfood.ca/en/Articles/Vitamins-and-Minerals/What-You-Need-to-Know-About-Potassium.aspx#.U8pytxbtL_V.

15.4.1 International standards and guidance (*The Codex Guideline Levels*)

International standards relating to the control of radioactivity in food and drinking water have been developed by the Food and Agricultural Organization of the United Nations (FAO), The International Atomic Energy Agency (IAEA) and the World Health Organization (WHO).

The webpage of the IAEA published on November 28, 2017 contains the article 'Natural radioactivity in Food: Experts Discuss Harmonising International Standards' which provides information about harmonising international guidance-acceptable levels of natural radioactivity in food.

The World Health Organization (WHO) has developed a framework, including guidance levels, for the management of both naturally occurring and man-made radionuclides in drinking water (WHO, 2011; https://www.iaea.org/newscenter/news/natural-radioactivity-in-food-experts-discuss-harmonizing-international-standards). The Joint FAO/WHO Codex Alimentarius Commission (2006), also published guideline levels applicable to radionuclides contained in foods traded internationally for human consumption, which have been contaminated following a nuclear or radiological emergency.

The International Atomic Energy Agency (IAEA) (2017) also worked on the development of principles for harmonised guidance on natural radioactivity in food in non-emergency situations. The International Consultative Group on Food Irradiation (ICGFI) (1999) collected and published available information about consumer attitudes to irradiated food.

15.5 Application of food irradiation

The application of ionising radiation to food, or food irradiation for short, is a technology that improves the safety and extends the shelf life of foods by reducing or eliminating microorganisms and insects. Like pasteurising milk and canning fruits and vegetables, irradiation can make food safer for the consumer. In the USA the Food and Drug Administration (FDA) is responsible for regulating the sources of radiation that are used to irradiate food. The FDA approves a source of radiation for use on foods only after it has determined that it is safe to irradiate the food. Useful information is available by the US FDA (2018: 'Food irradiation: what you need to know').

> Irradiation does not make food radioactive, compromise nutritional quality, or noticeably change the taste, texture, or appearance of food. In fact, any changes made by irradiation are so minimal that it is not easy to tell if a food has been irradiated.

Food irradiation applications are discussed in the World Nuclear Association's (WNA), 'Radioisotopes in Food & Agriculture' (updated April 2021). It is affirmed that irradiation can serve many purposes, the most important of which are listed below (US FDA, 2018):

- ► 'Prevention of foodborne illness – to effectively eliminate organisms that cause foodborne illness, such as *Salmonella* and *Escherichia coli* (*E. coli*).
- ► Preservation – to destroy or inactivate organisms that cause spoilage and decomposition and extend the shelf life of foods.
- ► Control of insects – to destroy insects in or on tropical fruits imported into the United States. Irradiation also decreases the need for other pest-control practices that may harm the fruit.
- ► Delay of sprouting and ripening – to inhibit sprouting (e.g. potatoes) and delay ripening of fruit to increase longevity.
- ► Sterilization – irradiation can be used to sterilize foods, which can then be stored for years without refrigeration. Sterilized foods are useful in hospitals for patients with severely impaired immune systems, such as patients with AIDS or undergoing chemotherapy. Foods that are sterilized by irradiation are exposed to substantially higher levels of treatment than those approved for general use'.

Food irradiation is the process of exposing food to a carefully controlled amount of energy in the form of high-speed particles or electromagnetic radiation. These particles and radiation occur widely in nature and are included within the energy that is continually reaching Earth from the sun. While the knowledge of how to produce them originated from research into nuclear energy many years ago, modern methods are available which are straightforward and safe (IFST, 2015).

The choice of irradiation method will depend on the material needing to be treated. Thus, to treat the surface or a thin layer of a food, one would usually choose beta particles (i.e. electrons). These are easy to produce electronically but they do not have deep penetrating power. To treat a bulky product such as an entire sack of spices, one would choose gamma rays or X-rays.

The energy (otherwise known as ionising radiation) penetrates the food and produces free radicals from the material through which it passes. Free radicals are highly reactive and very short-lived; so short-lived that they cannot be detected in water-containing food almost immediately after it has been irradiated.

Ionising radiation is effective because high-speed electrons, gamma rays and X-rays as well as the free radicals they produce denature sensitive cell material, importantly DNA (deoxyribonucleic acid) which is the largest molecule in the nucleus and also RNA (ribonucleic acid). DNA consists of a very long ladder twisted into a double helix. The backbone is composed of sugar and phosphate molecules while the rungs of the ladder are comprised of four nucleotide bases (cytosine, thymine, adenine and guanine), which are joined weakly in the middle by hydrogen bonds. Disruption of these weak hydrogen bonds prevents replication and causes cell death while exerting minimal effects on non-living tissue.

Living organisms deprived of intact DNA or RNA will cease to function. Thus, parasites such as tapeworms and disease-causing microorganisms, such as *Salmonella* species (both of which will occasionally be found in raw food) can be controlled or destroyed by irradiation. The irradiation dose depends on the treated object and the purpose of irradiation. Different irradiation doses required for achieving the listed purposes are presented in Table 15.3, together with the treated objects.

In addition to inhibiting spoilage, irradiation can delay ripening of fruits and vegetables to give them a longer shelf-life. Its ability to control pests and reduce required quarantine periods has been the principal factor behind many countries adopting food irradiation practices.

Table 15.3. Food irradiation applications.[1]

Irradiation dose	Purpose	Treated food
Low dose (up to 1 kGy)	Inhibition of sprouting	Potatoes, onions, garlic, ginger, yam
	Insect and parasite disinfestation	Cereals, fresh fruit, dried foods
	Delay ripening	Fresh fruit, vegetables
Medium dose (1-10 kGy)	Extend shelf-life	Fish, strawberries, mushrooms
	Halt spoilage, kill pathogens	Seafood, poultry, meat
High dose (10-50 kGy)	Industrial sterilisation	Meat, poultry, seafood, prepared foods
	Decontamination	Spices, etc.

[1] Source: World Nuclear Association (WNA), Radioisotopes in Food and Agriculture. Available at: https://www.world-nuclear.org/information-library/non-power-nuclear-applications.aspx.

As well as reducing spoilage after harvesting, increased use of food irradiation is driven by concerns about foodborne diseases and by growing international trade in foodstuffs which must meet stringent standards of quality. On their trips into space, astronauts eat foods preserved by irradiation.

While food irradiation has consistently been shown to be safe in clinical studies, consumer concerns have largely limited its use to imported spices and fruits in the USA and Europe. The relevant bodies in each jurisdiction stipulate that foods that have undergone irradiation must be labelled accordingly.

There are three sources of radiation approved for use on foods (GHI, 2020):
- ▶ Gamma rays that are emitted from radioactive forms of the element cobalt (Cobalt-60) or of the element caesium (Caesium-137). Gamma radiation is used routinely to sterilise medical, dental, and household products and is also used for the radiation treatment of cancer.
- ▶ X-rays that are produced by reflecting a high-energy stream of electrons off a target substance (usually one of the heavy metals) into food. X-rays are also widely used in medicine and industry to produce images of internal structures.
- ▶ Electron beams (or e-beams) which are similar to X-rays and are a stream of high-energy electrons propelled from an electron accelerator into food.

The Institute of Food Science and Technology (IFST, 2015), claimed that irradiation, carried out under conditions of Good Manufacturing Practice, is an effective, widely applicable food processing method judged to be safe on extensive available evidence and can reduce the risk of food poisoning, control food spoilage and extend the shelf-life of foods without any detriment to health and with minimal effect on nutritional or sensory quality. This view has been endorsed by international bodies such as the WHO, the FAO and *Codex Alimentarius*.

A number of applications for irradiation have been identified, aimed at improving safety and reducing food spoilage. According to IFST (2015), the application areas include:

- Grain and grain products: a low dose of irradiation (less than 1 kGy) for insect control, where a dose of 150-700 Gy is sufficient.
- Poultry and poultry products, including mechanically recovered meat, to reduce numbers of *Salmonella*, *Campylobacter* and other food poisoning bacteria: doses of up to 3 kGy (fresh) and up to 7 kGy (frozen) have been recommended. In 2012, the Food and Drug Administration (FDA) extended the maximum dosage for poultry in the USA to 4.5 kGy.
- Red meats, particularly hamburger meat, to reduce numbers of *E. coli* O157:H7 and other food-poisoning bacteria: doses of up to 4.5 kGy (fresh) and up to 7 kGy (frozen) have been recommended by IFST. The irradiation of meat in the USA was extended by the FDA in 2012 to cover unrefrigerated meat.
- Frogs' legs, especially in Belgium, France, the Netherlands and Finland: the same doses as for poultry products are applicable.
- Dried herbs and spices, to reduce levels of contaminating micro-organisms generally and to reduce or eliminate food poisoning bacteria in particular: doses up to 10 kGy have been recommended by IFST. Herbs and spices are the food materials most commonly irradiated. These raw agricultural products, grown and harvested by traditional methods are only processed by mild drying which does not reduce the level of microbes present. Alternative methods to reduce microbial numbers using chemicals, such as ethylene oxide and methyl bromide were applied but are now considered dangerous to humans and/or the environment. This has led to a large trade in steam pasteurised spices, but this process can result in flavour losses.
- Some seafood, in particular warm water shrimps/prawns and other shellfish, to improve their microbiological safety: doses up to 3 kGy have been recommended. Low doses (<3 kGy) eliminate 90-95% of spoilage organisms, resulting in an improvement in shelf-life and eliminating all vegetative bacterial pathogens. Shrimps in ice have a shelf life of 7 days; treating them with 1.5 kGy adds another ten days. 1 kGy eliminates both *E. coli* and *Vibrio* spp. in oysters without detracting from their raw quality. 20% of potential oyster consumers said they would be prepared to consume irradiated oysters now that their safety has been significantly enhanced. Oysters treated with 2 kGy have a shelf-life of 21 to 28 days under refrigeration, compared to 15 days for their non-irradiated counterparts. Vibrios, most common in crustaceans and bivalve molluscs, are very sensitive to irradiation and are reduced to below detectable levels with a treatment of only 300 Gy.

- Certain fruits and vegetables, in order to reduce the numbers of microorganisms particularly those that cause spoilage: doses of up to 2 kGy have been recommended. Irradiation of onions, garlic, mung beans and tamarind is commercially viable in Thailand. Irradiation is also useful in combating the rice weevil (*Sitophilus oryzae*) and the lesser grain borer (*Rhyzopertha dominica*). It is particularly effective against internal feeders. Only a few species are internal feeders, but larvae and pre-emergent pupae present the greatest challenge.
- The USA approved the use of irradiation of spinach and iceberg lettuce in 2008 on grounds of safety and shelf-life extension. This followed several cases of food poisoning attributed to *E. coli* contamination, including fatalities.
- Bulbs and tubers, such as potatoes and onions: doses of less than 1 kGy have been recommended by IFST, 2015.
- Cereals, grains and certain fruits, such as papaya and mango, as a quarantine measure, to kill insects: doses of 1 kGy are recommended by IFST, 2015.
- In South Africa 1,754 tonnes of herbs and spices were irradiated during 2004. The only irradiated fruit was the dried mango. At retail sale the term 'radurised' is used on the label to denote an irradiated product.
- High dose irradiation to produce sterile foods, such as ready meals, for special medical diets, emergency or space diets. These foods are irradiated by doses of 45 kGy to render the foods microbiologically sterile. The irradiation is carried out under frozen conditions to minimise adverse sensory effects.

More than 50 countries have given approval for over 60 products to be irradiated (ISGFI, 1999). The USA, China, the Netherlands, Belgium, Brazil, Thailand and Australia are among the leaders in adopting the technology.

EU legislation concerning food irradiation based on two Directives (framework Directive 1999/2/EC and implementing Directive 1999/3/EC) was adopted by the European Union in 1999 (EC, 1999a,b). The framework Directive permits irradiation providing it:

- is necessary;
- presents no hazard;
- is beneficial to consumers;
- is not substituting good manufacturing practice.

Currently, regulations on food irradiation in the European Union are not fully harmonised. Directive 1999/2/EC establishes a framework for controlling irradiated foods, their labelling and importation, while Directive 1999/3 establishes an initial positive list of foods that can be irradiated and traded freely between Member States.

The Official Journal of the EU, 2009, carries a table of Member States' authorisations of food and food ingredients that can be treated with ionising radiation. A list of foods and food ingredients authorised for irradiation by EU member states has been published (European Commission, nd – a,b).

The control of food irradiation is assigned to the European Union Reference Laboratories and the European Union Reference Centres which have been working under Regulation (EU) No 2017/625 (EC, 2017) on official controls since 29 April 2018.

15.6 Looking ahead

As the International Consultative Group on Food Irradiation (ICGFI, 1999), predicted, consumer awareness of food irradiation is increasing. In certain countries, labelled irradiated foods have become standard commodities. In others, irradiated foods are available to a limited number of consumers, and in some, irradiated foods are not permitted. Where irradiated foods are available, consumers have purchased them because of their satisfaction with product quality and safety.

Increased understanding by consumers and utilisation of irradiation by the food industry will increase consumer welfare by enhancing food safety by reducing foodborne pathogens, increasing the availability of a wide variety of nutritious, flavourful, high-quality fruits and vegetables, and reducing food spoilage. The majority of consumers respond positively to these benefits.

References

Akgoz, M., Gokmen, A. and Gokmen, I.G., 2000. Radioactive contamination on soil and vegetation at Eastern Black Sea Coast of Turkey, ten years after Chernobyl accident. In: Vosniakos, F.K., Cigna, A.A., Foster, P. and Vasilikiotis, G. (eds) Radiological impact assessment in the South-Mediterranean area. Technological Educational Institution of Thessaloniki, Thessaloniki, Greece, pp. 9-22.
Aladjadjiyan, A., 2006. Physical hazards in the agri-food chain. In: Luning, P.A., Devlieghere, F. and Verhé, R. (eds) Safety in the agri-food chain. Wageningen Academic Publishers, Wageningen, the Netherlands, pp. 209-222.

Bologa, A.S., Osvath, I. and Dovlete, C., 2000. Gamma radioactivity along the Romanian Coast of the Bleak Sea during 1983-1988. In: Vosniakos, F.K., Cigna, A.A., Foster, P. and Vasilikiotis, G. (eds). Radiological impact assessment in the South-Mediterranean area. Technological Educational Institution of Thessaloniki, Thessaloniki, Greece, pp. 29-46.

Codex Alimentarius Commission, 2006. Codex General Standard for Contaminants and Toxins in Food and Feed, Schedule 1 – Radionuclides, CODEX STAN 193-1995. Available at: http://www.fao.org/fileadmin/user_upload/livestockgov/documents/1_CXS_193e.pdf.

DiSavino, S., 2011, 8 April. Analysis: a month on, Japan nuclear crisis still scarring. International Business Times (Australia). Available at: https://www.reuters.com/article/us-nuclear-japan-month-idUSTRE7377I120110408.

Edwards, G., 2021. About radioactive bananas. Canadian Coalition for Nuclear Responsibility. Available at: http://www.ccnr.org/About_Radioactive_Bananas.pdf.

Egorov V.N., Polikarpov G.G. and Stokozov N.A., 2000. Inventory of 90Sr and 137Cs in the Bleak Sea before and after the Chernobyl NPP accident. In: Vosniakos, F.K., Cigna, A.A., Foster, P. and Vasilikiotis, G. (eds) Radiological impact assessment in the South-Mediterranean Area. Technological Educational Institution of Thessaloniki, Thessaloniki, Greece, pp. 113-138.

El-Naggar, H.A., 2000. The radioactivity in the environment of Egypt. In: Vosniakos, F.K., Cigna, A.A., Foster, P. and Vasilikiotis, G. (eds) Radiological impact assessment in the South-Mediterranean area. Technological Educational Institution of Thessaloniki, Thessaloniki, Greece, pp. 139-152.

European Commission (EC), 1999a. Directive 1999/2/EC of the European Parliament and of the Council of 22 February 1999 on the approximation of the laws of the Member States concerning foods and food ingredients treated with ionising radiation. Official Journal of the European Union L 66, 13.3.1999: 16-23. Available at: http://data.europa.eu/eli/dir/1999/2/oj.

European Commission (EC), 1999b. Directive 1999/3/EC of the European Parliament and of the Council of 22 February 1999 on the establishment of a Community list of foods and food ingredients treated with ionising radiation. Official Journal of the European Union L 66, 13.3.1999: 24-25. Available at: http://data.europa.eu/eli/dir/1999/3/oj.

European Commission (EC), n.d. – a. Food Irradiation. Available at: https://food.ec.europa.eu/safety/biological-safety/food-irradiation_en.

European Commission (EC), n.d. – b. Biological Safety. Available at: https://food.ec.europa.eu/safety/biological-safety_en.

European Commission (EC), 2017. Regulation (EU) 2017/625 of the European Parliament and of the Council of 15 March 2017 on official controls and other official activities performed to ensure the application of food and feed law, rules on animal health and welfare, plant health and plant protection products, amending Regulations (EC) No 999/2001, (EC) No 396/2005, (EC) No 1069/2009, (EC) No 1107/2009, (EU) No 1151/2012, (EU) No 652/2014, (EU) 2016/429 and (EU)

Applied food science

2016/2031 of the European Parliament and of the Council, Council Regulations (EC) No 1/2005 and (EC) No 1099/2009 and Council Directives 98/58/EC, 1999/74/ EC, 2007/43/EC, 2008/119/EC and 2008/120/EC, and repealing Regulations (EC) No 854/2004 and (EC) No 882/2004 of the European Parliament and of the Council, Council Directives 89/608/EEC, 89/662/EEC, 90/425/EEC, 91/496/EEC, 96/23/ EC, 96/93/EC and 97/78/EC and Council Decision 92/438/EEC (Official Controls Regulation). Official Journal of the European Union L 95, 7.4.2017: 1-142. Available at: http://data.europa.eu/eli/reg/2017/625/oj

Fackler, M. and Wald, M., 2011, 05 May. Life in limbo for Japanese near damaged nuclear plant. New York Times. Available at: https://www.nytimes.com/2011/05/02/world/ asia/02japan.html.

General Standard for Contaminants and Toxins in Food and Feed (GSCTFF), CODEX STAN 193-1995. Available at: http://www.ico.org/projects/Good-Hygiene-Practices/ cnt/cnt_en/sec_2/docs_2.1/Codex%20General%20Stan%20Tox.pdf.

Global Harmonization Initiative (GHI), Consensus Document on Food Irradiation: Discordant international regulations of food irradiation are a public health impediment and a barrier to global trade, 2018 revised 2020. Available at: https://www.globalharmonization.net/sites/default/files/pdf/GHI-Food-Irradiation_ October-2018_revised_09-2020.pdf.

Gorbanov, S., 2000. Agriculture. In: Vosniakos, F.K., Cigna, A.A., Foster, P. and Vasilikiotis, G. (eds) Radiological impact assessment in the South-Mediterranean area. Technological Educational Institution of Thessaloniki, Thessaloniki, Greece, pp. 247-296.

Henriksen, T., 2015. Radiation-and-health. Department of Physics, Oslo University, Norway. Available at: https://www.mn.uio.no/fysikk/english/services/knowledge/ radiation-and-health/chap06.pdf.

Institute of Food Science and Technology (IFST), 2015. The use of irradiation for food quality and safety. Available at: https://www.ifst.org/resources/information-statements/food-irradiation.

International atomic energy agency (IAEA), 2008. The International Nuclear and Radiological Event Scale, User manual. Available at: https://www-pub.iaea.org/ MTCD/Publications/PDF/INES2013web.pdf.

International Atomic Energy Agency (IAEA), 2017. Natural radioactivity in food: experts discuss harmonizing international standards. Available at: https://www.iaea. org/newscenter/news/natural-radioactivity-in-food-experts-discuss-harmonizing-international-standards.

International consultative group on food irradiation (ICGFI), 1999. Consumer attitudes and market response to irradiated food. ICGFI, Vienna, Austria. Available at: http:// www-naweb.iaea.org/nafa/fep/public/consume.pdf.

Klug, T., 2021, 24 April. Fact check: 5 myths about the Chernobyl disaster. Deutsche Welle. Available at: https://www.dw.com/en/fact-check-5-myths-about-the-chernobyl-nuclear-disaster/a-57314231.

Kritidis, P., Michaelidu, S., Kaestner, P., Vandecasteele, C., Vosniakos, F. and Misaelides, P., 2000. Survey of soil radioactivity in Cyprus. In: Vosniakos, F.K., Cigna, A.A., Foster, P. and Vasilikiotis, G. (eds) Radiological impact assessment in the South-Mediterranean area. Technological Educational Institution of Thessaloniki, Thessaloniki, pp. 311-320.

McCurry, J., 2021, 13 April. Fukushima: Japan announces it will dump contaminated water into sea. The Guardian, Available at: https://www.theguardian.com/environment/2021/apr/13/fukushima-japan-to-start-dumping-contaminated-water-pacific-ocean.

Miller, R., Drissner, J., Kamert, S., Kaminski, S., Klemt, E. and Zibold, G., 2000. Control and management of semi-natural ecosystems contaminated by caesium radionuclides from the Chernobyl fallout. In: Vosniakos, F.K., Cigna, A.A., Foster, P. and Vasilikiotis, G. (eds) Radiological impact assessment in the South-Mediterranean area. Technological Educational Institution of Thessaloniki, Thessaloniki, pp. 321-334.

Ne'erman, E., Butenko, V., Brenner, S. and Lavi, N., 2000. Radiocesium derived from the Chernobyl accident in the surface soils of Israel. In: Vosniakos, F.K., Cigna, A.A., Foster, P. and Vasilikiotis, G. (eds) Radiological impact assessment in the South-Mediterranean area. Technological Educational Institution of Thessaloniki, Thessaloniki, pp. 335-360.

Othman, I., 2000. The impact of the Chernobyl accident on Syria. In: Vosniakos, F.K., Cigna, A.A., Foster, P. and Vasilikiotis, G. (eds). Radiological impact assessment in the South-Mediterranean area. Technological Educational Institution of Thessaloniki, Thessaloniki, pp. 361-370.

Papucci, C., Delfanti, R. and Torricelli, L., 2000. In: Vosniakos, F.K., Cigna, A.A., Foster, P. and Vasilikiotis, G. (eds) Radiological impact assessment in the South-Mediterranean area. Technological Educational Institution of Thessaloniki, Thessaloniki, pp. 393-401.

US Food and Drug Administration (FDA), 2018. 'Food irradiation: what you need to know'. Available at: https://www.fda.gov/food/buy-store-serve-safe-food/food-irradiation-what-you-need-know.

Vosniakos, F.K., 2001, Studies on the radioactive transfer in air, soil, plants, foods. PhD Thesis. Sofia, Bulgaria.

Vosniakos, F.K., 2012. Radioactivity transfer in environment and food. Springer, Berlin, Heidelberg, Germany.

Vosniakos, F.K., Zoumakis, N.M. and Diomou, C., 2000. The level of 137-Cs in Greek soils one decade after the Chernobyl accident. In: Vosniakos, F.K., Cigna, A.A., Foster, P. and Vasilikiotis, G. (eds) Radiological impact assessment in the South-Mediterranean area. Technological Educational Institution of Thessaloniki, Thessaloniki, pp. 23-28.

World Health Organization (WHO), 2011. Nuclear accidents and radioactive contamination of foods, 30 March 2011. Available at: https://www.who.int/publications/m/item/nuclear-accidents-and-radioactive-contamination-of-foods.

World Nuclear Association (WNA), n.d. Radioisotopes in food and agriculture. Available at: https://www.world-nuclear.org/information-library/non-power-nuclear-applications.aspx.

Zehringer, M., 2016. Radioactivity in food: experiences of the food control authority of Basel-city since the Chernobyl accident. InTech Open, Rijeka, Croatia. https://doi.org/10.5772/62460

16. Dietary reference values

This is what you need to know about dietary reference values and how they should be applied

Janneke Verkaik-Kloosterman

National Institute for Public Health and the Environment (RIVM), P.O. Box 1, 3720 BA Bilthoven, the Netherlands; janneke.verkaik@rivm.nl

Abstract

A balanced diet provides an adequate intake of nutrients. This is important for maintaining health, as all nutrients have specific functions in the human body. The amount of a nutrient required to maintain an individual's health is the nutrient requirement. The requirement varies between nutrients and between individuals, and depends on, for instance, the person's age, gender or physiological status (like pregnancy). As the requirement of an individual is unknown, dietary reference values (DRV) are set at the population level, based on current scientific knowledge. DRV is an umbrella term for a set of nutrient-related reference values, with different origins and purposes. DRV are intended as guidance for optimal nutrition and as a benchmark to measure progression towards that standard. Traditionally the DRV were set as a reference amount to ensure sufficient intake to prevent nutritional deficiency diseases. Currently, the risk reduction of e.g. chronic diseases is also being discussed for inclusion in setting DRV. In addition to DRV for requirement, there is also a DRV identifying high intakes that could result in adverse health effects. The aim of this chapter is to offer the reader an introduction to the different types of DRV and how they are set. It will also be explained how the DRV can be applied correctly for an evaluation of the diet.

Bart Wernaart and Bernd van der Meulen (eds)
Applied food science
DOI: 10.3920/978-90-8686-933-6_16, © Janneke Verkaik-Kloosterman 2022

Key concepts

▶ Dietary reference values (DRV) are a set of reference values that can be used to assess and plan nutrient intakes of a healthy population.

▶ The average requirement (AR) is the habitual amount estimated to be sufficient to meet the requirement for half of the population; the other half of the population will require higher amounts.

▶ The population reference intake (PRI) is the habitual amount estimated to be sufficient to meet the requirement for a large part (97.5%) of the population. The population reference intake is calculated as the average requirement plus 2 SD (standard deviation).

▶ If there is not enough scientific data to set an average requirement and accompanying population reference intake, an adequate intake (AI) is set. It is assumed that the amount set as adequate intake is sufficient to meet the requirement of almost the whole population. The adequate intake has more uncertainties compared to the population reference intake.

▶ The tolerable upper intake level (UL) is the highest level of intake that is unlikely to pose adverse health effects in a population. It is not a recommended intake, but it is a high level of intake that is tolerated biologically without adverse health effects. Intakes above the UL may result in adverse health effects, but this depends on dose, duration of high intake, and individual sensibility, among other things.

▶ Habitual intake is the long-term average intake. Many dietary and nutrient recommendations are set in such a way that these should be met over time. This means that some daily fluctuation of the nutrient intake and foods is alright, if in the long term the intake meets the requirement.

Case 16.1 Dietary reference values worldwide

Worldwide, dietary reference values (DRV) are available. Some countries collaborate and set DRV together, like the Nordic countries (Iceland, Denmark, Finland, Sweden, Norway) and the German-speaking European countries D-A-CH (Germany, Austria, Switzerland). On the other hand, other countries set their own DRV or adopt DRV from scientific literature or other countries or organisations (like WHO or EFSA). Although the same scientific evidence is available and population groups are physiologically similar, DRV may vary from country to country (see Table 16.1 for some examples).

Reasons for differences may be the lack of a standard approach setting DRV, or scientific/expert judgement in the process of setting DRV, which may be influenced by cultural and regional factors. In addition, DRV are regularly updated, and new scientific evidence may alter the value of the DRV. Further differences in age categories, unit that the DRV is expressed in, and type of DRV set, may result in other DRV values.

Table 16.1. Overview of average requirement (population reference intake) or adequate intake (indicated with *) of some nutrients set by different countries/organisations for adult female.

Nutrient	World Health Organization (WHO, 2004)		EFSA (2019)		US Institute of Medicine (USA) (IoM, 1997, 1998, 2000a, 2001, 2011)	
	Age (yrs)	value	Age (yrs)	value	Age (yrs)	value
Vitamin A µg RE/d	19-65	270*	≥18	490 (650)	≥19	500 (700)[2]
Vitamin B1 (thiamine) mg/d	≥19	1.1*	≥18	0.072 (0.1) mg/MJ	19-50	0.9 (1.1)
Vitamin B2 (riboflavin) mg/d	≥19	1.1*	≥18	1.3 (1.6)	19-70	0.9 (1.1)
Vitamin B3 (niacin) mg NE/d	≥19	14*	≥18	1.3 (1.6) mg NE/MJ	≥19	11 (14)
Vitamin B6 mg/d	19-50	1.3*	≥18	1.3 (1.6)	19-50	1.1 (1.3)
Pantothenic acid mg/d	≥19	5.0*	≥18	5*	19-50	5*
Folate µg DFE/d	19-65	320 (400)		250	19-50	320 (400)
Vitamin B12 µg/d	19-65	2.0 (2.4)	≥18	4*	19-50	2 (2.4)
Vitamin C mg/d	19-65	45*	≥18	80 (95)	19-50	60 (75)
Vitamin D µg/d	19-50	5*	≥18	15*	19-50	10 (15)
Vitamin K µg/d	19-65	55*	≥18	70* (K$_1$)	≥19	90*
Calcium mg/d	19-menopause	1000*	18-24	860 (1000)	19-50	800 (1000)
Selenium µg/d	19-65	20.4 (26)	≥18	70*	19-50	45 (55)
Magnesium mg/d	19-65	220*	≥18	300*	19-30	255 (310)
Iron mg/d	≥18[3]	19.6-58.8*[4]	≥18 [b]	7 (16)	19-50	8.1 (18) [c]
Iodine µg/d	≥13	150*	≥18	150*	≥19	95 (150)

[1] DFE = dietary folate equivalent; NE = niacin equivalent; RE = retinol equivalent.

[2] In RAE = retinol activity equivalent.

[3] Premenopausal women only.

[4] Depends on bioavailability of dietary iron (range 5-15%).

16.1 Dietary reference values

In this section a short history of dietary reference values (DRV) is provided as well as the definition of the different reference values included in the DRV (largely based on EFSA, 2010; Institute of medicine, 2006; department of Health, 1991).

recommended daily allowance (RDA)

Until the 1990s a single reference value was set: the recommended intake or recommended daily allowance (RDA). The recommended intake was defined as the intake amount (more than) sufficient to meet the nutritional needs of virtually all healthy individuals in a population group. This resulted in misinterpretations when using the RDA in research as well as education, as it was thought that this was the amount recommended for each individual. In a later definition of the RDA it was framed as an average amount of a nutrient that should be provided to each person in a population group if the nutritional needs of virtually everyone in the group are to be met. It was thought that using the word average would make clear that this amount is not equal to what an individual should consume (NRC, 1989; Department of Health, 1991). But, even after redefinition of the RDA, it was still not understood that this reference value is not the same as the recommended intake for individuals or groups. Therefore, in the early 1990s, a new set of reference values was developed in the United Kingdom, describing the requirement distribution (Department of Health, 1991). In the mid-1990s a reference value for high intakes was also introduced by FAO/WHO (EFSA, 2006).

reference values

Dietary reference values (DRV) is an umbrella term for several reference values providing information on the requirement as well as the safety of intake. The precise nutrient requirement differs from person to person. However, an individual's nutrient requirement is unknown. In addition, scientific data is generally insufficient to describe the exact nutrient requirement distribution for a group of individuals. But with the assumption that the requirement distribution is normally distributed around the mean requirement and with a specific inter-individual variability, a population requirement distribution can be drawn (Figure 16.1). These DRV are usually set for a healthy, normal weight population group. In addition, separate DRV are often set by age, gender and life stage (e.g. pregnancy). As we can see in Case 16.1 and Table 16.1, DRV are adopted per country, by groups of countries, or by organisations. Therefore, depending on the interpretation of the available scientific evidence as well as cultural or regional factors, different DRVs are used worldwide. In this chapter the EFSA terms are used; synonyms used by some other organisations are provided below.

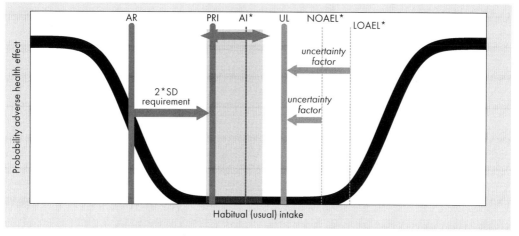

Figure 16.1. Schematic overview of the paradigm of dietary reference values and their relation to intake and probability of adverse health effects (Verkaik-Kloosterman, 2011). * AI, NOAEL, and LOAEL have dashed lines and are surrounded by a shaded area as they are not related exactly to requirement. AR: average requirement; PRI: population reference intake; AI: adequate intake; SD: standard deviation; UL: tolerable upper intake level; NOAEL: no observed adverse effect level; LOAEL: lowest observed adverse effect level.

average requirement The (estimated) average requirement (AR) is a value for daily intake estimated to meet the requirement of half the population group. This means that this value is (more than) sufficient for 50% of the population group, but still **population reference** inadequate for the other 50%. The population reference intake (PRI; also called **intake** recommended daily allowance) is derived from the AR and represents the minimum daily intake meeting the nutrient requirement of 97.5% of the population group. If a normal requirement distribution is assumed, the PRI is calculated as the AR plus 2 standard deviations of the AR (SD_{AR}). To estimate the SD_{AR} a coefficient of variation (%CV) is assumed, representing the between-person variation. From the %CV and the AR the SD_{AR} can be calculated; %CV = SD_{AR}/AR. Organisations setting DRV generally use a %CV between 10-25%. The %CV differs between nutrients as well as between organisations. The DRV provided in the case show that, for instance, for folate a %CV of 12.5 is used, while for vitamin B_{12} the %CV is 10% and for vitamin D 25% (%CV is calculated as $(((PRI-AR)/2)/AR) \times 100$).

For some nutrients it is known that the assumption of a normal requirement distribution is not valid. A well-known example is the iron requirement of women of childbearing age. Their iron needs are partly determined by menstrual losses. In Table 16.1 the AR and PRI for iron are respectively 7 and 16 mg/d for EFSA and 8.1 and 18 mg/d for Institute of Medicine (USA); the large difference between AR and PRI indicates a distribution with a tail to the right. If a similar tail would be expected at the left side of the distribution, the 2.5th percentile of the distribution (AR – 2SD) would be a negative value, and as such impossible. These menstrual losses are highly variable among women and show a skewed distribution. In addition, the menstrual losses are consistent within woman. As a result some women have much higher iron losses each month compared to others (NRC, 1986). Also, for other age categories some organisations do not assume a normal requirement distribution for iron, for instance children aged 1-8 years old (Australian Bureau of Statistics, 2015). In such situations, the PRI cannot be estimated as easily as described above. But the PRI can still be calculated from the AR, if the non-normal requirement distribution can be transformed to normality (Institute of Medicine, 2001). On the transformed scale the 50th and 97.5th percentile can be estimated as described earlier (+2SD) and these values can be transformed back to the normal scale to represent the AR and PRI. Another approach can be followed if the total requirement distribution is calculated by the summation of several specific requirement distributions representing parts of the requirement (e.g. growth). With Monte Carlo simulation, a large population of individuals is simulated for which a total requirement is calculated by drawing a value from each of the specific distributions and summing them. This results in a population's distribution of the total requirement, from which the 50th and 97.5th percentile can be calculated to represent the AR and PRI, respectively.

adequate intake

If scientific data is insufficient to quantify an AR and accompanying PRI, the adequate intake (AI, sometimes also referred to as population reference intake, although it is different from the PRI described above) is set as DRV. The AI is expected to meet or even exceed the nutrient amount required to fulfil the requirement of virtually the whole population group (PRI). An example of insufficient data is when no data on dose-response between health effects and intake is available. In such cases the current intake of a population with no signs of deficiency is used to set an AI. Other examples of insufficient data are a single experimental study, or a combination of data from different approaches that are not sufficient on their own to estimate an AR. There are differences and similarities between the AI and PRI. Both are expected to cover the requirement of nearly all subjects in a population group. But the AI is much less certain than the PRI, due to the larger degree of scientific judgement. As such, the AI should be applied with more care than the PRI. A similarity between AI and PRI is that for healthy individuals it is generally not established that intakes above the PRI or AI would have beneficial health effects.

tolerable upper intake level

The last DRV to be discussed is the tolerable upper intake level (UL). This value is not related to the requirement, but concerns safety. The UL is the highest level of daily intake that is unlikely to pose a risk of adverse health effects for virtually all in the population group, including sensitive groups. It is a high level of intake that can be biologically tolerated with high probability, but it is not a recommended intake level. Increased nutrient intake from nutrient-dense foods, e.g. food fortification and dietary supplements, gave rise to the need for a tolerable upper intake level (Institute of Medicine, 1994, 1997, 1998, 2006).

16.2 How DRV are set – average requirement and adequate intake

physiological requirement

To establish DRV for the requirement, information is needed on the physiological requirement and the dietary requirement of the nutrient. The physiological requirement of a nutrient is the rate at which a nutrient has to be provided to the body to support metabolism and maintenance of functions (EFSA, 2010). This varies between individuals due to (epi) genetic differences, age, and sex, among other things. In addition, part of the variation is caused by physiological state, like pregnancy or breastfeeding, as well as by individual variation in the response to stressors like infection and trauma. It is generally assumed that for most nutrients for a subpopulation with similar age, sex, and physiological state, the requirement distribution has a more or less normal shape. The human body is able to deal with (large) day-to-day variation in nutrient intake in the short and medium long term by making changes in some processes in the body. An example is modification of the absorption of the nutrient in the body or a change in the nutrient elimination. Another example is the mobilisation of nutrient body storage or the other way around with construction of body storage. Together with the physiological requirement, this capacity for metabolic adaptation should be considered when setting DRV.

The dietary requirement is the amount of a nutrient that needs to be provided by the diet to maintain health in a healthy individual (EFSA, 2010). A general assumption in setting DRV is that the dietary requirement for energy and all other nutrients is satisfied. There are several factors that are important to consider. For most nutrients, the diet is the only source. An exception is vitamin D, which can also be produced by the body if the skin is exposed to UV light. Furthermore, the nutrient can be present in diet preformed or as a precursor. In the case of vitamin A for instance, retinol is the preformed form and carotenoids are pro-vitamin A precursors. The physiological requirement differs from the dietary requirement, because factors like bioavailability and bio-efficacy play a role. Similar to the physiological requirement, the dietary requirement varies between individuals, but again the general assumption is that the distribution is normal for subgroups with similar age, sex and physiological state. However, this is not valid for all nutrients, for instance iron.

| criterion of adequacy | To set DRV, a criterion of adequacy has to be determined, for instance the risk of deficiency, the relationship between nutrient intake and a function in the body (e.g. status parameter, enzyme activity), or maintenance of the body storage pool or body composition (EFSA, 2010). These are the classic criteria, but nowadays (surrogate) clinical endpoints (mortality or morbidity) are also considered relevant. The criteria to be used in setting DRV is a scientific judgement, taking into account and weighting all the scientific evidence. |

criterion of adequacy

To set DRV, a criterion of adequacy has to be determined, for instance the risk of deficiency, the relationship between nutrient intake and a function in the body (e.g. status parameter, enzyme activity), or maintenance of the body storage pool or body composition (EFSA, 2010). These are the classic criteria, but nowadays (surrogate) clinical endpoints (mortality or morbidity) are also considered relevant. The criteria to be used in setting DRV is a scientific judgement, taking into account and weighting all the scientific evidence.

types of studies

Many types of studies could be considered when setting DRV, e.g. animal studies, case studies/reports, human feeding studies, randomised (clinical) trials, and observational studies. Each of these types of studies has its own (dis)advantages and limitations. The total scientific evidence from different types of studies is a basis for setting DRV. However, often the evidence is weighted by including, for example, the quality of the study type. Experimental animal studies are, for instance, often not applied in setting DRV, due among other things to the contrast of the control of the experiment versus the free-living situation of people. Furthermore, the dose and administration route of animal studies may differ from what is relevant for humans (Institute of Medicine, 2005). Tables 16.2-16.4 give an overview of the different types of scientific evidence used as a basis for setting DRV for vitamins by EFSA.

human feeding studies

Experimental human feeding studies have generated a lot of the knowledge on nutrient requirement for the prevention of deficiency (EFSA, 2010; Institute of Medicine, 2005). These studies provide valuable information on the relationship between the nutrient intake and biomarkers of health. Setting DRV with these types of data is called the classical approach. In some of these human feeding studies, subjects are confined, making it possible to closely control intake as well as activities. In addition, recurring (and complete) sampling of biological materials, like urine and faeces, is possible. In such nutrient balance studies, the situation in which input equals output is studied. It is assumed that the body is then saturated, based on the notion that the size of the body pool of the nutrient is appropriate and that the body pool will not change due to the experiment. Another type of human feeding study is the depletion-repletion study. In depletion-repletion studies, subjects are given a diet lacking in or with very low levels of a specific nutrient until clinical deficiency signs appear. Then the nutrient is added to the diet, and the lowest level resulting in a reversal of the clinical deficiency signs in all subjects is the minimum nutrient requirement. Generally, the study period of human experimental feeding studies is limited to days or weeks, so long-term effects are not known. Furthermore, due to the amount of control the results may not be representative for free-living subjects. Other limitations are the limited number and variety of subjects (often healthy young men), and generally, few different doses are included in such studies, often due to time and cost constraints.

Table 16.2. Overview of type of scientific basis used to set the DRV for requirement for adults ((≥18 years) for several vitamins by EFSA (EFSA, 2013, 2014a,b,c, 2015a,b,c, 2016a,b,c, 2017a,b; SCF, 1993).

	Approach	Publication date underlying scientific data	Type of DRV
Vitamin A	Factorial	1987	AR (PRI)
Vitamin B_1	Depletion-repletion	1942, 1943, 1965, 1966, 1970 & 1979	AR (PRI)
Vitamin B_2	Intake-excretion	1946, 1950, 1993, 2016	AR (PRI)
Vitamin B_3	Depletion-repletion	1956 & 1969	AR (PRI)
Pantothenic acid	Median/mean intake	European surveys 1996-2010	AI
Vitamin B_6	Depletion-repletion	1991 (older women)	Women: AR (PRI)
	Extrapolated from adult women		Men: AR (PRI)
Folate	Depletion-repletion	2000 (agrees with 1983, 1987)	AR (PRI)
Vitamin B_{12}	Combination observational and	Intervention: 2012	AI
	intervention (limited data)	Observational: 2010	
Vitamin C	Intake-excretion	1979, 1991 & 2013 (men)	AR (PRI)
	Extrapolated from adult men		AR (PRI)
Vitamin D	Intake-status relationship	35 trials (adult & children) 1990-2012	AI
Vitamin E	Intake-status relationship	Surveys in 8 EU countries 2000-2012	AI
Vitamin K	Intake-status relationship	1988	AI

Table 16.3. Overview of type of scientific basis used to set the DRV for requirement for infants (7-11 months) for several vitamins by EFSA (EFSA, 2013, 2014a,b,c, 2015a,b,c, 2016a,b,c, 2017a,b; SCF, 1993).

	Approach	Publication date underlying scientific data	Type of DRV
Vitamin A	Extrapolation from adults with correction growth factor		AR (PRI)
Vitamin B_1	Depletion-repletion	1942, 1943, 1965, 1966, 1970 & 1979	AR (PRI)
Vitamin B_2	Upward extrapolation from fully breastmilk fed infants 0-6 months	1980,1983, 1988, 1990, 1999, 2005, 2006	AI
Vitamin B_3	Identical to adults per MJ		AR (PRI)
Pantothenic acid	Upward extrapolation from fully breastmilk fed infants 0-6 months	1977, 1983, 1984, 2005, 1981	AI
Vitamin B_6	Average of upward (from 0-6 months) and downward (from adults) extrapolation	1985, 1989	AI
Folate	Upward extrapolation from fully breastmilk fed infants 0-6 months	1988, 1999, 2004, 2006, 2009, 2012	AI
Vitamin B_{12}	Extrapolation from adults with correction growth factor		AI
Vitamin C	3x amount known to prevent scurvy	1976	PRI
Vitamin D	Intake-status relationship	1986, 2001, 2012, 2013 & 2016	AI
Vitamin E	Upward extrapolation from fully breastmilk fed infants 0-6 months	2004, 2011	AI
Vitamin K	Identical to adults per kg body weight		AI

Table 16.4. Overview of type of scientific basis used to set the DRV for requirement for children (1-18 years) for several vitamins by EFSA (EFSA, 2013, 2014a,b,c, 2015a,b,c, 2016a,b,c, 2017a,b; SCF, 1993).

	Approach	Publication date underlying scientific data	Type of DRV
Vitamin A	Extrapolation from adults with correction growth factor		AR (PRI)
Vitamin B$_1$	Depletion-repletion	1942, 1943, 1965, 1966, 1970 & 1979	AR (PRI)
Vitamin B$_2$	Extrapolation from adults with correction growth factor		AR (PRI)
Vitamin B$_3$	Identical to adults per MJ		AR (PRI)
Pantothenic acid	Median/mean intake	European surveys 2003-2011	AI
Vitamin B$_6$	Extrapolation from adults with correction growth factor		AR (PRI)
Folate	Extrapolation from adults with correction growth factor		AR (PRI)
Vitamin B$_{12}$	Extrapolation from adults with correction growth factor		AI
Vitamin C	Extrapolation from adults with correction growth factor		AR (PRI)
Vitamin D	Intake-status relationship	35 trials (adult & children) 1990-2012	AI
Vitamin E	Intake-status relationship	Surveys in 8 EU countries 2000-2012	AI
Vitamin K	Identical to adults per kg body weight		AI

randomised clinical trial   Another type of experimental study is the randomised clinical trial (RCT) in which subjects are randomly allocated to a specific exposure (nutrient) (Institute of Medicine, 2005). In contrast to the above-described experimental human feeding study, the level of control is limited. If the sample size is large enough, the studied groups are expected to be similar with respect to known and unknown confounding factors that may be related to the disease risk. With a smaller sample size, matching on known confounding factors is applied to reduce the bias. Furthermore, due to some degree of control and randomness in exposure, part of the confounding present in studies in a free-living situation is reduced. Although RCTs are a kind of standard method studying the relationship between nutrient intake and health, there are limitations. For example, there may be selection bias, as subjects willing to participate may be a specific selection. Therefore, results may not be representative for the whole population. These studies demand a lot of effort from participants and maintaining adherence to the intervention may be difficult. In addition, the number of nutrients or nutrient combinations studied is generally limited and the amount of intake levels (dose) is also limited. As with experimental feeding studies, the study period is often relatively short, which may especially have an impact on the study of chronic health effects.

observational studies

Observational studies lack the controlled setting, but have the advantage of the free-living situation as a basis (Institute of Medicine, 2005). These types of studies are useful for finding an association between intake level or nutritional status and disease risk, though the causality of this association is difficult to obtain. However, this may be strengthened by consistency between different studies in diverse populations as well as by measuring exposure more objectively and taking into account potential confounding factors. There are some factors that make it difficult to study the association between nutrient intake or nutritional status and disease risk with observational studies. It is, for instance, possible that there is limited variation in intake in the study population, making it difficult to find a dose-response relationship. In addition, the diet of humans is a complex mix of foods and food components. The intake of several components from the human diet may be highly correlated. So, it may not be clear which of these highly correlated components is associated with the health effect nor whether it is associated with a mix of components rather than a single one. Furthermore, these types of studies often rely on self-reported food consumption, which may include systematic bias. However, with current techniques, biomarkers of exposure or disease are also included in these studies, and may reduce some of this bias.

balance studies

For many nutrients DRV are based on data from balance studies, biochemical indicators, or a factorial approach (EFSA, 2010; Institute of Medicine, 2005). With status or functional parameters (biochemical indicators), the amount of intake (dose) required to reach a specific value of these parameters (response) is studied. In the factorial approach, the various factors that determine the requirement for maintenance of a specified plasma level or body store are summed up. To be able to do this, measurements of the nutrient amount leaving the body via several pathways are required; and the additional amounts necessary for growth or during pregnancy or lactation also need to be estimated.

observed intake levels

If none of the above data is sufficient to set DRV, observed intake levels are used (EFSA, 2010). Sometimes, these values are supported by limited scientific evidence from the above-described study types, or by the observation of no deficiency signs in the population. For young infants the DRV is often based on the average intake of a nutrient by full-term born infants consuming exclusively human milk from healthy and well-nourished mothers (Institute of Medicine, 2005). It is not known to what extent the nutrient levels in human milk may exceed the requirement. However, due to ethical constraints this is difficult or maybe even impossible to study, as reducing the nutrient content of (human) milk may result in adverse health effects for the child in the long or short term.

extrapolation Extrapolation is also applied when setting DRV for specific life-stage groups. For instance, no or limited data is available for children, but a DRV is set for adults (Institute of Medicine, 2005). This DRV is then based on difference in body weight or body circumstances, or energy intake extrapolated. Assumptions about reference body weight and age categories will influence the actual DRV.

16.3 How DRV are set – tolerable upper intake level

The tolerable upper intake level (UL) often applies to chronic daily intakes and is generally related to total intake from foods and dietary supplements. For some nutrients there is an exception, and the UL is only set for e.g. intake from dietary supplements (e.g. magnesium) or intake of the synthetic (added) form (e.g. folic acid). This is because the underlying scientific studies only show adverse effects associated with intake from dietary supplements or a specific chemical form of the nutrient. The UL of most nutrients is not related to acute effects; this means that a single high intake will not immediately result in adverse health effects. There are however exceptions, as in the case of the UL for retinol for women of childbearing age or during pregnancy, as single high intakes are already associated with teratogenicity (i.e. defects in the embryo) (EFSA, 2006).

Nutrients can cause adverse effects at (chronic) high intakes. Due to differences in sensitivity between individuals, the UL is generally set as an intake level unlikely to pose a risk of adverse effects for most members of the general healthy population, including sensitive subjects. However, some subgroups may still be at risk due to extreme/distinct vulnerability (EFSA, 2006).

The framework for deriving an UL for nutrients is based on the risk assessment procedure for other food chemicals, but there are some differences (EFSA, 2006; FAO/WHO, 2006). In contrast to other food chemicals, nutrients are, within a dose range, essential for human health. And adverse health effects are generally not expected for nutrients until a specific threshold level is exceeded. This threshold differs between nutrients as well as between individuals. In addition, nutrients have a long history of safe consumption in a balanced diet. Furthermore, for some nutrients there is knowledge about chronic intake at levels higher than those obtained from normal diet, for instance from dietary supplements. Unlike for food chemicals, data about the adverse effects of high nutrient intake are more often from human studies. Also, homeostatic regulation of the body content is the case for several nutrients. This means that the body is capable of adapting or changing body processes in order to maintain the nutrient level in e.g. blood within a specific range. This will protect the body, at least temporarily, against too low or too high levels. Examples of these changes in body processes are adaptation of absorption, excretion or metabolic processes. Furthermore, body stores also play a role, as they can be used or made.

no observed adverse effect level The UL is derived from the no observed adverse effect level (NOAEL) or lowest observed adverse effect level (LOAEL) (Figure 16.1) (EFSA, 2006). In an ideal situation, a distribution of individual NOAELs or LOAELs (or thresholds) would be the basis for the UL. The UL would be situated at the low end of this theoretical distribution. Unfortunately, this data is not available. Due to ethical constraints, toxicological intervention studies with nutrients in humans are not feasible. Most data are therefore from observational studies in groups or individuals (e.g. case descriptions). The uncertainty is considered by taking into account uncertainty factors in deriving the UL from the NOAEL or LOAEL. The level of certainty determines the size of the uncertainty factor, and this factor is generally lower if the UL is derived from a NOAEL compared to a LOAEL. The size of the uncertainty factor is generally based on expert judgement taking into account study as well as nutrient-specific aspects. Compared to food chemical risk assessment, the uncertainty factors applied for nutrients are generally lower, because otherwise the UL would come below the references for requirement. ULs are often derived for separate age or life-stage groups, so factors influencing sensitivity to high intakes like physiological changes due to growth and maturation are considered.

identify the adverse health effects A first step in deriving the UL is to identify the adverse health effects that are caused by excessive intake of the nutrient (hazard identification) (EFSA, 2006). Good quality and extensive human studies are prioritised in identifying the hazard. However, other type of studies can also be used. All evidence from human, animal and *in vitro* studies is searched to study the likelihood of a nutrient resulting in adverse health effects. If observed effects are adverse, the decision is generally based on scientific (expert) judgement. Aspects like the causality of the relationship (e.g. consistency of study results, strength of association, dose-response, biological plausibility), the knowledge of mechanistic aspects underlying the adverse effect, the quality and completeness of data as well as identification of the highly sensitive sub-populations, are included in this judgement.

hazard characterization In the hazard characterization (next step) the dose-response association between the nutrient intake and adverse effect is determined (EFSA, 2006). From this association a NOAEL or LOAEL is derived for the critical endpoint. The critical endpoint is the most sensitive indicator of adverse effects of the nutrient, so the adverse effect (indicator) with the lowest NOAEL or LOAEL. Using the critical endpoint in deriving the UL ensures that it will also protect against the other (potential) adverse effects which are related to higher intakes.

There are several uncertainties that should be considered, for instance the extrapolation of study results to a general population, or extrapolation of results from animal studies to human, and also interindividual variation and extrapolation of results from studies with sub-chronic exposure to chronic exposure. This is accounted for by using uncertainty factors, which are based on scientific expert judgement. The NOAEL or LOAEL is divided by the uncertainty factor to derive the UL, so the UL value is lower than the NOAEL/LOAEL, unless the uncertainty factor equals 1. In general, the uncertainty factor rises with greater uncertainty; however, the uncertainty factors should be chosen in such a way that the nutritional needs are considered. It is unwarranted for the UL to be lower than the recommended intake.

UL for children

There is often insufficient data to determine the UL for children. For these groups the UL is generally extrapolated from the UL from adults based on difference in body size, physiology, metabolism, absorption, and excretion. In practice, the UL for children is often extrapolated from the UL for adults based on difference in body weight or body surface.

differences

Although the derivation of the UL has a scientific basis as well as a generic approach, scientific expert judgement is part of the process. This partly explains differences in the UL derived by different organisations. However, there are also other reasons for differences. For example, more recently evaluated ULs may be based on more scientific data then older ULs. As the scientific knowledge grows, regular evaluation of the UL is required to keep it up to date. Table 16.5 shows a comparison of the UL for zinc set by the Scientific Committee on Food (SCF, currently EFSA) and Institute of Medicine (IoM). Although these ULs were set only one year apart, the scientific data on which they are based, as well as some of the judgements, are different. In addition, Table 16.5 shows some quality indicators of the studies used to set the UL. All these studies have a limited number of subjects, a limited variety in life-stages, and a relatively short study period.

16.4 How to apply DRV for evaluation of dietary intake of a population group

Dietary reference values (DRV) can be used for different goals, e.g. dietary planning, dietary assessment, nutritional labelling, and fortification (Institute of Medicine, 2000b, 2003a,b). In addition, DRV are included in the nutritional information provided with foods and dietary supplements. The amount of nutrients is presented as proportion of a DRV. This information may help consumers in their dietary choices (Institute of Medicine, 2003b). Although DRV should not be interpreted as nutritional goals or recommendations for individuals, they are used for dietary planning for individuals as well as

Table 16.5. Comparison of derivation of tolerable upper intake level (UL) of zinc for adults by Scientific Committee on Food (SCF, predecessor of EFSA) (EFSA, 2006) and Institute of Medicine (IoM) (Institute of Medicine, 2001).

	SCF (2002)		IoM (2001)	
Critical effect	Copper status		Reduced copper status	
NOAEL	50 mg/d		-	
LOAEL	-		60 mg/d (from Yadrick (1989; supported by others)	
Studies included	Davis et al. (2000), Milne et al. (2001), Bonham et al. (2003a,b)		Prasad et al. (1978), Greger et al. (1978), Burke et al. (1981), Fischer et al. (1984), Festa et al. (1985), Samman and Roberts (1988), Yadrick et al. (1989), Boukaiba et al. (1993)	
Characteristics of included studies	Davis et al. (2000)	N: 25 ♀ Suppl zinc 90 days Metabolic study 53 mg/d zinc	Prasad et al. (1978)	N: 1 ♂ 26 yrs Zinc intake 150-200 mg/d
	Milne et al. (2001)	N: 21 ♀ Suppl zinc 90 days Metabolic study 53 mg/d zinc	Greger et al. (1978)	N: 14 ♀ 12-14 yrs 7.4 mg or 13.4 mg food
	Bonham et al. (2003a)	N: 19 ♂ Suppl zinc 14 weeks Zinc 30 mg/d supple & 10 mg/d diet	Burke et al. (1981)	N: 11 ♀♂ 56-83 yrs 7.8 mg or 23.26 mg fortified food
	Bonham et al. (2003b)	N: 19 ♂ Suppl zinc 14 weeks Zinc 30 mg/d supple & 10 mg/d diet	Fisher et al. (1984)	N: 26 ♂ adult Placebo or 50 mg (gluconate)
			Festa et al. (1985)	N: 9 ♂ 21-27 yrs 1.8, 4.0, 6.0, 8.0, or 18.5 mg food
			Samman and Roberts (1988)	N: ? ♀♂ 150 mg (sulphate)
			Yadrick et al. (1989)	N: 18 ♀ 25-40 yrs 50 mg (gluconate)
			Boukaiba et al. (1993)	N: 44 ♀♂ 73-106 yrs Placebo or 20 mg
Uncertainty factor	2 → small sample size, relatively short study period, rigidly controlled metabolic studies		1.5 → extrapolation LOAEL to NOAEL, reduced copper status is rare in humans, higher uncertainty factor not justified	
UL	25 mg/d		40 mg/d	

groups. The idea is that the habitual dietary intake is nutritionally adequate, so the probability of both inadequate and excessive intake is low (Institute of Medicine, 2003a). In addition, DRV are used in regulations for food fortification and dietary supplements (Institute of Medicine, 2003b); for instance, to set maximum fortification levels or maximum amounts in food supplements based on the dietary intake and UL. In another example, DRV can be used to set up fortification programmes with the aim of reaching adequate and safe habitual nutrient intakes in a population, e.g. for iodine. The evaluation of nutrient intake with DRV is an aim in itself but is also part of the dietary planning as well as fortification and dietary supplementation policy. In this section we explain how DRV can be correctly applied to evaluate the dietary intake of a population group. In the next section (16.5) the dietary assessment of individuals is described.

epidemiological association studies

Before evaluation with DRV is possible, information on the dietary intake is required. There are several methods for collecting this data. With often applied methods in epidemiological association studies (e.g. a food frequency questionnaire (FFQ)), dietary intake data is collected over a longer time-span, e.g. month or year, but with fewer details. Subjects are generally asked to fill in their average daily intake of specific food (groups) within the timespan studied. To limit participants' burden and costs, data is often collected from a limited number of foods or food groups. This data is sufficient to rank subjects from low to high consumers, however, the data is insufficient for estimating the absolute intake amount with precision. Therefore, dietary data collected in this way can be used in association studies, relating high or low intake to a specific health outcome, but are less useful in evaluation of the dietary intake with DRV.

dietary intake data

With other methods, like a 24-hour recall or food diary, dietary intake data are collected with a lot of detail. In addition, these methods use open questions to gain insight into all food and drinks consumed, whereas in a Food Frequency Questionnaire (FFQ) there is a limit to the number of foods for which the frequency (and amount) of consumption is collected. So, an FFQ will not be able to collect information on all food and drinks consumed. Furthermore, the timeframe of e.g. 24-hour recall and food diary is generally a relatively short period (e.g. a day), among other things because of the burden and cost to the participants. With the 24-hour recall, the dietary intake of a person will be (slightly) different when collected on different recall days; this is called within-person variation. In addition, when data is collected from different participants on the same day, the dietary intake of each person will be different. This is called between-person variation. Although the great detail in information is an advantage of e.g. 24-hour recall, one drawback is that, due to the short time period of data collection per person, these data contain both within- and between-person variation. This has an effect on the population's intake distribution as illustrated in Figure 16.2. The habitual intake distribution, only containing between-person variation, is more narrow than the intake distribution based on one 24-hour recall per person. The average of a repeated 24-hour recall reduces the within-person variation but for most nutrients will not result in the habitual intake distribution. That will require more replicates. How many will depend on the nutrient. With repeated 24-hour recalls and a statistical correction for within-person variation, it is possible to estimate the population's habitual intake distribution. At least two 24-hour recalls on non-consecutive days, for a large enough subgroup of the study population, are required to calculate the habitual intake distribution. This statistical correction contains several steps: (1) transformation of the observed data to a normal or symmetrical distribution, (2) within-person variation is estimated and removed on the transformed scale with as a result a shrunken distribution, and (3) this shrunken distribution is transformed back to the original scale. Several programs are available for this, e.g. SPADE, NCI, and MSM (Dekkers *et al.*, 2014; Harttig *et al.*, 2011; Tooze *et al.*, 2006).

Why is it important to have detailed and habitual dietary intake data? As explained above, with a detailed data collection the consumption of all food and drinks are (ideally) included. As such, the absolute intake estimated is expected to be closer to the true intake compared to the non-detailed methods. However, it is important to keep in mind that all these methods of collecting data on food consumption rely on memory and are self-administered (some are guided by an interviewer). Therefore, bias about the true intake occurs. Furthermore, it is important to evaluate habitual intake, as for most DRV the health effects included are based on long-term low or high intakes (chronic). This means that a single day with low or high intakes will not immediately result in negative health effects. But if this intake is more habitual, these health effects may appear. Looking at Figure 16.2, the proportion of a population with an intake below a specific cut-off value is influenced by the width of the intake distribution, especially at the tails of the distribution. This may for instance result in overestimation of the proportion with intakes below a specific cut-off value in the left tail of the distribution when it is based on 1 day 24-hour recall information, compared to the habitual intake distribution.

16.4.1 Evaluation of intakes using the average requirement

Carriquiry described how dietary reference values (DRV) for requirement can be applied correctly to estimate the (in)adequacy of nutrient intake of a population group (Figure 16.3) (Carriquiry, 1999). The easiest to perform is the average requirement (AR) cut-point approach. With this approach, the proportion of the population with habitual intake below the AR is calculated. Under certain conditions, this calculated proportion is an estimate of the

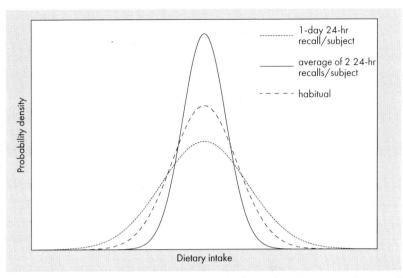

Figure 16.2. Comparison of population's intake distribution based on one 24-hour recall per subject, average of two 24-hour recalls per subject, and habitual intake.

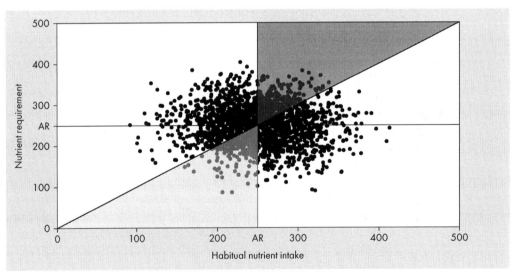

Figure 16.3. Visualisation of the average requirement cut-point approach. Black dots represent individuals with their specific habitual intake and personal requirement. Black diagonal line represents a situation in which the habitual intake of an individual is equal to the individual requirement. As such, for all black dots at the right of the diagonal line, the habitual intake exceeds their personal requirement. The proportion of black dots left of the diagonal line is the proportion of the population with inadequate intakes. As the number of individuals (black dots) in the light grey and dark grey triangle is similar, the proportion of the population below the AR is a good prediction of the proportion with intake left of the diagonal line. This is only valid under some conditions.

prevalence of inadequate intakes in the population. The first condition is that the intake and requirement are independent. This means that people with a higher requirement do not (automatically) have a higher intake of that nutrient. This condition is for instance not met for energy, as people with a higher energy demand will consume more energy. For other nutrients, it is generally assumed that this condition is met. The second condition is that the requirement distribution is symmetrical, not necessarily normal. For most nutrients a normal requirement distribution is assumed, so this condition is met. However, for iron this condition is not met, as the requirement distribution seems to be skewed for some population groups, e.g. women of childbearing age. The last condition is that the variance of the requirement distribution is relatively small compared to the variance of the habitual intake distribution. This means that the variation in habitual intake in a population is larger than the variation in individual requirements in this population. This condition is generally thought to be satisfied in a free-living population. However, this condition may not be met for institutionalised individuals, as their dietary consumption may be more similar between persons. Simulation studies in particular showed that the AR cut-point method provides a good approximation of the true prevalence of inadequate intake. An exception is when the estimated proportion with inadequate intakes is very low (<8-10%) or very high (>90-92%).

Beaton's full probability approach

In some situations when it is inappropriate to apply the AR cut-point approach, there are alternative approaches; for instance, Beaton's full probability approach and Monte Carlo simulation approach. Both approaches require detailed information on the shape of the requirement distribution, in contrast to the AR cut-point method, and the habitual intake distribution. Because the individual requirement is unknown, it is not possible to combine the requirement distribution with the intake distribution and calculate the proportion with intakes below their individual requirement. A first step in Beaton's full probability approach is computing a risk curve. This risk curve links each habitual intake level with the risk of adverse health effects due to low intake. This requires information on the shape of the requirement distribution. For most nutrients a normal distribution is assumed with a specified %CV. But for iron a distribution skewed to the right is derived for e.g. women of childbearing age, to take into account the large variation in menstruation blood losses. The risk curve can be computed from the cumulative requirement distribution. The prevalence of inadequate intake in a population is calculated by summing for each habitual intake the probability of having a specific habitual intake multiplied by the risk of inadequacy at that specific habitual intake. It should be noted that the risk curve contains several assumptions, due to limited information on the exact requirement distribution. Different assumptions could result in significantly different risk curves. Therefore, errors in these assumptions may result in biased estimates of the prevalence with inadequate intakes. Similar to the AR cut-point method, Beaton's full probability approach is meant to estimate the population's prevalence of inadequate intake and cannot identify people with inadequate intakes. Table 16.7 provides an example of the calculations in Beaton's full probability approach.

Monte Carlo simulation

Another probabilistic approach is Monte Carlo simulation. With Monte Carlo simulation for uncertain variables, values are randomly generated many times (iterations). Based on the possible values, a probability distribution can be generated for each of the uncertain variables. In the Monte Carlo simulation approach, draws from the population's requirement distribution and the population's habitual intake distribution are combined to identify those for which the habitual intake is below the drawn requirement (as the individual requirement is unknown). The proportion of draws with habitual intakes below the requirements is an estimate of the proportion with inadequate intakes in a population.

16.4.2 Evaluation of intakes using recommended daily allowance

Because the recommended daily allowance (RDA) is by definition an amount exceeding the requirement of a large part of the population, this value cannot be applied as cut-point evaluating dietary intake of a group. The proportion with intakes below the RDA will result in an overestimation of the prevalence with inadequate intakes.

Table 16.7. Calculation of Beaton's full probability approach (Australian Bureau of Statistics, 2015).

Percentile	Habitual intake population at percentile	Proportion (%) of the population	Proportion inadequacy for this percentile (= proportion of requirement above intake of this percentile)[1]	Prevalence inadequacy
0	HI_1	$S_1 = 1$	P_1	$R_1 = P_1 \times S_1$
1	HI_2	$S_2 = 1$	P_2	R_2
2	HI_3	$S_3 = 1$	P_3	R_3
3-10	HI_4	$S_4 = 7$	P_4	R_4
10-49	HI_5	$S_5 = 40$	P_5	R_5
50-79	HI_6	$S_{99} = 30$	P_6	R_6
80-99	HI_7	$S_{100} = 20$	P_7	R_7
Total prevalence inadequacy =				$\sum R_i$ over i=1 to 7

[1] Based on the requirement distribution of the population and represents the area under the curve.

Some evaluate dietary intake by comparing the median (mean) intake of a group with the RDA. The common assumption is that if the habitual median intake is at or above the RDA, the prevalence of inadequate intakes would be acceptably low. However, due to a larger variation in intake compared to requirement, the median intake should exceed the RDA to result in an acceptable low prevalence of inadequate intakes. How much greater the median intake should be depends on e.g. the variability in habitual intake relative to the variation in requirement.

As nutrients with an RDA also have an AR, it is better to apply the AR cut-point method or the approaches to evaluation with assumed requirement distributions as described earlier to estimate the prevalence of inadequacy.

16.4.3 Evaluation of intakes using adequate intake

In contrast to the AR, the AI cannot be used to quantify the prevalence of inadequate intakes in a population. It is however possible to apply the AI in a qualitative evaluation (Institute of Medicine, 2000b). It is generally considered that if the median habitual intake of a group is at or higher than the AI, there is a low prevalence of inadequate intakes. But if the median habitual intake is below the AI, no statement about the inadequacy can be made, because the habitual nutrient intake distribution is often not normally distributed but skewed to the right. The median and mean intake do not overlap. In evaluation with the AI, it is important to use the median intake and not the mean.

16.4.4 How is the tolerable upper intake level applied

As described earlier, the tolerable upper intake level (UL) is at the lower end of the distribution of the relationship of intake and adverse health effects (Carriquiry and Camaño-Garcia, 2006). As such, it is thought to be sufficiently low to protect virtually the whole population. Although in the opposite direction, this is a similar concept to the recommended daily allowance (RDA) and adequate intake (AI).

As the UL is the level of intake unlikely to pose adverse health effects for most people in a group, habitual intakes above the UL cannot be considered as unsafe. The actual risk depends on several factors including, dose, duration, and individual sensitivity. To be able to identify the health risks associated with habitual intakes above the UL, a dose-response relationship is required between the nutrient and critical end point. Unfortunately, this dose-response relationship is not available for most nutrients and most age and life-stage groups. Therefore, quantifying the risk associated with high intakes of a nutrient is not possible.

In many studies evaluating the dietary intake with the UL, the proportion of the population with intakes above the UL is reported. Often this is described as the proportion of people with excessive intakes. However, based on the definition of the UL, this is incorrect. Others define this proportion as people at risk of excessive intakes or potentially at risk of adverse effects. The other way around, focusing on the proportion of the population with intakes below the UL makes more sense in relation to the definition of the UL. This is the proportion considered very likely to have safe (no excessive) intakes.

Similar to evaluation with dietary reference values (DRV) for requirement, it is important to know if the critical effect is based on acute intakes or habitual intakes. In addition, it is important to include the total nutrient intake from all sources (e.g. dietary supplements, fortified foods, regular foods). However, some ULs are only set for intake from specific sources or for specific forms of a nutrient.

16.5 Applying dietary reference values to assess diet of an individual

Besides the use of dietary reference values (DRV) to evaluate the diet of population groups, there is often also a wish to have a reference against which to compare the diet of a specific person, for instance for nutritional counselling or education. Because an individual's nutrient requirement is unknown and also the habitual nutrient intake of an individual is unknown, it is impossible to assess whether a diet is sufficient to fulfil the requirements of that specific person. Biochemical indicators of nutritional intake or status (e.g. blood levels or urinary excretion), or clinical signs of deficiency or other health effects could provide information on the adequacy, but this is more invasive and often health effects or changes in biochemical indicators appear after long-term too low intakes.

recommended daily allowance Often it is suggested that the recommended daily allowance (RDA) could be used as a goal for individuals. As this amount assures adequate intake in the largest part (97.5%) of the population, it is suggested that with an intake this high most individuals will in any case have sufficient intake. However, it is important to acknowledge that for most people a less high intake will also be adequate. So, most people will have more than adequate intakes with intakes at the RDA.

In 2000 the Institute of Medicine proposed a statistical approach to quantitatively assess an individual's diet and decide on the adequacy or excessiveness of **average requirement** nutrient intake (Institute of Medicine, 2000b). It is proposed to use the average requirement (AR) for this assessment and not the RDA, and to include a person's mean observed intake over several days as a best estimate of the habitual intake. As will be explained in Section 16.6, the number of days required to have a valid estimate of the habitual intake varies between nutrients. The difference between the observed intake and AR together with their standard deviation (SD)s are used in this approach to conclude the adequacy of the diet of a specific person. If this value is large and positive, it is likely that the nutrient intake is adequate. By contrast, if this value is large and negative, it is likely that the nutrient intake is inadequate. In between these outcomes there is uncertainty about the (in)adequacy of the nutrient intake. With which value of this difference could it be concluded with some degree of confidence that the person's (unobservable) habitual intake will exceed the (unobservable) actual requirement of that person? This requires the SD of the difference between the observed intake and AR. This SD depends on the number of days with information on dietary intake, the SD of the requirement, and the within-person SD of the intake. This latter could be taken from a large food consumption survey in a similar group of people to the individual in this assessment. The ratio of the difference between the observed intake and AR and its SD can be used to determine the probability that the intake is below or above the requirement. In Table 16.8 and 16.9 this is illustrated with a fictitious example. This approach is only valid for nutrients with a normal distribution of the habitual intake as well as a normal distribution of the requirement. Currently, no method is available if these conditions are not met.

For nutrients with an AI and/or UL, a qualitative approach can be applied, similar to the dietary assessment of groups. A mean intake of an individual (over many days) is likely to be non-excessive if it remains below the UL. With a mean intake at or above the UL, no statement about the excessiveness can be provided. The reverse is true for the evaluation with the AI. If the mean intake of an individual is at or above the AI, there is a low risk of inadequate intakes, whereas no statement could be made if the mean intake of a person remains below the AI. The confidence of these conclusions decreases with a smaller number of days on which the mean intake is based.

Due to all the uncertainties, the outcomes of these methods should be interpreted as guidance rather than the exact truth. The bigger picture of an individual should be taken into account, including characteristics (e.g. age, gender) and health status.

Table 16.8. Overview of parameters to assess adequacy of individual's diet.

Variable	Value
Gender	woman
Age	35 years
Nutrient	calcium
AR [ref EFSA]	750 mg/d
$\%CV_{AR}$	10%
SD_{AR}	75 mg
RDA	950 mg/d
Number of days dietary intake	3
Mean intake	940
Day-to-day SD_{intake}	325
D=Mean intake − AR	940-750=190
V_{req}=variance requirement = $(SD_{AR})^2$	$(75)^2$=5,625
V_{intake}=variance intake = $(SD_{intake})^2$	$(325)^2$=105,625
d=$SD_{(mean\ intake\ -\ AR)}$=$\sqrt{(V_{req} + (V_{intake}/days))}$	$\sqrt{(5,625 + (105,625/3))}$=202
D/d = Ratio (mean intake − AR) / $SD_{(mean\ intake\ -\ AR)}$	190/202=0.94
Conclusion	~85% confidence that calcium intake is adequate for this person[a]

[1] Based on standard z-score table; 0.94 is close to 1, 1 has a probability of correct conclusion of 0.84 (Table 16.9).

Table 16.9. Probability of correct conclusion based on standard z-score table.

D/d (z-score)	Probability correct conclusion (precision 0)	Conclusion, habitual intake is
>2	0.98	Adequate
>1.5	0.93	Adequate
>1	0.84	Adequate
0	0.50	(In)Adequate
<-1	0.84	Inadequate
<-1.5	0.93	Inadequate
<-2	0.98	Inadequate

16.6 General considerations using DRV in the evaluation of dietary intake

There are several considerations when using dietary reference values (DRV) in the evaluation of the dietary intake of groups or individuals, which are important in the interpretation of the results. Below some of these considerations are briefly introduced.

Due to the uncertainties in intake estimation, nutrient composition of food, DRV, etc., the evaluation of dietary intake with DRV should be interpreted as a first signal. If evaluation with the AR results in a high estimated proportion of the population with inadequate intakes, the next step is to verify this with nutritional status research or research on clinical signs of (too) low intakes. A general rule of thumb is that an estimated proportion with inadequate intakes >10% may imply a public health issue (Institute of Medicine, 2000b) and requires additional research. The outcome of this additional research may support the results of the dietary assessment. In that case, a plan could be drawn up as to how to increase the nutrient intake in a (sub)population, for instance by nutritional education to shift the dietary pattern, food fortification to increase the nutrient content of (specific) foods, or the advice to consume dietary supplements with a specified amount of the nutrient. But it is also possible that there is a contradiction between the results of the dietary assessment and the addition research on nutritional status or clinical signs. Then there is no need to improve the nutrient intake, but for future dietary assessment it is important to find out what caused the contradiction. There are several options worth studying: quality of the food composition and dietary supplement composition databases for that nutrient, issues in collecting food consumption data that may result in underestimation of the nutrient, and further also the uncertainties in the DRV. As discussed earlier, there is uncertainty in DRV, partly due to lack of data and expert judgement. Additional research to improve the (basis of the) DRV may be required.

DRV are set for different age-gender or life-stage groups, so a difference in requirement due, for instance, to growth and body size are included. All approaches to evaluate dietary intake from groups with DRV should be applied for the appropriate age-gender or life-stage groups, otherwise over- or underestimation of the proportion with potential inadequate or excessive intakes will occur. For children, for example, the DRV is often extrapolated based on average or mid-body weight of the age-class. As such, one can imagine that the DRV for a specific age unit in that age-class would be different due to different bodyweight. For instance, if we take the age-class 1-3-year olds, the DRV are often based on the body weight of 2-year olds. So, the DRV might be an overestimation for 1-year olds and an underestimation of 3-year olds. But as long as the whole age range is present in the study population, the idea is that the proportion with inadequate intakes will average over the whole age-class. If the study population only contains 2- and 3-year olds, the proportion with inadequate intakes may be underestimated. This is not only the case for the DRV for which the proportion below or above is estimated (AR and UL), but also for the qualitative evaluation with the AI.

Besides age-gender class or life-stage the study population should be representative for the population of interest. Aspects like socio-economic status, educational level, region, following a specific eating habit (e.g. vegan) are important. This is also important for the DRV, which should be set for a population group similar to the population of interest or under study. For some nutrients, e.g. iron and zinc, bioavailability differs based on the dietary pattern. Therefore, subjects with higher intake of plant-based foods (iron) or phytate (zinc) will require a higher intake to have a similar intake to those with a higher intake of animal-based foods or less phytate, so their DRV will be higher.

For the comparison with DRV the habitual intake is of interest. For groups it is generally accepted that the habitual intake distribution can be estimated based on at least two independent replicate measurements per subject in combination with a statistical procedure to remove the within-person variation. To reduce participants' burden and costs, this procedure is also performed in populations of which only a subset has duplicate measurements. It is then important that this subgroup is of a sufficient size. The exact size needed is unknown and not thoroughly studied, but there is some rule of thumb that repeated measurements of 50-100 subjects are required (Dodd *et al.*, 2006). More research is required to better understand the sample size of the replicate measurements. For studies with only single measurements per subject, it is proposed to use estimates of within-person : total variance from other studies with comparable study population and apply this (Luo *et al.*, 2021; Verkaik-Kloosterman *et al.*, 2019). Although this is technically feasible, the true ratio remains unknown, and the use of different ratios will create different results. Therefore, it is recommended to conduct a sensitivity analysis, applying a range of ratios to study this effect. In addition, it is recommended that for studies with replicate data, this ratio is presented together with the other results. So, there will be insight into the variation in this ratio between study populations and nutrients, which is valuable for studies with only single measurements. For the estimation of habitual intake without statistical correction for within-person variation, the average intake over multiple days could be a good estimate. However, it is important to know that the number of days required to have a good estimate of the habitual intake of an individual varies between nutrients. The smaller the ratio within : between-person variation, the fewer number of days are required to estimate the nutrient intake with some accuracy. For vitamin A for instance, the number of days required is high, but fewer days are required for carbohydrates (Chun and Davis, 2012). This is also true for the estimate of the habitual intake of an individual.

This chapter provided an overview of the current methods for deriving DRVs and the current evaluation approaches. Regular evaluation of the DRV is required to take into account the growing knowledge on nutrition and health. The evaluation of the intake with DRV is an ongoing process as the dietary intake may vary from time to time. In addition, new techniques and methodologies are being developed and these may change the DRV, intake estimate and/or evaluation procedure.

References

Australian Bureau of Statistics, 2015. Beaton's full probability approach for iron. Available at: https://www.abs.gov.au/AUSSTATS/abs@.nsf/Lookup/4363.0.55.001 Chapter6510312011-13.

Bonham, M., O'Connor, J.M., McAnena, L.B., Walsh, P.M., Stephen Downes, C., Hannigan, B.M. and Strain, J.J., 2003a. Zinc supplementation has no effect on lipoprotein metabolism, hemostasis, and putative indices of copper status in healthy men. Biological Trace Element Research 93: 75-86. https://doi.org/10.1385/BTER:93:1-3:75

Bonham, M., O'Connor, J.M., Alexander, H.D., Coulter, J., Walsh, P.M., McAnena, L.B., Downes, C.S., Hannigan, B.M. and Strain, J.J., 2003b. Zinc supplementation has no effect on circulating levels of peripheral blood leucocytes and lymphocyte subsets in healthy adult men. British Journal of Nutrition 89: 695-703. https://doi.org/10.1079/BJN2003826

Boukaïba, N., Flament, C., Acher, S., Chappuis, P., Piau, A., Fusselier, M., Dardenne, M. and Lemonnier, D., 1993. A physiological amount of zinc supplementation: effects on nutritional, lipid, and thymic status in an elderly population. American Journal of Clinical Nutrition 57: 566-572. https://doi.org/10.1093/ajcn/57.4.566

Burke, D.M., DeMicco, F.J., Taper, L.J. and Ritchey, S.J., 1981. Copper and zinc utilization in elderly adults. Journal of Gerontology 36: 558-563. https://doi.org/10.1093/geronj/36.5.558

Carriquiry, A.L. and Camaño-Garcia, G., 2006. Evaluation of dietary intake data using the tolerable upper intake levels. Journal of Nutrition 136: 507s-513s. https://doi.org/10.1093/jn/136.2.507S

Carriquiry, A.L., 1999. Assessing the prevalence of nutrient inadequacy. Public Health Nutrition 2: 23-33. https://doi.org/10.1017/s1368980099000038

Chun, O. and Davis, C., 2012. Variation in nutrient intakes and required number of days for assessing usual nutrient intake among different populations. Journal of Nutritional Disorders & Therapy 2: 1-4.

Davis, C.D., Milne, D.B. and Nielsen, F.H., 2000. Changes in dietary zinc and copper affect zinc-status indicators of postmenopausal women, notably, extracellular superoxide dismutase and amyloid precursor proteins. American Journal of Clinical Nutrition 71: 781-788.

Dekkers, A.L., Verkaik-Kloosterman, J., van Rossum, C.T. and Ocké, M.C., 2014. SPADE, a new statistical program to estimate habitual dietary intake from multiple food sources and dietary supplements. Journal of Nutrition 144: 2083-2091. https://doi.org/10.3945/jn.114.191288

Department of Health, 1991. Dietary reference values for food energy and nutrients for the United Kingdom. Her Majesty's Stationery Office, London, UK.

Dodd, K.W., Guenther, P.M., Freedman, L.S., Subar, A.F., Kipnis, V., Midthune, D., Tooze, J.A. and Krebs-Smith, S.M., 2006. Statistical methods for estimating usual intake of nutrients and foods: a review of the theory. Journal of the American Dietetic Association 106: 1640-1650. https://doi.org/10.1016/j.jada.2006.07.011

European Food Safety Authority (EFSA), 2006. Tolerable upper intake levels for vitamins and minerals. Available at: https://www.efsa.europa.eu/sites/default/files/efsa_rep/blobserver_assets/ndatolerableuil.pdf

European Food Safety Authority (EFSA), 2010. Scientific opinion on principles for deriving and applying Dietary Reference Values. EFSA Journal 8: 1458.

European Food Safety Authority (EFSA), 2013. Scientific opinion on dietary reference values for vitamin C. EFSA Journal 11: 3418.

European Food Safety Authority (EFSA), 2014a. Scientific opinion on dietary reference values for folate. EFSA Journal 12: 3893.

European Food Safety Authority (EFSA), 2014b. Scientific opinion on dietary reference values for niacin. EFSA Journal 12: 3759.

European Food Safety Authority (EFSA), 2014c. Scientific Opinion on Dietary Reference Values for pantothenic acid. EFSA Journal 12: 3581.

European Food Safety Authority (EFSA), 2015a. Scientific opinion on Dietary Reference Values for cobalamin (vitamin B12). EFSA Journal 13: 4150.

European Food Safety Authority (EFSA), 2015b. Scientific opinion on dietary reference values for vitamin A. EFSA Journal 13: 4028.

European Food Safety Authority (EFSA), 2015c. Scientific Opinion on Dietary Reference Values for vitamin E as α-tocopherol. EFSA Journal 13: 4149.

European Food Safety Authority (EFSA), 2016a. Dietary reference values for vitamin B6. EFSA Journal 14: e04485.

European Food Safety Authority (EFSA), 2016b. Dietary reference values for vitamin D. EFSA Journal 14: e04547.

European Food Safety Authority (EFSA), 2016c. Dietary reference values for thiamin. EFSA Journal 14: e04653.

European Food Safety Authority (EFSA), 2017a. Dietary reference values for vitamin K. EFSA Journal 15: e04780.

European Food Safety Authority (EFSA), 2017b. Dietary reference values for riboflavin. EFSA Journal 15: e04919.

European Food Safety Authority (EFSA), 2019. DRV Finder. Available at: https://multimedia.efsa.europa.eu/drvs/index.htm.

Festa, M., Anderson, H.L., Dowdy, R.P. and Ellersieck, M.R., 1985. Effect of zinc intake on copper excretion and retention in men. American Journal of Clinical Nutrition 41: 285-292.

Fischer, P.W., Giroux, A. and L'Abbé, M.R., 1984. Effect of zinc supplementation on copper status in adult man. American Journal of Clinical Nutrition 40: 743-746. https://doi.org/10.1093/ajcn/40.4.743.

Food and Agriculture Organization of the United Nations / World health organization (FAO/WHO), 2006. A model for establishing upper levels of intake for nutrients and related substances. Report of a joint FAO/WHO technical workshop on nutrient risk assessment WHO headquarters, Geneva, Switzerland 2-6 May 2005, FAO/WHO, Geneva, Switzerland.

Greger, J.L., Baligar, P., Abernathy, R.P., Bennett, O.A. and Peterson, T., 1978. Calcium, magnesium, phosphorus, copper, and manganese balance in adolescent females. American Journal of Clinical Nutrition 31: 117-121. https://doi.org/10.1093/ajcn/31.1.117

Harttig, U., Haubrock, J., Knüppel, S. and Boeing, H., 2011. The MSM program: web-based statistics package for estimating usual dietary intake using the Multiple Source Method. European Journal of Clinical Nutrition 65: S87-S91.

Institute of Medicine, 1994. How should the recommended dietary allowances be revised? Washington, DC, USA.

Institute of Medicine, 1997. Dietary reference intakes for calcium, phosphorus, magnesium, vitamin D, and fluoride. Washington, DC, USA. https://doi.org/10.17226/5776

Institute of Medicine, 1998. Dietary reference intakes for thiamin, riboflavin, niacin, vitamin B6, folate, vitamin B12, pantothenic acid, biotin, and choline. Washington, DC, USA. https://doi.org/10.17226/6015

Institute of Medicine, 2000a. Dietary reference intakes for vitamin C, vitamin E, selenium, and carotenoids. Washington, DC. https://doi.org/10.17226/9810

Institute of Medicine, 2000b. Dietary reference intakes: applications in dietary assessment. Washington, DC, USA. https://doi.org/10.17226/9956

Institute of Medicine, 2001. Dietary reference intakes for vitamin A, vitamin K, arsenic, boron, chromium, copper, iodine, iron, manganese, molybdenum, nickel, silicon, vanadium, and zinc. Washington, DC, USA. https://doi.org/10.17226/10026

Institute of Medicine, 2003a. Dietary reference intakes: applications in dietary planning. Washington, DC, USA. https://doi.org/10.17226/10609

Institute of Medicine, 2003b. Dietary reference intakes: guiding principles for nutrition labeling and fortification. Washington, DC, USA. https://doi.org/10.17226/10872

Institute of Medicine, 2005. Dietary reference intakes for water, potassium, sodium, chloride, and sulfate. Washington, DC. https://doi.org/10.17226/10925

Institute of Medicine, 2006. Dietary reference intakes: the essential guide to nutrient requirements. Washington, DC, USA.

Institute of Medicine, 2011. Dietary reference intakes for calcium and vitamin D. Washington, DC, USA. https://doi.org/10.17226/13050

Luo, H., Dodd, K.W., Arnold, C.D. and Engle-Stone, R., 2021. Introduction to the SIMPLE macro, a tool to increase the accessibility of 24-hour dietary recall analysis and modeling. Journal of Nutrition 151: 1329-1340. https://doi.org/10.1093/jn/nxaa440

Milne, D.B., Davis, C.D. and Nielsen, F.H., 2001. Low dietary zinc alters indices of copper function and status in post-menopausal women. Nutrition 17: 701-708.

National Research Council (NRC), 1986. Nutrient adequacy: assessment using food consumption surveys. NRC Subcommittee on Criteria for Dietary Evaluation, Washington, DC, USA.

National Research Council (NRC), 1989. Food and nutrition board recommended dietary allowances, 10th edition. National Academy Press, Washington, DC, USA.

Prasad, A.S., Brewer, G.J., Schoomaker, E.B. and Rabbani, P., 1978. Hypocupremia induced by zinc therapy in adults. JAMA 240: 2166-2168. https://doi.org/10.1001/jama.1978.03290200044019

Samman, S. and Roberts, D.C.K., 1988. The effect of zinc supplements on lipoproteins and copper status. Atherosclerosis 70: 247-252. https://doi.org/10.1016/0021-9150(88)90175-X

Scientific Committee for Food (SCF), 1993. Nutrient and energy intakes for the European Community. Reports of the Scientific Committee for Food, 31st Series. Food-Science and Technique, Luxembourg, Luxembourg.

Tooze, J.A., Midthune, D., Dodd, K.W., Freedman, L.S., Krebs-Smith, S.M., Subar, A.F., Guenther, P.M., Carroll, R.J. and Kipnis, V., 2006. A new statistical method for estimating the usual intake of episodically consumed foods with application to their distribution. Journal of the American Dietetic Association 106: 1575-1587.

Verkaik-Kloosterman, J., 2011. Estimation of micronutrient intake distributions: development of methods to support food and nutrition policy making. Wageningen University and Research, Wageningen, the Netherlands.

Verkaik-Kloosterman, J., Dekkers, A.L.M., de Borst, M.H. and Bakker, S.J.L., 2019. Estimation of the salt intake distribution of Dutch kidney transplant recipients using 24-h urinary sodium excretion: the potential of external within-person variance. American Journal of Clinical Nutrition 110: 641-651. https://doi.org/10.1093/ajcn/nqz134

World Health Organization (WHO), 2004. Vitamin and mineral requirements in human nutrition. WHO, Geneva, Switzerland.

Yadrick, M.K., Kenney, M.A. and Winterfeldt, E.A., 1989. Iron, copper, and zinc status: response to supplementation with zinc or zinc and iron in adult females. American Journal of Clinical Nutrition 49: 145-150. https://doi.org/10.1093/ajcn/49.1.145

17. Nutrition and health

This is what you need to know about how nutrition research efforts help prove the health effects of foods

Alie de Boer

Food Claims Centre Venlo, Faculty of Science and Engineering, Maastricht University, Nassaustraat 36, 5911 BV Venlo, the Netherlands; a.deboer@maastrichtuniversity.nl

Abstract

Since the 1980s, functional foods have been developed as products that deliver benefits beyond their normal nutritional value. Increased welfare has allowed consumers to buy food products not merely to prevent hunger, but also to maintain or improve their health. At the same time, important technological and scientific advances have allowed for the development of and research into healthy food products. Today, nutrition and health are key topics in the development of foods, both for the general public as well as for targeted groups that benefit from specific developments. But what does it mean when a product is proved to be healthy? What kind of scientific evidence is needed to be able to claim such health effects? Following an introduction to the terminology used to define 'healthy food products', I describe how nutritional science and pharmacology have developed over the last decades, and how this has influenced the definition of health. And whereas from a regulatory perspective, health effects of food (components) need to be substantiated by well-controlled trials, scientific evidence providing insights into nutrition can come from various types of studies. Finally, this chapter highlights potential developments in designing research for substantiating health effects, such as N-of-1 trials for personalised nutrition.

Bart Wernaart and Bernd van der Meulen (eds)
Applied food science
DOI: 10.3920/978-90-8686-933-6_17, © Alie de Boer 2022

Key concepts

- ▶ Functional foods are (processed) food products that provide health benefits beyond the nutritional effects of their 'regular' counterparts, due to the ingredients that are added or removed from these products.
- ▶ Botanicals are substances or extracts derived from plants, algae, fungi, or lichens and are often a mixture of compounds.
- ▶ Nutrition claims are regulated statements or other representations stating, implying, or suggesting the (beneficial) nutritional content of a food.
- ▶ Health claims are regulated statements or other forms of representation that state, imply, or suggest that a nutrient or other food constituent has certain health benefits.
- ▶ 'Ability to adapt' is a term used to define health, referring to the ability of a healthy person to respond to (external) challenges within the 'normal' or 'healthy' homeostatic range.
- ▶ Products, services and/or advice (or a combination of these) that offer the personalisation of individuals' nutrition based on their needs and preferences, using demographic and lifestyle-related phenotypic and/ or genotypic information is named personalised nutrition.
- ▶ Randomised controlled trials (RCTs) are experimental studies in which two groups of (mostly healthy) volunteers are exposed to a compound of interest or a placebo, under controlled circumstances, after which biomarkers of interest are measured in these volunteers. This study design is considered to provide the highest quality evidence to establish causal relationships between nutrition and health.
- ▶ N-of-1 or single-subject trials are trials in which a health outcome or behaviour is measured repeatedly in one (healthy) participant.

Case 17.1. Kale as superfood.

Since the early 2000s, kale has become an increasingly popular crop for farmers to grow and consumers to buy (Šamec *et al.*, 2018). This seems to have been triggered by the fact that marketeers have labelled kale as a 'superfood': a seemingly healthy product, rich in nutrients and – supposedly – supporting health. But the term 'superfood' has also been used to highlight numerous other products, including but not limited to blueberries, goji berries, chia seeds, and wheatgrass (Van den Driessche *et al.*, 2018). Every few months, a new food product seems to be hyped up using the term 'superfood'. Even though different sources of evidence might show that these products contain a variety of nutrients, or suggest that some of the components affect mechanisms that are thought to be relevant in health effects (such as antioxidant status), there is no conclusive evidence to suggest that any of these products are more important than other foods normally included in food-based dietary guidelines (Harvard T.H. Chan School of Public Health, 2021; Van den Driessche *et al.*, 2018). The term 'superfood' is therefore often considered an 'empty' marketing term to stimulate sales of a product, without allowing consumers to benefit from any actual substantiated health effect following consumption of these products. These cases show that whilst scientific information about nutrition can originate from different sources, it is important to critically assess whether this scientific evidence is firstly of good quality and secondly, can also be translated to human health.

17.1 Introduction

For centuries, the main purpose of food has been to provide sufficient nutrients and prevent undernourishment (Afman and Müller, 2006; Georgiou *et al.*, 2011). But in today's society, food and nutrition play a different role: next to satisfying hunger and providing necessary macro- and micronutrients, nutrition is increasingly used to support and sometimes even optimise health (De Boer, 2015; Witkamp, 2021). Economic and technological advancements since the 1950s, in particular advancements in food manufacturing processes, have allowed for the development of new foods that could play a role beyond merely providing nutrients. These products are known as functional foods (Jones and Jew, 2007; Katan and Roos, 2004). Whilst epidemiological studies show the important role of nutrition in maintaining health and reducing the risk of developing specific diseases, maintaining a healthy diet (e.g. high in fruits, vegetables, unrefined grains and low in saturated fats and sodium) is known to be challenging for consumers (Katan and Roos, 2004). The development of functional foods was therefore considered a response to the consumer demand for foods that could support their health and quality of life (Jones and Jew, 2007; Katan and Roos, 2004). The increased consumer interest in the potential health-enhancing effects of foods, together with an ageing population that many countries were witnessing, created the opportunity for food businesses to develop new food products aimed at maintaining or improving health (Georgiou *et al.*, 2011; Gulati and Berry Ottaway, 2006; Menrad, 2003; Weenen *et al.*, 2013).

Key to the marketing of these products is that the benefit of consuming the product can be communicated to consumers (de Boer, 2021). In the European Union, this can only be done in the form of a nutrition claim or a health claim[52]. A nutrition claim describes the nutritional content of a product, for example

nutrition claim

---

[52] Regulation (EC) No 1924/2006 of the European Parliament and of the Council of 20 December 2006 on nutrition and health claims made on foods. Official Journal of the European Union L 404: 9-25. *Consolidated version 13 December 2014.*

health claims

describing that a product is high in fibre or that it is 'light' (because it has reduced calories, reduced carbohydrates, or reduced fat content). With health claims, food producers can go one step further: a health claim describes the beneficial effect of consuming the ingredient on maintaining health or reducing a specific risk factor in disease development. Whereas function claims (the first type of health claim) describe the role that a nutrient or ingredient has in regular bodily processes, the latter type of health claims, risk reduction claims, link the intake of an ingredient to reducing a specific risk factor in the development of a disease. An example of such a risk reduction claim is the claim about plant stanols in margarine, which lower LDL cholesterol. Lowering LDL cholesterol has been associated with decreasing the risk of developing cardiovascular diseases.

scientific evidence

**pre-market
authorisation**

Globally, such advertisements are often regulated: in different countries and jurisdictions, such claims are required to be based on scientific evidence (De Boer and Bast, 2015a; Domínguez Díaz et al., 2020). In the EU, nutrition and health claims can only be used on a product once they are authorised by the European Commission.[53] This pre-market authorisation decision is mainly based upon the opinion of the European Food Safety Authority (EFSA) on the scientific substantiation submitted for such a claim (De Boer, 2021; EC, 2006; Lenssen et al., 2018). But whilst upon the entry into force of the European nutrition and health claim regulation in 2006, over 40,000 putative health claims were submitted to the European Commission, fifteen years later a meagre 265 claims were authorised (De Boer, 2021; De Boer et al., 2014). Next to discussions about the clarity of the procedures in place (De Boer and Bast, 2015b; Khedkar et al., 2016; Lenssen et al., 2018), this has also been attributed to debates about what constitutes scientific substantiation for health effects (De Boer, 2021; Witkamp, 2021). In this chapter, I will provide an overview of methods used in nutritional sciences to study health and health benefits.

17.2 The broad category of foods

terminology

Before being able to address which methods are used to measure health benefits of foods and nutrition, it is important to gain an overview of the terminology used in scientific literature as well as popular media to describe the actual compounds, ingredients and products that are considered 'health enhancing' components of food. Around the 1980s, the food industry actively started to develop and commercialise functional foods: foods that deliver benefits beyond their normal nutritional value (Georgiou et al., 2011; Katan and Roos, 2004;

---

[53] Regulation (EC) No 1924/2006 of the European Parliament and of the Council of 20 December 2006 on nutrition and health claims made on foods. Official Journal of the European Union L 404: 9-25. *Consolidated version 13 December 2014.*

Siró *et al.*, 2008). Subsequent to the introduction of functional foods, other food products that focused on health were also brought to market, including for example fortified foods, food supplements and nutraceuticals (Gulati and Berry Ottaway, 2006). These products generally have in common that they are foods with actions that go beyond normal nutritional properties of food consumption, due to an ingredient they do not contain (because it is removed in the production process) or due to the addition of an ingredient. Substances that are added to elicit a health effect following consumption of the specific product are considered active ingredients, and when they come from nature

bioactives these components are mostly referred to as 'bioactives' or 'bioactive substances' (Biesalski *et al.*, 2009). The marketing of foods for their specific health benefits is nowadays not necessarily restricted to products that have undergone processing; also, the functionality of conventional products or specific nutrients that they contain are increasingly highlighted (De Boer, 2015; Proestos, 2018; Van den Driessche *et al.*, 2018). These products are often referred to as

superfoods 'superfoods' (Proestos, 2018) and include for example tomatoes containing lycopene (Van Steenwijk *et al.*, 2020), or blueberries in which the anthocyanin content is associated with health benefits (Kalt *et al.*, 2020). Considering that all nutrients serve specific functions in the body, it has simultaneously been questioned in literature whether foods can even be considered *not functional*, and the different terms therefore have been challenged (Scrinis, 2008). Still, in scientific literature describing the effects of nutrition and healthy diets, different terms are often used interchangeably. A snapshot of terminology used, and their associated meanings, are indicated in Table 17.1.

These terms again highlight that the aim of consuming a food product may have changed for specific foods, from merely serving to support the nutritional intake to improving health. Still, the definitions used in Table 17.1 are different from terminology that is in legislation restricted to medicinal products: products that can be used in the prevention, treatment, or cure of diseases; or claim to do so.[54] From a regulatory perspective, the claimed effect of a product determines whether the product is regulated as a medicinal product or as a food product. This highly influences which terminology can be used to promote a food product for the health benefits it is suggested to elicit (De Boer, 2021; Röttger-Wirtz and de Boer, 2021).

---

[54] Directive 2001/83/EC of the European Parliament and of the Council of 6 November 2001 on the Community code relating to medicinal products for human use. Official Journal of the European Union L 311: 76-128. *Consolidated version 25 May 2021.*

Table 17.1. Terminology for 'health-enhancing' food products (adapted from De Boer, 2015).

Term	Definition
Bioactive	Component in food or drug which is able to modulate a biological process, originating from nature (Biesalski *et al.*, 2009; Mateo Anson *et al.*, 2012).
Cosmoceuticals	Commonly defined as cosmetic products with bioactive substances which elicit positive cosmetic effects (Harrison-Dunn, 2015).
Food for special medical purposes	Food used under medical supervision which is specifically processed or formulated and intended for the dietary management of patients, for partial or exclusive feeding of patients with limited capacity to use ordinary food or specific nutrients or with other medically determined nutrient requirements.[1]
Food supplement	Concentrated source of a nutrient or other nutritional or physiological active substances aimed to supplement the normal diet, marketed in dose form.[2] In jurisdictions outside of the EU often defined as 'dietary supplement' (Lenssen *et al.*, 2019).
Functional foods	Also known as 'health foods'. Commonly defined as food products which provide health benefits beyond normal nutritional effects due to biologically active components (European Food Information Council, 2015; Katan and Roos, 2004).
Health-enhancing product	Commonly defined as a component/ingredient or food which can boost health.
Nutraceuticals	Commonly defined as food products with bioactive substances which elicit positive, drug-like effects on health (Harrison-Dunn, 2015).
Nutricosmetics	Commonly defined as food products with bioactive substances which elicit positive cosmetic effects (Harrison-Dunn, 2015).
Superfoods	Marketing term for food products with supposedly high amounts of healthy nutrients, claiming to lead to health benefits. Mostly referring to conventional, whole foods that are known to have healthy constituents (Proestos, 2018; Voedingscentrum, 2015).

[1] Regulation (EU) No 609/2013 of the European Parliament and of the Council of 12 June 2013 on food intended for infants and young children, food for special medical purposes, and total diet replacement for weight control. Official Journal of the European Union L 181: 35-56. *Consolidated version 28 April 2021.*

[2] Directive 2002/46/EC of the European Parliament and of the Council of 10 June 2002 on the approximation of the laws of the Member States relating to food supplements. Official Journal of the European Union L 183: 51-75. *Consolidated version 20 March 2021.*

17.3 The development of nutritional sciences and pharmacology

The strong relationship between food and pharmaceuticals is exemplified by the well-known phrase that has been attributed to Hippocrates, in 500 BC: '*Let food be your medicine and medicine be your food. Only nature heals, provided it is given the opportunity.*' (Georgiou *et al.*, 2011; Witkamp and van Norren, 2018). Where in traditional medicine this connection between food and medicinal products is still well acknowledged, both specialisations have developed separately in Western societies (De Boer, 2015). Whilst nutrition **nutritional research** and nutritional research originally focussed on the safety of food and on providing sufficient nutrients to prevent undernourishment, pharmacology developed as the study of the effects on organs or body functions elicited by natural or synthetically derived active compounds, focussing on single components and single targets to elicit these effects (Georgiou *et al.*, 2011; Witkamp, 2021). However, the increasing abilities to process foods, newly generated insights into adverse effects of overconsumption, and the sparked interest of consumers in health have transformed the nutritional sciences into a multi-disciplinary field, that includes research areas such as metabolomics, epidemiology, behavioural sciences and psychology (Afman and Müller, 2006; Allison *et al.*, 2015; Georgiou *et al.*, 2011). One of the current trends in nutritional research is to study the effects of specific food components with the so-called 'omics' or 'system biology' approach, thereby combining data generated by analysing effects on gene-expression patterns (transcriptomics), protein-expression patterns (proteomics) and on metabolite profiles (metabolomics) (Afman and Müller, 2006; Van Ommen *et al.*, 2009; Van Ommen and Stierum, 2002). By combining such measurements on different levels of complex biological systems, a holistic picture of disease processes and effects of active components can be generated (Lindon *et al.*, 2006; Mollet and Rowland, 2002; Nicholson *et al.*, 2007; Van Ommen *et al.*, 2009; Van Ommen and Stierum, 2002). These data can be used in new strategies such as personalised nutrition and personalised medicine, when the individual genetic makeup is taken into account with advising or administering nutrition and medicinal products (Ginsburg, 2001; Lindon *et al.*, 2006; Sikalidis, 2018; Van Ommen *et al.*, 2009). These advances have also resulted in new insights into the potential effects of individual components from the diet, and their combined effects.

pharmacology   In pharmacology, much effort has been put into understanding and developing single mechanisms that can be targeted with selective acting agents (Bast and Hanekamp, 2013). Such selectivity should lead to optimally effective medicinal products that would provide few side effects (Pierce, 2012). It is however increasingly recognised that seemingly selective compounds also actually work through multiple, non-specific effects that follow from ingestion (Pierce, 2012). Pharmacological research is therefore also shifting from studying single, highly selective and potent drugs with a single medicinal product to research into pathways and multi-target pharmacology and the use of 'dirty drugs' affecting more sites of action, thereby lowering multiple risk factors for diseases (Eussen *et al.*, 2011; Georgiou *et al.*, 2011). Whilst nutrition research has shifted towards pharmacology, pharmacology is also being seen to increasingly rely on expertise gained from nutrition research (De Boer, 2015). This is for example shown in the application of insights gained into metabolic regulation of different processes, such as bioavailability, in pharmacology (Georgiou *et al.*, 2011). At the same time, knowledge about the complexity of pathological disturbances helps provide an understanding of the fate of nutritional components, by applying pharmacodynamic and pharmacokinetic knowledge to nutritional sciences (Georgiou *et al.*, 2011; Witkamp and van Norren, 2018).

As previously described in the literature by e.g. Georgiou *et al.* (2011), Mark-Herbert (2004) and Witkamp (2021), the increased emphasis on the health-enhancing capacities of foods and food ingredients shows how the fields of nutrition and pharmacology are growing ever closer together. In addition to the aims of consuming products, and maintaining and improving health, the boundaries between the concepts of health and disease are also thought to be blurring (Georgiou *et al.*, 2011; Witkamp, 2021). This will be further elaborated upon in the next section.

17.4. Defining and measuring health effects

health, definition

Methods for measuring health effects are greatly influenced by how we define health. Following the Second World War, the World Health Organization (WHO) in 1946 defined health as 'a state of complete physical, mental and social well-being, and not merely the absence of disease or infirmity' (WHO, 2020). This definition was considered an important shift away from defining health merely as the 'absence' of diseases or pathologies (The Lancet [editorial], 2009). Still, the definition is highly debated in scientific literature today: the definition is still considered to approach health as an 'absolute' state of well-being and it has been suggested that it is an ill-fit with the rising incidence of chronic diseases (Huber *et al.*, 2011; The Lancet [editorial], 2009; Witkamp, 2021). As described by Witkamp (2021), the definition is also described as focusing too much on the present status (Witkamp, 2021). Scientific methods have allowed for the development of new insights into what is considered a 'healthy' versus 'diseased' state (Van Ommen *et al.*, 2009; Witkamp, 2021). These new measurements and the understanding of homeostasis, the maintenance of a steady state in the human body to which many physiological processes contribute, originate among other things from studying how nutrients affect gene expression (nutrigenomics) (Elliott *et al.*, 2007; Van Ommen and Stierum, 2002). This has led different researchers to suggest defining health as 'the ability to adapt', resilience, or phenotypic flexibility: the capability of the human body to return to a homeostatic state that follows upon specific challenges (Stroeve *et al.*, 2015; Van Ommen *et al.*, 2009; Witkamp, 2021).

When health and disease are considered static states, analyses are aimed at identifying how specific compounds such as foods or pharmaceuticals affect these states. This would imply that a clinical endpoint needs to be measured, a 'characteristic [...] that reflects how someone feels, functions or survives' (Boobis *et al.*, 2013; Witkamp, 2021). As it is often impossible to measure such endpoints

biomarkers

directly in research (for example, due to the time it takes after an intervention for the clinical endpoint to develop), surrogate endpoints or biomarkers can be measured as an indicator to signal the development of a disease (Boobis *et al.*, 2013; Stroeve *et al.*, 2015; Witkamp, 2021). Such surrogate measurements can be taken from blood, urine, or other body fluid analyses, but data generated through genomics and metabolomics, for example, can also serve as biomarkers (Boobis *et al.*, 2013). Biomarkers can be used to identify exposure to foods (biomarkers of exposure, e.g. serum concentrations), whether such exposure results in a biological response (are normal bodily functions affected), or provide insights into whether a substance can improve health or reduce risk factors for disease (as intermediate endpoints) (Boobis *et al.*, 2013; Witkamp, 2021).

randomised clinical
trials

The effect of exposure to a food ingredient or component can be estimated through randomised clinical trials, RCTs. Based on evidence-based medicinal practices, RCTs are considered the golden standard for establishing a causal relationship between consumption of a food (ingredient) and the biological effect (Blumberg *et al.*, 2010). In a RCT to study the health effects of nutrition, two groups of volunteers are exposed to exactly the same meal or diet, with the only difference being the consumption of the compound of interest (the intervention) or a placebo to this intervention. Researchers expect that this compound affects a specific clinical endpoint or biomarker of interest. Volunteers are randomly assigned to receiving the intervention or the placebo, preferably without either the participant or researcher knowing this, making it a double-blind study. This randomisation should allow for both groups to be relatively similar, which should minimise the risk of bias (Hariton and Locascio, 2018). The intervention group and placebo group undergo the same procedures, and in the final analysis researchers examine whether this compound of interest indeed had an effect on their outcome of interest. If a double-blind RCT shows a difference between the intervention group and the placebo group on the biomarker or endpoint of interest, this is attributed to the only difference in volunteers: the treatment of receiving (or not receiving) the product of interest, the intervention. As this is a highly controlled setting, an RCT is considered to have the lowest risk of bias, thus leading to the highest quality of evidence about how a compound can affect the outcome of interest (Willett, 1998).

The fact that RCTs are considered the gold standard in health research has also affected their role in the substantiation of health claims in the European Union. Regulation 1924/2006, the Nutrition and Health Claim Regulation, specifies that claims need to be substantiated with scientific evidence. At the request of the European Commission, the European Food Safety Authority EFSA conducts a scientific assessment of the submitted health claim dossier. EFSA Panel members critically review whether the submitted scientific evidence indeed proves a causal relationship between a well-characterised ingredient from food and a well-defined beneficial physiological effect (De Boer, 2021; Lenssen *et al.*, 2018). The requirements for health claim dossiers are further defined in Commission Regulation (EC) No 353/2008[55] and in different EFSA guidance documents that provide either general administrative scientific specifications (EFSA, 2021), or specific scientific guidance for different health effects such as those related to the functioning of the nervous system (EFSA, 2012a) or physical performance (EFSA, 2012b). EFSA's guidance documents were the result of various scientific research projects into which scientific evidence should be required for supporting health benefits of foods (Aggett *et al.*, 2005; Boobis *et al.*, 2013; De Boer *et al.*, 2014; Diplock *et al.*, 1999). This has resulted in a requirement for the 'highest possible standard' of evidence (Luján and Todt, 2020). The development of scientific opinions on the dossiers that are submitted by industry have also been described by Panel members as a 'huge peer review process' (De Boer and Bast, 2015b). Even though this is not further specified anywhere in legislation or guidance documents, in practice the need to substantiate the causal link between consumption of a compound and a health benefit is often translated into needing two independently conducted, well-designed RCTs in which – upon ingestion of this food or food ingredients – significant differences are measured in one relevant biomarker. Proving the health benefits of consuming foods is, however, often challenging for various reasons: nutrition is known to result in multiple, but subtle, effects; the biomarkers that are used may be more appropriate for reflecting a state of disease rather than health processes; and it may require a long timeframe to establish the actual impact of nutrients (Blumberg *et al.*, 2010; De Boer *et al.*, 2014; Stroeve *et al.*, 2015; Weseler and Bast, 2012; Witkamp, 2021). The low number of authorised health claims has therefore also been (partially) explained in the literature to be due to the reliance on RCTs to establish health effects of nutrition (De Boer, 2021; Lenssen *et al.*, 2018; Todt and Luján, 2021).

human intervention studies Trials, or human intervention studies, are considered the type of experimental study design that has the best ability to establish causal relationships between the intake of substances and biomarkers of health. RCTs are therefore found at the top of hierarchy of evidence pyramids, listing different study designs based on the relative strength of the evidence produced by the study design (Dahlgren Memorial Library, 2021; Willett, 1998). Only meta-analyses and systematic reviews, studies in which the results of multiple original experimental studies are grouped and critically reviewed, are considered to provide a higher strength of evidence. Most hierarchies of evidence include these studies at the

---

[55] Commission Regulation (EC) No 353/2008 of 18 April 2008 establishing the implementation of rules for applications for authorisation of health claims as provided for in Article 15 of Regulation (EC) No 1924/2006 of the European Parliament and of the Council. Official Journal of the European Union L 109:. 11-16. *Consolidated version 21 December 2009.*

top of the pyramid, even though these types of studies cannot be considered experimental study designs as such, but rather provide a 'lens' with which to group and analyse individual studies (Murad *et al.*, 2016). Since its introduction in 1979 by the Canadian Task Force on the Periodic Health Examination (Canadian Task Force on the Periodic Health Examination, 1979), authors from different fields have suggested adaptions to both the content (including or excluding specific study designs) and the shape of the hierarchy (Burns *et al.*, 2011; Milano, 2015; Murad *et al.*, 2016; Tugwell and Knottnerus, 2015). The utility of the hierarchy of evidence, and especially the emphasis on RCTs, has however been criticised for nutrition research (Satija *et al.*, 2015). The complex nature of nutrition research and dietary intake may not be sufficiently well resembled in RCTs, and prospective, observational studies may therefore be more appropriate and feasible for studying the effects of nutrition on health (Satija *et al.*, 2015).

In Table 17.2, I provide an overview of study designs and types of evidence that may be appropriate in identifying relevant information for nutrition research purposes. For all study designs listed, there is not one study design that should be relied on completely (Tugwell and Knottnerus, 2015). It is rather the combination of different sources of evidence that can provide in-depth insights into what effects substances have and through what mechanisms these substances cause these health benefits. This is also reflected in scientific guidance documents of EFSA for health claims, in which it is stipulated that published and unpublished studies should be included in the dossier (EFSA, 2021). A systematic and transparent search must be conducted in the published literature, to identify relevant studies on the relationship that is put forward in the dossier. These studies need to be reported upon, to show how to support the health benefit of the substance of interest (EFSA, 2021).

Study designs that are traditionally lower in this hierarchy are observational studies: non-experimental studies in which no intervention is made to the normal (dietary) patterns, but in which observations are made to relate intake to certain health outcomes (Bouter *et al.*, 2005). This can be done prospectively ('looking ahead'), by letting people collect data over a certain period of time, or retrospectively ('looking back'), asking people to reflect on their intake and linking this to their health status at a certain point in time. A prospective cohort study is an example of an observational study design, in which a group of people is followed for a long period of time. At the start of the study, participants are free of the disease in which the researchers are interested. By means of questionnaires and analyses of biomarkers over time, the occurrence of a specific health outcome or disease can be related to exposure to a specific compound from, for example, the diet (Boushey *et al.*, 2006).

Table 17.2. Study designs in nutrition research.

Study design	Methodological quality requirements and considerations	Considerations for nutritional studies
Meta-analyses & systematic reviews	• Accounting for publication bias. • Including high-quality studies of interest.	Allows for critically grouping and filtering well-designed studies to obtain more certainty about associations studied.
Randomised controlled trials (RCTs)	Reduce the risk for selection bias, information bias and confounding through: • Sufficient power (sample size) to detect effect of intervention. • Randomised. • Placebo-controlled. • Double blind execution (researchers and study subjects blinded to intervention or placebo).	RCTs allow for participants to be randomly assigned to groups, which should nullify the source of measured and unmeasured confounding. They may be appropriate for studying single nutrients and clearly defined substances, but are less suitable for studying dietary intake, as this exposure is complex, and different interactions and synergies may occur within the diet across different components (Satija et al., 2015).
Prospective observational studies (e.g. cohort studies)	• Accounting for information bias by selecting reliable exposure measurement instruments. • Accounting for selection bias (only including people in which specific health outcomes of interest have not yet been established). • Accounting for confounding by identifying and (statistically) adjusting for all relevant confounders.	Statistical analysis of obtained data can provide insights into the relative risk: what is the difference in probability of getting the outcome of interest after the exposure. Studying dietary behaviour in a population in an uncontrolled environment most resembles daily consumption. Although many factors are uncontrolled for, compared to a RCT, this design is often used in nutritional science for this reason (Satija et al., 2015).
Retrospective observational studies (e.g. case-control studies)	• Accounting for recall bias and other information bias. • Accounting for selection bias. • Accounting for confounding.	Recalling dietary information is known to be complex. Needing to report this retrospectively increases the risk for recall bias. Also, cases are selected and do not occur in the population like in a prospective study. This means that the true risk in the population may be different. Therefore, it is not the relative risk but the adjusted measure, called the odds ratio, that should be used (Fitzmaurice, 2000).

Retrospective observational studies without control groups (e.g. case series or ecological studies)	• Hypothesis-generating studies. Each study design has specific risks of biases and validity problems that need to be considered.	Hypothesis-generating studies. Such descriptive studies provide insights into potential associations between intake and health, that need to be tested further.
	• Case series (multiple case reports or case studies) are used to describe different individual cases (or patients), for which a researcher aims to establish a pattern of exposure (Bouter et al., 2005).	In the 1990s, ecological studies have provided insights into how diet affects the risk of cancer, among other things. These were, however, indirect measures. Additionally, the diet is complex and multidimensional (and not merely one exposure). Therefore, the associations found needed to be tested further (Lagiou et al., 2002).
	• Ecological studies describe observations in which groups of people are followed. Different from other designs, findings focus on the full group and not individuals (Mackenbach, 1995).	
Expert opinions	Also known as eminence-based texts.	Can provide insights based on expert knowledge but can never be considered sufficient to establish a causal relationship. Mostly used for generating hypotheses.
Basic research	When conducting in vitro laboratory research, animal studies (including efficacy studies) and other types of fundamental research, good laboratory practices (GLP) and other guidelines should be adhered to, to foster the quality of the outcomes established in these studies.	Fundamental research is important for establishing the mechanism of action by which a substance can cause a beneficial effect on human health.

One step below prospective studies are retrospective observational studies. A case-control study is an example of a retrospective observational study, in which cases and controls are compared: the outcome status is already known, which allows for selecting a group with this outcome (the cases) and a group without this outcome (the controls), that is otherwise comparable to the group of cases. Subsequently, questionnaires such as food frequency questionnaires can be used to identify whether there has been a difference in, for example, intake or exposure to specific nutrients or compounds (Boushey et al., 2006; Bruemmer et al., 2009). Retrospective observational studies are at higher risk for recall bias or information bias (Blumberg et al., 2010; Song and Chung, 2010). Prospective studies into the effects of nutrition on health require following large numbers of volunteers for a long period of time before effects are found, even on disease processes that are common in the population (Willett, 1998). Due to these risks of bias, the lack of control over the exposure, and the limited effect sizes that can be detected, the strength of the evidence collected through such studies is considered lower than from a controlled trial (Song and Chung, 2010; Trepanowski and Ioannidis, 2018; Willett, 1998). Even though these epidemiological studies have shown associations between diets and health outcomes, the methodological concerns have often given rise to criticism (Hall, 2020).

animal studies

Next to experimental and observational human studies, other types of study also play an important role in researching the effects of nutrients on health and disease. In laboratory-based animal studies, experimental animals are fed a specific diet containing or omitting the nutrient of interest, after which the changes in health are observed by researchers. To identify such health effects, various animal models have been developed to mimic specific disease processes, which allows for studying how nutrients (similarly to medicinal products) affect the disease onset or progression. In animal studies, the diet and therefore the exposure to the nutrients can be tightly controlled and these types of experiments have greatly contributed to our understanding of nutrients affecting health and disease detailing, for example, the mechanisms that are affected by specific dietary compounds (Baker, 2008; Hunter, 1998). Such studies are considered supporting evidence in health claim substantiation processes (Aggett *et al.*, 2005; EFSA, 2021). Benefit assessments are, however, mostly conducted in humans rather than experimental animals (Boobis *et al.*, 2013). Similarly, *in vitro* cell and molecular studies and modelling studies can support health claim applications by providing insights into the mechanism of action of a compound. *In vitro* experiments can, for example, provide detailed insights into the kinetics of substances or analyse single organ effects of separate compounds (Boobis *et al.*, 2013; Van den Abbeele *et al.*, 2013). Computational experiments, *in silico* research, can support the exploration and analysis of data from new methods such as nutrigenomics (Martinez-Mayorga and Montes, 2016). Research findings that have not been confirmed in humans can, however, not be considered sufficient in establishing the causal relationship between the intake of a food (ingredient) and a beneficial physiological effect for health claims in Europe (Boobis *et al.*, 2013; De Boer, 2021).

17.5 Looking ahead: the future of establishing nutritional health effects

So far, applications for new function health claims (Article 13.5 claims) in the EU seem to rely mainly on well-designed randomised controlled trials in humans that focus on single effects of single nutrients. Guidance documents for health claims stipulate that the highest quality scientific evidence is required to support such causal relationships, even though different forms of evidence are described to be potentially suitable (EFSA, 2021). Nutrition science is, however, moving forward in designing new methods better aligned with the definition of health as the ability to adapt and the pleiotropic, more subtle effects that can be expected from food and food ingredients (Weseler and Bast, 2012; Witkamp, 2021; Witkamp and Van Norren, 2018).

resilience-based
measurement
methods

Two prominent advances in nutrition to measure health benefits are firstly the development of resilience-based measurement methods, and secondly the advances in personalised nutrition strategies. Originating from the ability to adapt as a more suitable definition for health, various researchers have started studying the health effects of nutrition by employing challenge-based studies. In these studies, the homeostatic state of healthy volunteers is temporarily challenged by, for example, consuming a large amount of glucose in the oral glucose tolerance test, or the more sophisticated nutritional challenge PhenFlex, a milkshake-like product rich in energy, fats, protein and sugar (Stroeve *et al.*, 2015). Following such a challenge, it is measured how fast the healthy volunteer goes back to their steady state, to gain insights into how resilient the phenotype is to such stressors (Van Ommen *et al.*, 2014; Witkamp, 2021). Measuring the nutritional phenotype, both molecular and physiological stress responses, allowing for better measuring of actual health benefits instead of measuring single biomarkers of disease development, is made possible with -omics techniques such as nutrigenomics, epigenomics, metabolomics and microbiomics; nutrigenetics (how genes affect the response to dietary intake); the development of multibiomarker panels; and the integration of data sources through big data, machine learning and Artificial Intelligence (AI) approaches (Adams *et al.*, 2020; Elliott *et al.*, 2007; Ordovas *et al.*, 2018; Stroeve *et al.*, 2015).

personalised
nutrition

These challenge-based methods and measurement of the nutritional phenotype also allow for the analysis of the effects of foods and food ingredients on a more personal level (Schork and Goetz, 2017; Sikalidis, 2018). The field of personalised nutrition is gaining a lot of interest in research, industry and popular media (Röttger-Wirtz and De Boer, 2021), but so far, limited successful personalised interventions have been seen (Schork and Goetz, 2017). To progress in the development of personalised nutrition products and advice, in addition to data analysis techniques as described above, researchers have renewed their interest in conducting single-subject or N-of-1 trials (Potter *et al.*, 2021; Schork and Goetz, 2017; Witkamp, 2021). Single-subject trials were already used in clinical practice back in the early 1950s, to individualise strategies to deal with pain in patients. By collecting a large volume of information, by conducting several measurements within a study subject, these studies are able to identify the response of one individual to a specific intervention, whereas regular RCTs provide more general insights based on less data (Schork and Goetz, 2017). Tracking individuals over time provides insights into within-person variability, whilst meta-analyses of such N-of-1 trials could be used to analyse whether the findings are consistent over a larger group of individuals (Adams *et al.*, 2020;

Potter *et al.*, 2021). The wide accessibility of tools to generate lifestyle-based information nowadays, for example wearable devices and smartphone applications, is expected to boost the use of N-of-1 trials for studying personalised nutrition interventions (Adams *et al.*, 2020; Potter *et al.*, 2021). In addition to these methodological developments, other strategies are also being developed to optimise nutrition research on identifying health effects. This includes, but is not limited to, studying subjects remotely or in real time, facilitating domiciled feeding studies, or combining different sources of evidence by establishing dietary index scores (Hall, 2020; Shivappa *et al.*, 2014; Witkamp, 2021).

17.6 Conclusions

This chapter highlights the fact that information about the health benefits of nutrition can be generated from different types of experimental and observational studies. Currently however, randomised controlled trials are considered essential for substantiating a health claim in the European Union, and thus establishing the causal relationship between food (ingredient) intake and a beneficial physiological effect. Data obtained through other study designs can be supportive in health claim authorisation requests but are considered less or not sufficiently conclusive in supporting a claim. The use of RCTs in nutrition research seems well-aligned with the pharmaceutical approach of testing single effects of single compounds. There is, however, some debate as to whether this is the best method for analysing and establishing the health effects (versus the influence on biomarkers of disease) of nutrients and other food components (De Boer *et al.*, 2014; Van Ommen *et al.*, 2009; Witkamp, 2021; Witkamp and van Norren, 2018). Different researchers have called for the definition of health to be adapted to mean 'the ability to adapt', focusing on resilience within the homeostatic range of healthy volunteers (Huber *et al.*, 2011; The Lancet [editorial], 2009; Van Ommen *et al.*, 2009; Witkamp, 2021). This should also stimulate the measurement of other biomarkers of health (instead of biomarkers of disease) in research and could allow for studies in which combined insights into multiple effects are identified to reflect the pleiotropic effect of nutrition (De Boer *et al.*, 2014; Stroeve *et al.*, 2015; Weseler and Bast, 2012). Whilst a range of methodologies are already available to study health effects of nutrition, it is essential to continue the development of these methodologies to improve the insights that can be generated through experimental and observational studies. This will allow for more detailed insights into the actual health benefits of consuming certain foods and food ingredients, and potentially allow for more personalised strategies to optimise and support health with nutrition.

References

Adams, S.H., Anthony, J.C., Carvajal, R., Chae, L., Khoo, C.S.H., Latulippe, M.E., Matusheski, N. V., McClung, H.L., Rozga, M., Schmid, C.H., Wopereis, S. and Yan, W., 2020. Perspective: guiding principles for the implementation of personalized nutrition approaches that benefit health and function. Advances in Nutrition 11: 25-34.

Afman, L. and Müller, M., 2006. Nutrigenomics: from molecular nutrition to prevention of disease. Journal of the American Dietetic Association 106: 569-576.

Aggett, P.J., Antoine, J.-M., Danone, G., Asp, N.-G., Bellisle, F., Contor, L., Cummings, J.H., Howlett, J., Müller, D.J.G., Persin, C., Pijls, L.T.J., Rechkemmer, G., Tuijtelaars, S. and Verhagen, H., 2005. PASSCLAIM* consensus on criteria contents. European Journal of Nutrition 44: i5-i30. https://doi.org/10.1007/s00394-005-1104-3

Allison, D.B., Bassaganya-Riera, J., Burlingame, B., Brown, A.W., le Coutre, J., Dickson, S.L., van Eden, W., Garssen, J., Hontecillas, R., Khoo, C.S.H., Knorr, D., Kussmann, M., Magistretti, P.J., Mehta, T., Meule, A., Rychlik, M. and Vögele, C., 2015. Goals in nutrition science 2015-2020. Frontiers in Nutrition 2: 26.

Baker, D.H., 2008. Animal models in nutrition research. Journal of Nutrition 138: 391-396.

Bast, A. and Hanekamp, J.C., 2013. Chemicals and health – thought for food. Dose-Response 11: 295-300.

Biesalski, H.-K., Dragsted, L.O., Elmadfa, I., Grossklaus, R., Müller, M., Schrenk, D., Walter, P. and Weber, P., 2009. Bioactive compounds: definition and assessment of activity. Nutrition 25: 1202-1205.

Blumberg, J., Heaney, R.P., Huncharek, M., Scholl, T., Stampfer, M., Vieth, R., Weaver, C.M. and Zeisel, S.H., 2010. Evidence-based criteria in the nutritional context. Nutrition Reviews 68: 478-484.

Boobis, A., Chiodini, A., Hoekstra, J., Lagiou, P., Przyrembel, H., Schlatter, J., Schütte, K., Verhagen, H. and Watzl, B., 2013. Critical appraisal of the assessment of benefits and risks for foods, 'BRAFO Consensus Working Group.' Food and Chemical Toxicology 55: 659-675.

Boushey, C., Harris, J., Bruemmer, B., Archer, S.L. and Van Horn, L., 2006. Publishing Nutrition research: a review of study design, statistical analyses, and other key elements of manuscript preparation, Part 1. Journal of the American Dietetic Association 106: 89-96.

Bouter, L.M., Van Dongen, M.C.J.M. and Zielhuis, G.A., 2005. Onderzoeksopzet. Hoofdstuk 4. In: Bouter, L.M., Van Dongen, M.C.J.M. and Zielhuis, G.A. (eds) Epidemiologisch onderzoek – opzet en interpretatie. 5e herziene druk. Bohn Stafleu van Loghum, Houten, the Netherlands.

Bruemmer, B., Harris, J., Gleason, P., Boushey, C.J., Sheean, P.M., Archer, S. and Van Horn, L., 2009. Publishing nutrition research: a review of epidemiologic methods. Journal of the American Dietetic Association 109: 1728-1737.

Burns, P.B., Rohrich, R.J. and Chung, K.C., 2011. The levels of evidence and their role in evidence-based medicine. Plastic and Reconstructive Surgery 128(1): 305-310. https://doi.org10.1097/PRS.0b013e318219c171

Canadian Task Force on the Periodic Health Examination, 1979. The periodic health examination. Canadian Medical Association Journal 121(9): 1193-1254.

Dahlgren Memorial Library, 2021. Guides: Evidence-Based Medicine Resource Guide: Clinical Questions, PICO, & Study Designs. Available at: https://guides.dml.georgetown.edu/ebm/ebmclinicalquestions.

De Boer, A., 2015. Interactions between nutrition and medicine in effect and law. MultiCopy Parkstad, Maastricht, the Netherlands.

De Boer, A., 2021. Fifteen years of regulating nutrition and health claims in Europe: the past, the present and the future. Nutrients 13: 1725.

De Boer, A. and Bast, A., 2015a. International legislation on nutrition and health claims. Food Policy 55: 61-70. https://doi.org/10.1016/j.foodpol.2015.06.002

De Boer, A. and Bast, A., 2015b. Stakeholders' perception of the nutrition and health claim regulation. International Journal of Food Sciences and Nutrition 66: 321-328. https://doi.org/10.3109/09637486.2014.986071

De Boer, A., Vos, E. and Bast, A., 2014. Implementation of the nutrition and health claim regulation – The case of antioxidants. Regulatory Toxicology and Pharmacology 68: 475-487. https://doi.org/10.1016/j.yrtph.2014.01.014

Diplock, A., Aggett, P., Ashwell, M., Bornet, F., Fern, E. and Roberfroid, M., 1999. Scientific concepts of functional foods in Europe: consensus document. British Journal of Nutrition 81: S1-S27.

Domínguez Díaz, L., Fernández-Ruiz, V. and Cámara, M., 2020. An international regulatory review of food health-related claims in functional food products labelling. Journal of Functional Foods 68: 103896.

Elliott, R., Pico, C., Dommels, Y., Wybranska, I., Hesketh, J. and Keijer, J., 2007. Nutrigenomic approaches for benefit-risk analysis of foods and food components: defining markers of health. British Journal of Nutrition 98: 1095-1100.

European Commission (EC), 2006. Regulation (EC) No 1924/2006 of the European Parliament and of the Council of 20 December 2006 on nutrition and health claims made on foods. Official Journal of the European Union L 404: 9-25.

European Food Information Council, 2015. Functional Foods.

European Food Safety Authority (EFSA), 2012a. Guidance on the scientific requirements for health claims related to functions of the nervous system, including psychological functions. EFSA Journal 10: 2816.

European Food Safety Authority (EFSA), 2012b. Guidance on the scientific requirements for health claims related to physical performance. EFSA Journal 10: 2817.

European Food Safety Authority (EFSA), 2021. General scientific guidance for stakeholders on health claim applications (Revision 1). EFSA 19: 6553. https://doi.org/10.2903/j.efsa.2021.6553

Eussen, S.R.B.M., Verhagen, H., Klungel, O.H., Garssen, J., van Loveren, H., van Kranen, H.J. and Rompelberg, C.J.M., 2011. Functional foods and dietary supplements: Products at the interface between pharma and nutrition. European Journal of Pharmacology 668: S2-S9.

Fitzmaurice, G., 2000. Some aspects of interpretation of the odds ratio. Nutrition 16(6): 462-463. https://doi.org/10.1016/S0899-9007(00)00286-0

Georgiou, N.A., Garssen, J. and Witkamp, R.F., 2011. Pharma-nutrition interface: The gap is narrowing. European Journal of Pharmacology 651: 1-8.

Ginsburg, G., 2001. Personalized medicine: revolutionizing drug discovery and patient care. Trends in Biotechnology 19: 491-496.

Gulati, O.P. and Berry Ottaway, P., 2006. Legislation relating to nutraceuticals in the European Union with a particular focus on botanical-sourced products. Toxicology 221: 75-87.

Hall, K.D., 2020. Challenges of human nutrition research. Science 367: 1298-1300.

Hariton, E. and Locascio, J.J., 2018. Randomised controlled trials – the gold standard for effectiveness research. BJOG 125: 1716.

Harrison-Dunn, A.-R., 2015. In the eye of the beholder: where does the EU stand on beauty claims? Nutraingredients.com, 15 March 2015. Available at: https://www.nutraingredients.com/Article/2015/03/18/Where-does-the-EU-stand-on-beauty-claims#

Harvard T.H. Chan School of Public Health, 2021. Superfoods or Superhype? | The Nutrition Source. Available at: https://www.hsph.harvard.edu/nutritionsource/superfoods/.

Huber, M., André Knottnerus, J., Green, L., Van Der Horst, H., Jadad, A.R., Kromhout, D., Leonard, B., Lorig, K., Loureiro, M.I., Van Der Meer, J.W.M., Schnabel, P., Smith, R., Van Weel, C. and Smid, H., 2011. How should we define health? BMJ 343: d4163. https://doi.org/10.1136/bmj.d4163

Hunter, D., 1998. Biochemical indicators of dietary intake. In: Willett, W. (ed.) Nutritional epidemiology. Oxford University Press, New York, NJ, USA, pp. 174-243.

Jones, P.J. and Jew, S., 2007. Functional food development: concept to reality. Trends in Food Science & Technology 18: 387-390.

Kalt, W., Cassidy, A., Howard, L.R., Krikorian, R., Stull, A.J., Tremblay, F. and Zamora-Ros, R., 2020. Recent research on the health benefits of blueberries and their anthocyanins. Advances in Nutrition 11: 224-236.

Katan, M.B. and Roos, N.M., 2004. Promises and problems of functional foods. Critical Reviews in Food Science and Nutrition 44: 369-377.

Khedkar, S., Ciliberti, S. and Bröring, S., 2016. The EU health claims regulation: implications for innovation in the EU food sector. British Food Journal 118: 2647-2665.

Lagiou, P., Trichopoulou, A. and Trichopoulous, D., 2002. Nutritional epidemiology of cancer: accomplishments and prospects. Proceedings of the Nutrition Society 61: 217-222.

Lenssen, K.G.M., Bast, A. and De Boer, A., 2018. Clarifying the health claim assessment procedure of EFSA will benefit functional food innovation. Journal of Functional Foods 47: 386-396. https://doi.org/10.1016/j.jff.2018.05.047

Lenssen, K.G.M., Bast, A. and De Boer, A., 2019. International perspectives on substantiating the efficacy of herbal dietary supplements and herbal medicines through evidence on traditional use. Comprehensive Reviews in Food Science and Food Safety 18: 910-922.

Lindon, J.C., Holmes, E. and Nicholson, J.K., 2006. Metabonomics techniques and applications to pharmaceutical research & development. Pharmaceutical Research 23: 1075-1088.

Luján, J.L. and Todt, O., 2020. Standards of evidence and causality in regulatory science: Risk and benefit assessment. Studies in History and Philosophy of Science Part A 80: 82-89.

Mackenbach, J.P., 1995. Public health epidemiology. Journal of Epidemiology and Community Health 49: 333.

Mark-Herbert, C., 2004. Innovation of a new product category – functional foods. Technovation 24: 713-719.

Martinez-Mayorga, K. and Montes, C.P., 2016. Role of nutrition in epigenetics and recent advances of *in silico* studies. In: Medina-Franco, J.L. (ed.) Epi-informatics: discovery and development of small molecule epigenetic drugs and probes. Academic Press, Amsterdam, the Netherlands, pp. 385-397.

Mateo Anson, N., Hemery, Y.M., Bast, A. and Haenen, G.R.M.M., 2012. Optimizing the bioactive potential of wheat bran by processing. Food & Function 3: 362-375.

Menrad, K., 2003. Market and marketing of functional food in Europe. Journal of Food Engineering 56: 181-188.

Milano, G., 2015. The hierarchy of the evidence-based medicine pyramid: classification beyond ranking. Joints 3: 101. https://doi.org/10.11138/jts/2015.3.3.101

Mollet, B. and Rowland, I., 2002. Functional foods: at the frontier between food and pharma. Current Opinion in Biotechnology 13: 483-485.

Murad, M.H., Asi, N., Alsawas, M. and Alahdab, F., 2016. New evidence pyramid. Evidence Based Medicine 21: 125. https://doi.org/10.1136/ebmed-2016-110401

Nicholson, J.K., Holmes, E. and Lindon, J.C., 2007. Metabonomics and metabolomics techniques and their applications in mammalian systems. In: Lindon, J.C., Nicholson, J.K. and Holmes, E. (eds) The handbook of metabonomics and metabolomics. Elsevier, Amsterdam, the Netherlands, pp. 1-34.

Ordovas, J.M., Ferguson, L.R., Tai, E.S. and Mathers, J.C., 2018. Personalised nutrition and health. BMJ 361: k2173. https://doi.org/10.1136/bmj.k2173

Pierce, G., 2012. Should we clean up the reputation of 'dirty drugs'? Canadian Journal of Physiology and Pharmacology 90: 1333-1334.

Potter, T., Vieira, R. and De Roos, B., 2021. Perspective: application of N-of-1 methods in personalized nutrition research. Advances in Nutrition 12: 579-589.

Proestos, C., 2018. Superfoods: recent data on their role in the prevention of diseases. Current Research in Nutrition and Food Science 6: 576-593.

Röttger-Wirtz, S. and De Boer, A., 2021. Personalised nutrition: the EU's fragmented legal landscape and the overlooked implications of EU Food Law. European Journal of Risk Regulation 12: 212-235.

Šamec, D., Urlić, B. and Salopek-Sondi, B., 2018. Kale (*Brassica oleracea* var. *acephala*) as a superfood: Review of the scientific evidence behind the statement. Critical Reviews in Food Science and Nutrition 59: 2411-2422.

Satija, A., Yu, E., Willet, W.C. and Hu, F.B., 2015. Understanding nutritional epidemiology and its role in policy. Advances in Nutrition 6(1): 5-18. https://doi.org/10.3945/an.114.007492

Schork, N.J. and Goetz, L.H., 2017. Single-subject studies in translational nutrition research. Annual Reviews 37: 395-422.

Scrinis, G., 2008. Functional foods or functionally marketed foods? A critique of, and alternatives to, the category of 'functional foods.' Public Health Nutrition 11: 541-545.

Shivappa, N., Steck, S.E., Hurley, T.G., Hussey, J.R. and Hébert, J.R., 2014. Designing and developing a literature-derived, population-based dietary inflammatory index. Public Health Nutrition 17: 1689-1696.

Sikalidis, A.K., 2018. From food for survival to food for personalized optimal health: a historical perspective of how food and nutrition gave rise to nutrigenomics. Journal of the American College of Nutrition 38: 84-95.

Siró, I., Kápolna, E., Kápolna, B. and Lugasi, A., 2008. Functional food. Product development, marketing and consumer acceptance – a review. Appetite 51: 456-67.

Song, J.W. and Chung, K.C., 2010. Observational studies: cohort and case-control studies. Plastic and Reconstructive Surgery 126: 2234.

Stroeve, J.H.M., Van Wietmarschen, H., Kremer, B.H.A., van Ommen, B. and Wopereis, S., 2015. Phenotypic flexibility as a measure of health: the optimal nutritional stress response test. Genes and Nutrition 10: 13.

The Lancet [editorial], 2009. What is health? The ability to adapt. The Lancet 373: 781.

Todt, O. and Luján, J.L., 2021. Rationality in context: regulatory science and the best scientific method: Science, Technology & Human Values Science 47(5): 1086-1108. https://doi.org/10.1177/01622439211027639

Trepanowski, J.F. and Ioannidis, J.P.A., 2018. Perspective: limiting dependence on nonrandomized studies and improving randomized trials in human nutrition research: why and how. Advances in Nutrition 9: 367-377.

Tugwell, P. and Knottnerus, J.A., 2015. Is the 'evidence-pyramid' now dead? Journal of Clinical Epidemiology 68: 1247-1250. https://doi.org/10.1016/j.jclinepi.2015.10.001

Van den Abbeele, P., Venema, K., Van De Wiele, T., Verstraete, W. and Possemiers, S., 2013. Different human gut models reveal the distinct fermentation patterns of arabinoxylan versus inulin. Journal of Agricultural and Food Chemistry 61: 9819-9827.

Van den Driessche, J.J., Plat, J. and Mensink, R.P., 2018. Effects of superfoods on risk factors of metabolic syndrome: a systematic review of human intervention trials. Food and Function 9: 1944-1966.

Van Ommen, B. and Stierum, R., 2002. Nutrigenomics: exploiting systems biology in the nutrition and health arena. Current Opinion in Biotechnology 13: 517-521.

Van Ommen, B., Keijer, J., Heil, S.G. and Kaput, J., 2009. Challenging homeostasis to define biomarkers for nutrition related health. Molecular Nutrition and Food Research 53: 795-804.

Van Ommen, B., Van der Greef, J., Ordovas, J.M. and Daniel, H., 2014. Phenotypic flexibility as key factor in the human nutrition and health relationship. Genes and Nutrition 9: 1-9.

Van Steenwijk, H.P., Bast, A. and De Boer, A., 2020. The role of circulating lycopene in low-grade chronic inflammation: a systematic review of the literature. Molecules 25: 4378. https://doi.org/10.3390/molecules25194378

Voedingscentrum, 2015. Superfoods. Voedingscentrum, The Hague, the Netherlands. Available at: https://www.voedingscentrum.nl/encyclopedie/superfoods.aspx

Weenen, T.C., Ramezanpour, B., Pronker, E.S., Commandeur, H. and Claassen, E., 2013. Food-pharma convergence in medical nutrition – best of both worlds? PLoS ONE 8: e82609.

Weseler, A.R. and Bast, A., 2012. Pleiotropic-acting nutrients require integrative investigational approaches: the example of flavonoids. Journal of Agricultural and Food Chemistry 60: 8941-8946.

Willett, W., 1998. Overview of nutritional epidemiology. In: Willett, W. (ed.) Nutritional epidemiology. Oxford University Press, New York, NY, USA, pp. 1-17.

Witkamp, R.F. and Van Norren, K., 2018. Let thy food be thy medicine ... when possible. European Journal of Pharmacology 836: 102-114.

Witkamp, R.F., 2021. Nutrition to optimise human health – how to obtain physiological substantiation? Nutrients 13: 2155.

World Health Organization (WHO), 2020. Constitution of the World Health Organization. WHO, Geneva, Switzerland.

18. Nudging nutrition

This is what you need to know about nudging food choices

Anastasia Vugts[1][*] and Remco Havermans[2]

[1]HAS University of Applied Sciences, Spoorstraat 62, 5911 KJ Venlo, the Netherlands; Maastricht University Campus Venlo, P.O. Box 8, 5900 AA Venlo, the Netherlands; a.vugts@has.nl [2]Maastricht University Campus Venlo, P.O. Box 8, 5900 AA Venlo, the Netherlands; r.havermans@maastrichtuniversity.nl

Abstract

Globally, dietary patterns are rapidly shifting toward a diet that is associated with various noncommunicable diseases (NCDs) such as diabetes, cardiovascular disease, and certain forms of cancer. This dietary pattern is characterised by the excess intake of highly processed, high-fat, high-sugar, and salt-rich foods at the expense of healthier less calorie-dense foods, like fresh fruits and vegetables, legumes, nuts, and seeds. Many public health interventions and policies have been developed and applied but have not successfully improved healthy food choices yet. A little over a decade ago, nudging was introduced as a new tool to change people's behaviour for their own or societal benefit by exploiting people's psychological biases and heuristics. The advantage of this approach is that a nudge is generally a very subtle form of manipulation that does not demand much cognitive effort and can be easily avoided by the so-called nudgee, thus making it less invasive than most policy regulations and preserving the nudgee's freedom of choice. In this chapter, we explain the need for changing poor dietary behaviours, define nudging, and discuss how and to what degree healthy nutrition can be nudged. Furthermore, we will address the question of whether all nudges truly preserve autonomy. We conclude that when it comes to promoting healthy nutrition, nudging complements but does not obviate the need for more intrusive interventions and regulations.

Bart Wernaart and Bernd van der Meulen (eds)
Applied food science
DOI: 10.3920/978-90-8686-933-6_18, © Anastasia Vugts and Remco Havermans 2022

Key concepts

- ▶ Nudging is a technique rooted in behavioural science for subtly influencing people's decisions.
- ▶ Nudges are changes in the choice architecture, the context in which people make choices.
- ▶ Nudging is a form of libertarian paternalism – *libertarian* in the sense that it preserves autonomy, but *paternalistic* because an external choice architect defines the desired outcome.
- ▶ Manipulation is when an individual is purposefully directed to act in a certain way (intentional or not).
- ▶ Autonomy is an important human need comprising the need for freedom of choice, agency, and self-constitution.
- ▶ Boosts are variations on nudging, taking into account the practical and ethical shortcomings of nudges by enhancing choice capacities and agency.

Case 18.1. Effective nudging?

In 2018, an undergraduate student at Maastricht University Campus Venlo conducted a study at a local homework assistance facility. She examined whether offering fruit (i.e. apples) at the facility would help the attending teenagers to eat more fruit. Across the different rooms at the facility, the undergraduate student placed bowls of apples in plain sight (or not) and/or within reach (or not) for a few consecutive weeks, hypothesising that the readily available (that is, clearly visible *and* proximate) apples would be consumed more often. No clear, statistically significant effect of this 'positioning' nudge was found though. If anything, some of the teenagers attending the facility seemed keener to throw apples at each other rather than eat them. Nudging can be used to promote healthy nutrition (e.g. increasing the consumption of fruit), but there are various types of nudges and, clearly, not every nudge is as effective as another.

18.1 Public health nutrition and nudging

The human diet has been (and is) subject to change over time. For millennia, mankind depended on gathering, scavenging, and hunting for sustenance. Innovations, such as agriculture and industrialisation, however, had a profound impact on our diet. Early agricultural settlements would rely on a small variety of staple crops and would be plagued by famines when yields happened to be low or when harvests (for whatever reason) failed completely. The occurrence of these famines receded with the introduction of various technological innovations (e.g. new farming methods and tools, as well as developments in food processing and packaging). The nutrition transition further accelerated after World War II, having both positive and negative health ramifications. The most important transition involved a change toward higher food security and increased food safety. However, the increase in food security and safety was matched by a shift in the composition of the modern diet. This 'Western diet' is characterised by the consumption of highly processed foods and meals low in fibre but high in saturated fats, sugar, or salt (Popkin, 1993, 1998, 2017).

Barry M. Popkin (1993) was careful to point out that the rate of dietary change varies in different global regions, depending on economic and demographic factors. Furthermore, this notion implies that certain countries or regions could be in different stages of the nutrition transition. Popkin (1998) also noted that, particularly in richer countries, there now seems to be another stage of dietary change impelled by a sense of urgency considering the health costs associated with a lifestyle of excessive junk food intake. The primary diet-related health consequence is excessive weight gain leading to obesity, which is an important risk factor for injuries, disabling osteoarthritis, hypertension, and a range of non-communicable diseases (NCDs), such as type 2 diabetes, cardiovascular disease, and various cancers (see e.g. Swinburn et al., 2011). In cancer treatment, obesity is also associated with poorer outcomes, perhaps due to lowered treatment tolerance (Slawinski et al., 2020). More recently, obesity has been found to be a prominent risk factor for increased susceptibility to and severity of COVID-19 (Kwok et al., 2020).

the obesity transition

Before 1975, hardly any country in the world had an adult obesity prevalence of more than 5%. But this changed rapidly from 1975 onward. The year 1975 thus marks the start of what Lindsay M. Jaacks and colleagues (2019) call the obesity transition, when an increase in obesity prevalence was first noted in several large countries. This is what they refer to as stage 1. The initial increase in obesity is first observed among richer citizens, notably women. In the second stage, obesity rises dramatically in all layers of society, in women, men and children. In the third stage, the difference between obesity prevalence in different layers of society becomes smaller and even becomes more prominent in citizens with a lower socio-economic background. The difference in prevalence between men and women also disappears. However, the total increase of the obesity prevalence may plateau in stage 3. This third stage is now observed in several European countries. Jaacks and colleagues predict that the fourth stage of the obesity transition will entail the decrease in prevalence of obesity, but they note that too few signs of this putative stage have been observed yet. No country or region has reported a decrease in the prevalence of obesity. This can be ascribed to a lack of comprehensive policies, and Jaacks et al. (2019, p. 231) are completely correct when pointing out that this represents 'one of the biggest population health failures of our time'.

ban on junk food

Clearly, national governments are reluctant to take decisive action to curb the obesity epidemic. Why? Well, one decisive policy could be a ban on junk food, food that has no nutritional value other than energy and is typically very high in fat and sugar or salt. As far as we are aware, no government has ever banned junk food, presumably as this measure would rob citizens of (1) individual freedom of choice and (2) access to foods they enjoy and prefer. Furthermore, banning junk food would cause financial pain to companies producing and selling these foods and hence lead to a loss of jobs, potentially threatening the livelihood of many families. This does not mean that policymakers are out of options in promoting healthy nutrition. An alternative form of population health policy that does not include coercive mandates and economic incentives concerns nudging.

nudging

In their book 'Nudge: improving health, wealth, and happiness', Richard Thaler and Cass Sunstein (2008) introduce the reader to Carolyn, a director of school food services. She and her friend Adam decide to run some experiments at the schools that Carolyn caters for. They want to find out whether by arranging and displaying certain food items differently (e.g. positioning healthier foods at eye level) they might be able to change consumption of these items accordingly. Adam expects to see a big impact and he is right. Depending on the arrangement of items in the school cafeterias, the sales of certain foods increase or decrease by as much as 25%.

The vignette describing Carolyn and Adam's experiments is wholly fictional but is used by Thaler and Sunstein (2008) to introduce the reader to their big idea: *nudging*. What, for example, might Carolyn do when she discovers that she can direct consumers to make a certain choice without changing menu options? Of course, she might implement that wisdom to increase sales and become rich(er). But that is not what Thaler and Sunstein have in mind for Carolyn. Carolyn is what Thaler and Sunstein term a *choice architect*. She is responsible for organising the school cafeterias in which students make their lunch decisions. As the organisation of food items influences students' lunch choices, Carolyn has the power to promote choices that may benefit the school students' health. In other words, Carolyn can nudge.

nudge policy

A nudge is a policy implemented by a choice architect and only qualifies as a nudge when it can be labelled as a form of *libertarian paternalism*. A nudge policy is libertarian in the sense that it preserves individuals' freedom of choice, and it is paternalistic in the sense that a choice architect (either a public or private party) defines what choice options are beneficial. Any health policy that is neither a mandate nor involves economic incentives, but predictably influences an individual's choice behaviour, and is cheap and easy to avoid/

ignore, counts as a nudge. Or as Thaler and Sunstein (2008: 25) put it when referring again to their vignette of Carolyn and Adam: 'Putting the fruit at eye level counts as a nudge. Banning junk food does not.'

Early in 2011, Andrew Hanks ran a study in a school cafeteria. This cafeteria had two identical lunch lines. Hanks changed one of these into a healthier convenience line, where students would only find the healthier options of the total menu on offer in this cafeteria. He found that students overall, after the introduction of the healthy convenience line, purchased nearly 20% more healthier lunch items. However, they did not consume more healthier lunch items. In other words, this school cafeteria nudge led to increased sales of healthier lunch items but also to an increased waste of these foods (Hanks *et al.*, 2012, see also Case 18.1). Effectively nudging healthy food choices may be as easy and effective as Thaler and Sunstein (2008) originally envisioned, and then may still only impart limited health benefits. Nudges are easy, but getting the desired result – help, not harm – is hard and requires thought and testing.

In this chapter, we aim to address several questions. First, how and what kind of policies have been developed to nudge nutrition? Second, how effective are these nudges at promoting healthy food choices and consumption? Third, are some nudges more libertarian (or paternalistic) than others, and if so, is that something a choice architect should worry about? Fourth, how acceptable are nudges to individuals? And fifth, what will the future hold for healthy food nudging or other policies that aim to influence nutritional choices?

18.2 Nudging healthy nutrition – a taxonomy

Behavioural insights about how to influence human decision-making – including food choices – are nothing new. Take, for example, the greengrocer at the marketplace who shouts 'pay two, get three!', the big supermarket with its coupons for extra cheap food, or the social media influencer touting the health benefits of avocado. They all apply, knowingly or unknowingly, proven psychological tactics that influence consumers' decisions about what food to buy. Consumer food choices are often made very fast without much thought. Daniel Kahneman (2011) famously distinguishes between two ways or modes of thinking: a 'system 1' thinking that requires little cognitive effort is fast but error-prone, and is used for day-to-day decisions such as deciding what to eat; a 'system 2' thinking that is slow, effortful but reliable, and is used for important decisions such as deciding whether to switch careers and become a chef. Richard Thaler and Cass Sunstein (2008) described various nudges that mostly exploit our cognitive biases and heuristics, or system 1 thinking.

system 1 thinking

Nudges were suggested by Thaler and Sunstein (2008) to influence human decision-making relating to a broad spectrum of areas, including for example finance, sustainability, climate, and health. In all those domains, a nudge ought to help and not harm. We need a nudge when we have to make difficult and rare decisions, when we don't fully understand our options and don't know what consequences our choices can have, or when a choice promises immediate and certain benefits versus an uncertain risk for longer term harm. Nudging is embraced by policymakers and behavioural scientists as a promising new way for promoting population health (Cadario and Chandon, 2020; Hummel and Maedche, 2019). Studies on the efficacy of nudging within the area of population health have focused mainly on nutrition (Ledderer *et al.*, 2020). Food choices are difficult choices as junk food options are highly palatable and preferred (offering immediate reward versus an uncertain possibility for a long-term adverse health outcome), and nutrition information on food products is often comprehensive and hence relies heavily on consumer understanding, background knowledge on nutrition, and health consciousness. In light of these considerations and the ongoing obesity epidemic, Thaler and Sunstein would presumably agree that when it comes to promoting healthy food choices and intake, we need a nudge.

taxonomies of nudging

Several taxonomies of nudging (or choice architecture) techniques have been suggested. In this chapter, when describing different types of nudges, we will refer to the taxonomy proposed by Münscher *et al.* (2016) (see also Hollands *et al.*, 2013). They state that there are three main categories of nudges that guide people's decisions: (1) information, (2) structure, and (3) assistance nudges. Within each of these categories, several types of nudges can be discerned.

information nudges

Information nudges refer to the presentation of information that may influence decision-making. These are nudges that simplify information, make information more visible, or inform the nudgee of a certain social norm. The use of front-of-pack nutrition labels/logos, or prominently displaying calorie information on menus are examples of such nudges. Messages that convey a descriptive social norm have also been applied.

structure nudges

Structure nudges refer to the (re-)arrangement of options. This includes the re-organisation of lunch items in a school cafeteria (e.g. Hanks *et al.* 2013), but also making whole wheat bread the default option in a sandwich choice situation (Van Kleef *et al.*, 2018), or increasing the amount of effort required to choose less healthy foods (see

assistance nudges

Cadario and Chandon, 2020). Assistance nudges aim to help individuals to follow through on their intentions, such as reminding them of healthy food options or helping them to forget about less healthy options. For example, positioning the less healthy menu options in the centre of the list of dishes prevents these unhealthy options from benefiting from primacy or recency effects when the customer is deciding what to eat (see e.g. Dayan and Bar-Hillel, 2011).

18.2.1 Methods of determining nudging efficacy and acceptability

In late 2020, a question that was on a lot of people's minds was whether the newly developed COVID-19 vaccines were any good. In November 2020, an interim analysis of the mRNA-based vaccine developed and produced by BioNTech and Pfizer, showed high efficacy (see Polack *et al.*, 2020). By then, the research into this vaccine had entered phase III. Phase III clinical trials are for testing the effectiveness of a new treatment (e.g. a drug or a vaccine) with a randomised controlled trial (RCT). An RCT is a scientific experiment that involves comparing at least two groups of participants: the experimental group receiving the new treatment of interest versus a control group that does not receive the new treatment (but the standard treatment, or a placebo treatment, or no treatment). The control group is necessary to establish whether the experimental treatment leads to a relatively favourable outcome. Statisticians William G. Cochran and Gertrude M. Cox (1950: 14) were careful to point out that '... if a new drug is to be tested in some ward of a hospital, the recovery rate in the ward before the drug was introduced is not a satisfactory control'. Indeed, there may be lots of reasons other than the new drug to explain why patients are now suddenly more likely to recover much faster in the ward. Randomly allocating patients to either an experimental group or a control group is necessary, with randomisation acting as a 'precaution against disturbances that may or may not occur and that may or may not be serious if they do occur' (Cochran and Cox, 1950: 8). In other words, failure to randomise can lead to unintended biases for or against the experimental treatment. As it is impossible to know all the variables that might moderate treatment effects, it is wise to randomise. Fernando P. Polack and colleagues (2020) reported that within the context of an ongoing multinational RCT, more than 40,000 participants (16 years and older) had already been randomly assigned to either receive two doses of the vaccine BNT162b2 or a placebo. At least 7 days after the second injection with this vaccine, a total of 170 participants developed COVID-19. Only 8 of these 170 participants had received the vaccine. Clearly, this vaccine is highly effective. Two doses of the vaccine lead to a marked reduction in the number of COVID-19 cases.

RCTs are considered the gold standard in medical research. But really, in any area of scientific research, randomised experimental designs are used (and required) to establish whether X (e.g. a COVID-19 vaccine) leads to Y (e.g. a reduction in COVID-19 cases). Randomised experiments can similarly be used to establish the efficacy of nudges. Does nudge X lead to healthier food choices Y? In the following sections we explore and qualify the empirical evidence for the efficacy of nudges designed to promote healthier eating behaviours. But merely establishing the utility of an intervention is not enough.

If a large proportion of individuals does not accept a policy, medical treatment, or vaccine, the intended beneficial outcome of these interventions becomes uncertain. Therefore, we will examine the evidence for the acceptability of various health-promoting nudges and discuss these findings within the framework of ethical concerns surrounding libertarian paternalism.

Case 18.2. Jamie's School Dinners.

In 2010, following the success of the documentary series *Jamie's School Dinners* from 2005, UK-based chef Jamie Oliver aimed to improve US school lunch programmes in a second season of the original series. In one of the episodes Jamie shows US school children how chicken nuggets are made. In the demonstration Jamie Oliver puts different parts of a chicken (including skin) in a blender, turning it into pink slime. Eventually he dips the pink slime in batter and breadcrumbs and starts frying the nuggets. The children seem disgusted, but when Jamie asks the children who would eat this, all of them raise their hand. This dismays Jamie who expected these children to refuse eating nuggets just like the UK children from the first season. Clearly, 'nudges' that seem to work in the UK may not be as effective in the US.

18.2.2 Nudging efficacy

Even though the concept of nudging (or at least the terminology and framework) is still young, it has managed to generate a lot of research interest allowing for meta-analyses and systematic reviews. In their recent review of the efficacy of healthy eating nudges, Romain Cadario and Pierre Chandon (2020) found no fewer than 11 meta-analyses already published before 2018. These analyses taken together suggest a relatively weak to moderate effect of nudges aiming to motivate healthier eating.

Cadario and Chandon (2020), in their meta-analysis, focused on field studies only, incorporating 299 effect sizes for healthy eating nudges. Importantly, none of these field studies qualified as an RCT. Some studies did not have a control condition, and if there was a control condition, randomisation (if at all) was not on the participant level but on the level of restaurants or cafeterias where the interventions were implemented. Overall, healthy eating nudges do seem to have an effect, albeit small. The effect size for a nudge intervention, however, depends on the type of nudge, with behavioural nudges being most effective. Behavioural nudge interventions (as Cadario and Chandon call them) are in the taxonomy of Münscher *et al.* (2016) mostly structure nudges; for example, interventions that aim to make the healthy choice more convenient or the unhealthy choice more cumbersome. Cadario and Chandon also found that nudges that discourage unhealthy food choices tend to garner larger effect sizes than do nudge interventions that aim to encourage healthy food choices. Oddly, regardless of nudge type and whether the nudge promotes healthy choices or decreases unhealthy food choices, effect sizes were markedly larger for US studies. Clearly, nudge efficacy differs between individuals but also between cultures. As demonstrated in Case 18.2, nudge interventions developed in one country may not always transfer to other countries.

We briefly searched the research literature for meta-analyses and systematic reviews on the efficacy of healthy eating nudges, in a similar manner to Cadario and Chandon (2020) and managed to identify 12 more reviews published between 2018 and 2021. Some of these reviews address the efficacy of healthy eating nudges within a wider discussion of the effectiveness of nudging (e.g. Hummel and Maedche, 2019; Ledderer *et al.*, 2020), or as a category of different behavioural nutrition interventions (Fergus *et al.*, 2021; Shimizu *et al.*, 2021). Other reviews focus more narrowly on one specific type of intervention, such as health warning labels (Clarke *et al.*, 2020) or front-of-pack nutrition labels (An *et al.*, 2021). Some researchers focused their review on specific settings, such as (1) online food ordering systems (Wyse *et al.*, 2021); (2) school cafeterias (Metcalfe *et al.*, 2021); (3) retail (Fergus *et al.*, 2021). Or reviews focused on specific population segments, such as (1) adolescents (Metcalfe *et al.*, 2021); (2) adults with low socio-economic status (Harbers *et al.*, 2020); (3) young adults (Shimizu *et al.*, 2021). Overall, all these review studies conclude that most published studies concerning healthy eating nudge interventions report positive outcomes concerning food choice, food purchase, or food intake (see e.g. Fergus *et al.*, 2021; Tørris and Mobekk, 2019; Vecchio and Cavallo, 2019). But effect sizes are often small (Harbers *et al.*, 2020; Hummel and Maedche, 2019; Ledderer *et al.*, 2020), sometimes evidence for beneficial effects is mixed (see An *et al.*, 2021; Bauer and Reisch, 2019; Metcalfe *et al.*, 2020), effect size magnitudes tend to vary a lot between studies, and the methodological quality of the reviewed studies is often very poor (see also Bauer and Reisch, 2019; Fergus *et al.*, 2021; Harbers *et al.*, 2020). The latter observation is nothing new but still disappointing. Clearly, the long-term acknowledgment of major study quality issues concerning nudging intervention studies has not led to improved quality of the studies being published. This is a serious problem, as the results of these studies may be confounded to the extent that even the weak beneficial effects of healthy eating nudges reported in these reviews represent an overestimation.

A meta-analysis cannot repair poorly designed studies complicating interpretation of the meta-analysis when it mostly includes these types of studies. One notable exception is the meta-analysis considering the effect of health warning labels on alcohol or food selection (Clarke *et al.*, 2020). Many of the included studies (though still relatively few in absolute terms) were well-controlled laboratory or online studies, and these studies taken together suggest that labels warning consumers of potential adverse health outcomes if consuming that product (in text, and sometimes accompanied with images) can have a major effect on product choice. This is promising but it still needs to be established whether such a large effect will transfer to naturalistic settings and whether that effect proves stable over the longer term.

When gauging the efficacy of healthy eating nudges, the conclusion here is necessarily tentative. Healthy eating nudges do seem to be effective. But the effect is likely to be small, especially when the nudge intervention comprises little more than a text message label (e.g. calorie/nutrition label, social norm message, with the notable exception of a warning label).

Case 18.3. The very hungry caterpillar.

In the children's book *The very hungry caterpillar* by the late Eric Carle, a day in the life of a caterpillar is described. The caterpillar wakes up hungry and starts eating and doesn't stop. At the end of the day the caterpillar anthropomorphically continues eating various products typical for human consumption (such as sausages, ice cream, and birthday cake). When the day is over the now big and fat caterpillar feels a bit nauseous. All ends well when the caterpillar eventually starts building a cocoon and turns into a beautiful butterfly. When Eric Carle passed away in 2021, an obituary in the Smithsonian Magazine referred to a 2015 interview for the Paris Review where Carle supposedly claimed to have fought bitterly with the publisher over the stomach-ache scene in his classic children's book. Having to include that scene felt wrong to Carle, who argued that the caterpillar just had a big appetite, which we all have sometimes, and that there's nothing shameful about it. This reference went viral on social media as it resonates (partly) with the problem of fat shaming and weight stigma that many people with overweight or obesity endure. But the Paris Review interview is fake. It is a satirical piece, ironically ridiculing concerns about weight stigma. This particular case is nonetheless interesting as it shows how 'well-being' can mean different things to different people.

18.2.3 Nudging acceptability

In addition to studies on the effectiveness of nudges, the acceptance of nudges has been empirically explored as well. Will the public accept nudges towards healthier food choices and if so why (not)? In an international survey, Lucia A. Reisch and Cass R. Sunstein (2016) presented participants with 15 nudge items/interventions and asked them to indicate whether they considered the hypothetical nudge to be acceptable. The less accepted nudges seemed to be related to the degree of being aware of the nudge. More implicit nudges were deemed less acceptable, suggesting that awareness and transparency influence people's experienced autonomy regarding nudges. More recently, Jonas Wachner, Marieke A. Adriaanse and Denise T.D. de Ridder (2020) explored the potential threat of nudges to autonomy in a series of studies. Participants were presented with hypothetical nudges to investigate the expected effect on autonomy in a way that resembles conversations about nudging in real-life settings in which public policy is debated. They found different results for the expected effect on autonomy depending on the type of nudge: whereas the default nudge had a high expected impact on autonomy, no such results were obtained for social norms nudges.

Cadario and Chandon (2019) conducted an online study, asking participants to rate their approval of seven different healthy eating nudges, such as descriptive nutrition labels (information nudge), salience enhancements to highlight the availability of healthy food options (assistance nudge), and enhancing convenience of healthier food choices and intake by repositioning healthy foods or decreasing portion size of unhealthy foods (structure nudges). Participants were moderately supportive of these nudges, but approval varied significantly between the types of nudges with the structure nudges rated as much less acceptable than information nudges. Note that this result is diametrically opposed to the actual relative efficacy of these nudges (see Cadario and Chandon, 2020). Cadario and Chandon (2019) thus suggest that there seems to be a trade-off between approval and effectiveness of healthy eating nudges. We believe a similar trade-off likely exists between nudges that foster healthy choices (more acceptable but less effective) versus nudges that discourage unhealthy food choices (less acceptable but more effective), as the unhealthy choice is very often the preferred choice. Importantly, simply increasing transparency when applying nudges will not automatically increase people's acceptance of nudging. System 2 nudges like salience enhancements were not particularly liked. Cadario and Chandon (2019) thus speculate that educating (not just informing) the public on the goals and relative efficacy of healthy eating nudges might increase acceptance of those nudges that are most effective (i.e. structure category nudges).

18.3 Ethics of nudging

As defined previously, a nudge is only a nudge when it qualifies as a form of libertarian paternalism. A nudge does not coerce the individual to make a certain choice, and yet it does promote making choices that are deemed beneficial by the choice architect. That latter 'paternalistic' aspect implies that certain nudges might still be thought of as undermining individual autonomy

(Selinger and Whyte, 2011; Hansen and Jespersen, 2013). In fact, one of the biggest ethical concerns (see also Chapter 3; Wernaart, 2022) about nudging is that, by exploiting people's cognitive heuristics and tendency for irrational thinking, it threatens autonomous decision-making. A counterargument is offered by Sunstein (2014) who argues that nudging is inevitable since we are already being subtly manipulated all the time. In his view, whether one should nudge is not the right question. The true moral question is in what direction choice architects should steer individuals. The contention here then is that the beneficial impact of a nudge on health and well-being legitimises any (limited) influence it has on personal autonomy. Ledderer and colleagues (2020) observantly point out that what counts as 'well-being' may well vary between individuals. It certainly varies between cultures and changes over time as illustrated in Case 18.3. Jennifer Hecht (2009) argued that Western culture values productivity and longevity over (and often at the expense of) short-term enjoyment and intense euphoric experiences, almost obsessively so. One can argue that this 'obsession' with a productive and long life is reflected by the types of public health interventions designed and tested, including healthy food nudges.

Apart from the question of whether nudging health is always morally justified, ethical discussions around nudges often conflate various conceptions of nudging and autonomy, and this has led to conceptual confusion (Engelen and Nys, 2020). Insights from a previous study indicate that ethical concerns about nudging and autonomy can be grouped into three dominant autonomy conceptualisations in nudging discussions: (1) freedom of choice, (2) agency, and (3) self-constitution (Vugts *et al.*, 2018). The impact of nudges on each aspect of autonomy is as follows. Nudges that influence freedom of choice influence what you do by promoting navigability but with the potential threat of coercion. Nudges that impact autonomy as agency influence what you think, enhancing your deliberative capacities but with the potential threat of manipulation. Lastly, nudges could have an influence on a person's self-constitution; such nudges could promote authenticity but with the potential negative consequences of indoctrination. Conceivably, the acceptability of nudges is largely determined by the degree to which one believes nudging violates one's autonomy: that is, one's freedom of choice, one's agency, or one's self-constitution.

autonomy   Whether nudges affect autonomous decision-making (and if so, in what way and to what extent) is a matter of ongoing ethical debate (Bovens, 2009; Hausman and Welch, 2010; Sunstein, 2015). These discussions tend to confuse (rather than enlighten) matters, in part because autonomy is a very broad notion referring to various concepts such as self-determination, independency, liberty, free will, authorship, freedom of choice, and so on. More recently, however, scholars have invested effort in viewing nudging in the context of more narrowly defined ideas revolving around autonomy.

Schmidt and Engelen (2020) describe some of the typical worries about nudging and autonomy while simultaneously distinguishing four dimensions of autonomy. The first dimension, freedom of choice, has to do with having options. Opponents and proponents of nudging think differently about how easily people can resist a nudge and to what extent options (and hence choice) are limited by nudges. The second dimension, volitional autonomy, involves the idea that actions should be in line with people's own preferences and desires. The worry here is that people are no longer making their own authentic choices. But proponents argue that it is written in the definition of a nudge that they should be 'easy and cheap to avoid' (Thaler and Sunstein, 2008: 6). The third dimension, rational agency, is all about the way in which we make decisions. The concern here is that many nudges exploit people's cognitive biases and heuristics, or 'System 1' thinking. This argument can be countered by responding that irrationality is inevitable: that is, no individual is suddenly 'liberated' from 'System 1' thinking in the absence of nudges. Cognitive biases are just as much part of the daily human experience as rational decision-making. Moreover, nudges come in various shapes and forms. Some exploit 'System 1' thinking, some depend on more rational and deliberate 'System 2' thinking. The fourth and last dimension of autonomy involves the absence of domination. The worry of some critics is that nudges could be wrongly used by the government to control citizens. This is a valid concern, but nefarious government leaders have all sorts of tools to control citizens. The worry also mistakenly views nudging as a new and singular technique. Indeed, many policies at all levels of society disqualify as nudges solely because they do not foster individual health and well-being. Furthermore, nudges are not necessarily less transparent than more traditional forms of public policy.

Sunstein (2015), when addressing concerns about autonomy, often compares nudges to a GPS in one's car for navigation purposes. Nudges send people in the right direction but of course if people decide to take another route they are free to do so. One could argue, as Sunstein does, that the GPS example implies that the nudgee chooses (autonomously) to be nudged. It is an amusing analogy but isn't really fitting. A more realistic comparison would be as follows. You are shopping in a supermarket and out of nowhere a voice tells you to 'go left here!'. A nudge comes without warning whereas a GPS is a known or even desired feature of a car. A consumer can choose to use the GPS or have a GPS installed in his/her car. Rarely, if ever, is the consumer asked whether s/he would like to be nudged. The idea of libertarian paternalism indicates that nudges ought to be viewed as policy measures or interventions that are mindful of a person's autonomy with the intent of preserving that autonomy as much as possible. The latter conclusion acknowledges that nudges often infringe autonomy, though only to a degree. Acknowledging such infringement, however, does not eliminate the ethical question of whether that infringement is justified.

dignity

On some of the covers of the book *Nudge* two graphically designed elephants are shown with one bigger elephant nudging a little (baby) elephant. This image, strong as it is, highlights the next ethical concern about nudging. Adults don't like to be treated like children. Instead, adults should be free to live their lives as they deem fit, taking into account personal values and goals. But the relationship between a nudger and a nudgee seems to resemble that of a parent and a child since the nudger seems to know what is best for the nudgee. For example, a nudger could, based on behavioural insights, decide that in a certain company cafeteria, employees now have to wait an additional 15 minutes before being served deep fried foods (e.g. fries). The nudger knows that such a waiting time likely results in less unhealthy choices (e.g. consumption of fries) in the cafeteria. Employees are likely to react with frustration – can't they decide for themselves whether or not they want to eat fries for lunch?

Schmidt and Engelen (2020) argue that the concern that nudges may infantilise nudgees (and hence undermine their human dignity) is an exaggeration of concerns regarding the presumed effect of nudges on different aspects/ dimensions of a person's autonomy. Certain nudges or forms of manipulation could be made transparent or could even be announced beforehand. This way people have the chance to think about their decisions and weigh their options. If the nudger in our example communicates the 15-minute waiting nudge beforehand and explains to the employees why s/he believes this nudge is important, the employees are treated as adults. They may still not like that nudge, but it would be hard to maintain that implementing the nudge in this manner is an affront to their dignity.

Illicit ends

Relatedly, nudges are always designed towards a certain end or goal, such as better health through healthier eating, more exercise, less alcohol consumption and smoking, etc. As mentioned in Schmidt and Engelen (2020), excessive forms of paternalism or a government imposing their values and goals on citizens, could be a potential consequence of implementing nudges. A hypothetical example of such a situation would be a nationally implemented anti-obesity health warning; putting pictures of morbidly obese people on products that are considered unhealthy (e.g. a hamburger, chocolate bars). Such a nudge could work (as has been shown by the anti-smoking campaigns using the same approach; Allemano, 2012). It could help people to be more acutely aware of the potential consequences of their dietary choices and thus prevent population-wide excessive weight gain. But effectiveness aside, such a nudge could have a negative impact on exactly the people being used to nudge the population to be more health conscious. Stigma is already a huge problem when it comes to obesity (Puhl and Heuer, 2010). Such a nudge would result in even more weight stigma and bias that likely exacerbates unhealthy eating behaviours in obese individuals. In an ideal situation, nudges would fit perfectly with the ends people have in mind themselves (because they fully align with their own wishes and goals). It is questionable whether such personalised nudging would be even possible. But it is clearly important for choice architects to be at least mindful of the goals, values, and aspirations of the nudgees and to ensure that the intended nudge serves their well-being.

learning effects

The final ethical concern we want to highlight here is an ethical question that results from an empirical one: If you take a nudge away, will its effect remain? Say, for example, a salient nudge (putting fruit at eye level) is taken away and the fruits are placed on the regular shelves again, will the nudge have a long-term effect on the behaviour of the nudgees? Will people still buy the same amount of fruit (assuming nudge efficacy) as if the nudge was still in place? This is an empirical question about the long-term effectiveness of nutrition nudges. Ethically, this empirical question connects with the above-mentioned concerns about dignity and autonomy. If nudges don't work in the longer term, then nudges should remain in place. However, people may then feel as if they are not being allowed to show they have learned from their 'mistakes'. If, however, nudges do work in the longer term, a nudge at some point in time can be removed. But then people may still feel as if the choices they now make were once altered to the wishes of someone else, much akin to indoctrination.

Proponents of nudging (Sunstein, 2014) counter the above arguments by saying that nudges could help people make the choices they would consciously like to make, but, without a nudge, are unable to make (because of all sorts of cognitive biases and tendencies). Nudges then serve as behavioural tools supporting the individual in his/her more general self-actualisation. Nudging then neither denies people's rights to learn from mistakes nor can it ever be perceived as a form of indoctrination. In conclusion, this is how proponents of nudging defend nudges against ethical criticisms: whenever a nudge excessively and unnecessarily undermines autonomy, is overly paternalistic, or not in line with the goals and well-being of the nudgee, it isn't a nudge. It is an effective strategy, but at the same time underscores the very importance of ethical consideration of any nudge before implementing it in the choice architecture.

18.4 Looking ahead, or the case for nudging, pricing, and boosting

As reviewed in this chapter, nudges are cheap and easy ways to try and direct individual's behaviour without having to resort to overly paternalistic, effortful, expensive, or coercive policies. In the context of promoting healthy nutrition, various nudges can be effective, but the benefits are likely to be small. As described in the sections and cases above, nudging does not always work as intended and may, occasionally, even backfire. The generally positive effects of nudges seem much too small to be able to curb the nearly global obesity epidemic and its associated risks for NCDs. Indeed, when comparing the potential effects of calorie labelling versus calorie taxes on students' lunch choices, taxing calories (elevating prices for more energy-dense food items) had a clear and much stronger effect on these choices than did labelling (Giesen *et al.*, 2011; see also Sacks *et al.*, 2011). We thus agree with Christine Tørris and Hilde Mobekk (2019) who stated that nudging '... alone will not solve the worldwide health challenges caused by poor health choices'.

From a pragmatic standpoint, healthy food nudging interventions do not preclude the implementation of additional policies that are much less libertarian, like the implementation of a calorie or sugar tax on foods. Recent studies on the joint efficacy of these measures show promising results. For example, Stuber *et al.* (2021) investigated the effect of nudges and pricing strategies (subsidies for healthy foods and taxes on unhealthy foods) on food choice in a virtual supermarket study and found the combination of these strategies (pricing and nudging) to be most efficacious in promoting healthier dietary purchase behaviour. This combined effect, though, was particularly apparent for dairy products and grains (for both encouraging healthier options and discouraging unhealthy options), but not for discouraging unhealthy beverages and snack purchases.

In a large survey study, more than 5,000 participants rated their acceptance of various policies aimed at reducing sugar intake. The suggested policies included taxation of sugary products but also nudges such as nutrition labels, and portion size reductions. Nutrition labels were most accepted, much more so than an intrusive sugar tax or portion size reduction (Hagmann *et al.*, 2018). Désirée Hagmann and colleagues conclude that as the effectiveness of a government policy strongly depends on its acceptance, the low acceptance of taxation (especially among individuals whose health status would benefit the most) should be taken into account by policymakers. But low acceptance alone, of course, does not invalidate a given policy, such as taxation. Furthermore, pricing strategies may also encompass subsidising healthy foods, which is generally more valued/accepted than taxing unhealthy foods (Waterlander *et al.*, 2010). For population-wide policies aimed at promoting healthier dietary patterns, a combination of nudging and pricing interventions appears particularly beneficial.

boosting

Apart from nudging and pricing, policymakers have yet another option for promoting healthier dietary patterns: boosting (Hertwig and Grüne-Yanoff, 2017). Boosts, likes nudges, are noncoercive interventions, but unlike nudges aim to foster individuals' competences to make their own choices. Nudges, especially structure nudges, change the behaviour of a consumer by manipulating the external choice architecture. With boosts, the primary target is the *cognitive* architecture of the individual consumer. The individual changes behaviour when s/he changes his/her mind. Where nudge interventions often exploit 'system 1' thinking, boosts deliberately try to harness 'system 2' thinking, which presumably would lead to longer lasting changes in behaviour. As Hertwig and Grüne-Yanoff note, once a nudge intervention is reverted, so is the behaviour within that choice architecture. But a change in external context will not necessarily lead to reversal of the individual's *cognitive* architecture. For example, if an individual has learned to value the consumption of ample fruits and vegetables, that value and the motivation to purchase lots of fruits and vegetables is not lost when the local supermarket changes the layout of its aisles.

Hertwig and Grüne-Yanoff (2017) point out that boosts can be education programmes but do not have to be an extensive intervention or campaign. A good example of a simple boost is the intervention designed by König and Renner (2019). In a series of three studies, participants were regularly prompted to select/eat a colourful lunch. Across studies, colourful meals included more fruits and vegetables. Simple reminders help individuals to select a more colourful and hence healthier meal more consciously. One could argue that boosts like the above prompts are still nudges that would perfectly fit within the nudging taxonomy that Münscher *et al.* (2016) has proposed (i.e. assistance nudge). In fact, Sunstein (2016) has considered boosts to be particularly attractive nudges, precisely because these interventions foster agency. The distinction between boosting and nudging seems fuzzier than Hertwig and Grüne-Yanoff (2017) have claimed, although discriminating between the route of establishing behaviour change (through altering external environment, or changing individual intentions) is meaningful. Again, from a pragmatic point of view, boosting and nudging are not mutually exclusive. Future research on how these two non-coercive types of interventions combined may be able to promote healthy food choices and intake would be particularly interesting. As far as we know, no such research has been published yet.

18.5 Conclusions

Nudging nutrition provides policymakers, or choice architects, with the option to use simple, cheap, noncoercive techniques to increase individuals' healthy food choices and intake. Not all nudges are very effective at doing so. Calorie labelling and descriptive nutrition labels in particular seem to exert very little effect on food choice and intake. More intrusive structure nudges, such as re-positioning food options to encourage healthy choices and discourage unhealthy choices, are much more effective but also much less accepted. People clearly value their autonomy, and infringements of that autonomy (whether by nudges, boosts, pricing, or mandates) are rarely considered acceptable, especially so when intrusive interventions aim to decrease unhealthy (but preferred) food choices.

Of course, one may question whether trying to direct individuals to consume healthier foods is morally justifiable in whatever form, but that poses a philosophical question that is far beyond the scope of this chapter. In summary, nudges are not simply a stand-in for more intrusive population-wide policies, but they are not superfluous either. Nudging is empowering. That is, many individuals have at some level (ranging from the household level to school classroom, to public parks, to a whole city or even country) the capacity to be a choice architect. That nudging power is limited and tackling the obesity epidemic may thus require more than a nudge. The current challenge is to find out how many more potentially intrusive or even coercive interventions and policies need to be implemented in order to enter the final stage of the nutrition and obesity transition: increased healthy eating and decreased prevalence of obesity.

References

Alemanno, A., 2012. Nudging smokers the behavioural turn of tobacco risk regulation. European Journal of Risk Regulation 3: 32-42. https://doi.org/10.1017/S1867299X00001781

An, R., Shi, Y., Shen, J., Bullard, T., Liu, G., Yang, Q., Chen, N. and Cao, L., 2021. Effect of front-of-package nutrition labeling on food purchases: a systematic review. Public Health 191: 59-67. https://doi.org/10.1016/j.puhe.2020.06.035

Bauer, J.M. and Reisch, L. A., 2019. Behavioural insights and (un)healthy dietary choices: a review of current evidence. Journal of Consumer Policy 42: 3-45. https://doi.org/10.1007/s10603-018-9387-y

Bovens, L., 2009. The ethics of nudge. In: Grüne-Yanoff, T. and Hansson, S.O. (eds) Preference change. Springer, Dordrecht, the Netherlands, pp. 207-219.

Cadario, R. and Chandon, P., 2019. Viewpoint: effectiveness or consumer acceptance? Tradeoffs in selecting healthy eating nudges. Food Policy 85: 1-6. https://doi.org/10.1016/j.foodpol.2019.04.002

Cadario, R. and Chandon, P., 2020. Which healthy eating nudges work best? A meta-analysis of field experiments. Marketing Science 39: 465-486. https://doi.org/10.1287/mksc.2018.1128

Clarke, N., Pechey, E., Kosīte, D., König, L.M., Mantzari, E., Blackwell, A.K.M., Marteau, T.M. and Hollands, G.J., 2020. Impact of health warning labels on selection and consumption of food and alcohol products: systematic review with meta-analysis. Health Psychology Review 15: 430-453. https://doi.org/10.1080/17437199.2020.1780147

Cochran, W.G. and Cox, G.M., 1950. Experimental designs. John Wiley, New York, NY, USA.

Dayan, E. and Bar-Hillel, M., 2011. Nudge to nobesity II: menu positions influence food orders. Judgment and Decision Making 6: 333-342.

Engelen, B. and Nys, T., 2020. Nudging and autonomy: analyzing and alleviating the worries. Review of Philosophy and Psychology 11: 137-156. https://doi.org/10.1007/s13164-019-00450-z

Fergus, L., Seals, K. and Holston, D., 2021. Nutrition interventions in low-income rural and urban retail environments: a systematic review. Journal of the Academy of Nutrition and Dietetics 121: 1087-1114. https://doi.org/10.1016/j.jand.2020.12.018

Giesen, J.C.A.H., Havermans, R.C., Nederkoorn, C. and Jansen, A., 2012. Impulsivity in the supermarket. Responses to calorie taxes and subsidies in healthy weight undergraduates. Appetite 58: 6-10. https://doi.org/10.1016/j.appet.2011.09.026

Hagmann, D., Siegrist, M. and Hartmann, C., 2018. Taxes, labels, or nudges? Public acceptance of various interventions designed to reduce sugar intake. Food Policy 79: 156-165. https://doi.org/10.1016/j.foodpol.2018.06.008

Hanks, A.S., Just, D.R., Smith, L.E. and Wansink, B., 2012. Healthy convenience: nudging students toward healthier choices in the lunchroom. Journal of Public Health 34: 370-376. https://doi.org/10.1093/pubmed/fds003

Hansen, P.G. and Jespersen, A.M., 2013. Nudge and the manipulation of choice: a framework for the responsible use of the nudge approach to behaviour change in public policy. European Journal of Risk Regulation 2013: 3-28.

Harbers, M.C., Beulens, J.W.J., Rutters, F., Boer, F. De, Gillebaart, M., Sluijs, I. and Van Der Schouw, Y.T., 2020. The effects of nudges on purchases, food choice, and energy intake or content of purchases in real-life food purchasing environments: a systematic review and evidence synthesis. Nutrition Journal 19: 103. https://doi.org/10.1186/s12937-020-00623-y

Hausman, D.M. and Welch, B., 2010. Debate: to nudge or not to nudge. Journal of Political Philosophy 18: 123-136. https://doi.org/10.1111/j.1467-9760.2009.00351.x

Hecht, J.M.,2009. The Happiness Myth. Harper Collins, New York, NY, USA.

Hertwig, R. and Grüne-Yanoff, T., 2017. Nudging and boosting: steering or empowering good decisions. Perspectives on Psychological Science 12: 973-986. https://doi.org/10.1177/1745691617702496

Hollands, G.J., Shemilt, I., Marteau, T.M., Jebb, S.A., Kelly, M.P., Nakamura, R., Suhrcke, M. and Ogilvie, D., 2013. Altering micro-environments to change population health behaviour: towards an evidence base for choice architecture interventions. BMC Public Health 13: 1218. https://doi.org/10.1186/1471-2458-13-1218

Hummel, D. and Maedche, A., 2019. How effective is nudging? A quantitative review on the effect sizes and limits of empirical nudging studies. Journal of Behavioral and Experimental Economics 80: 47-58. https://doi.org/10.1016/j.socec.2019.03.005

Jaacks, L.M., Vandevijvere, S., Pan, A., Mcgowan, C.J., Wallace, C., Imamura, F., Mozaffarian, D. and Swinburn, B., 2019. The obesity transition: stages of the global epidemic. The Lancet: Diabetes and Endocrinology 7: 231-240. https://doi.org/10.1016/S2213-8587(19)30026-9

Kahneman, D., 2011. Thinking, fast and slow. Farrar, Straus and Giroux, New York, NY, USA.

König, L.M. and Renner, B., 2019. Boosting healthy food choices by meal colour variety: results from two experiments and a just-in-time ecological momentary intervention. BMC Public Health 19: 975. https://doi.org/10.1186/s12889-019-7306-z

Kwok, S., Adam, S., Ho, J.H., Iqbal, Z., Turkington, P., Razvi, S., Le, C.W., Handrean, R. and Syed, A.A., 2020. Obesity: a critical risk factor in the COVID-19 pandemic. Clinical Obesity 10: e12403. https://doi.org/10.1111/cob.12403

Ledderer, L., Kjaer, M., Madsen, E.K., Busch, J. and Fage-Butler, A., 2020. Nudging in public health lifestyle interventions: a systematic literature review and metasynthesis. Health Education and Behavior 47: 749-764. https://doi.org/10.1177/1090198120931788

Metcalfe, J.J., Ellison, B., Hamdi, N., Richardson, R. and Prescott, M.P., 2020. A systematic review of school meal nudge interventions to improve youth food behaviors. International Journal of Behavioral Nutrition and Physical Activity 17: 77. https://doi.org/10.1186/s12966-020-00983-y

Münscher, R., Vetter, M. and Scheuerle, T., 2016. A review and taxonomy of choice architecture techniques. Journal of Behavioral Decision Making 29: 511-524. https://doi.org/10.1002/bdm.1897

Polack, F.P., Thomas, S.J., Kitchin, N., Absalon, J., Gurtman, A., Lockhart, S., Perez, J.L., Marc, G.P., Moreira, E.D., Zerbini, C., Bailey, R., Swanson, K.A., Roychoudhury, S., Koury, K., Li, P., Kalina, W.V, Cooper, D., Frenck, R.W., Hammitt, L.L., Türeci, Ö., Nell, H., Schaefer, A., Ünal, S., Tresnan, D.B., Mather, S., Dormitzer, P.R., Sahin, U., Jansen, K.U. and Gruber, W.C., 2020. Safety and efficacy of the BNT162b2 mRNA Covid-19 vaccine. New England Journal of Medicine 383: 2603-2615. https://doi.org/10.1056/NEJMoa2034577

Popkin, B.M., 1993. Nutritional patterns and transitions. Population and Development Review 19: 138-157. https://doi.org/10.2307/2938388

Popkin, B.M., 1998. The nutrition transition and its health implications in lower-income countries. Public Health Nutrition 1: 5-21. https://doi.org/10.1079/phn19980004

Popkin, B.M., 2017. Relationship between shifts in food system dynamics and acceleration of the global nutrition transition. Nutrition Reviews 75: 73-82. https://doi.org/10.1093/nutrit/nuw064

Puhl, R.M. and Heuer, C.A., 2010. Obesity stigma: important considerations for public health. American Journal of Public Health 100: 1019-1028. https://doi.org/10.2105/AJPH.2009.159491

Reisch, L.A. and Sunstein, C.R., 2016. Do Europeans like nudges? Judgment and Decision Making 11: 310-325.

Sacks, G., Veerman, J.L., Moodie, M. and Swinburn, B., 2011. 'Traffic-light' nutrition labelling and 'junk-food' tax: a modelled comparison of cost-effectiveness for obesity prevention. International Journal of Obesity 35: 1001-1009. https://doi.org/10.1038/ijo.2010.228

Schmidt, A.T. and Engelen, B., 2020. The ethics of nudging: an overview. Philosophy Compass 15: e12658. https://doi.org/10.1111/phc3.12658

Selinger, E. and Whyte, K., 2011. Is There a right way to nudge? The practice and ethics of choice architecture. Sociology Compass 5: 923-935. https://doi.org/10.1111/j.1751-9020.2011.00413.x

Shimizu, R., Rodwin, A.H. and Munson, M.R., 2021. A systematic review of psychosocial nutrition interventions for young adults. Journal of Nutrition Education and Behavior 53: 316-335. https://doi.org/10.1016/j.jneb.2021.01.002

Slawinski, C.G.V., Barriuso, J., Guo, H. and Renehan, A.G., 2020. Obesity and cancer treatment outcomes: interpreting the complex evidence statement of search strategies used and sources of information. Clinical Oncology 32: 591-608. https://doi.org/10.1016/j.clon.2020.05.004

Stuber, J.M., Hoenink, J.C., Beulens, J.W.J., Mackenbach, J.D. and Lakerveld, J., 2021. Shifting toward a healthier dietary pattern through nudging and pricing strategies: A secondary analysis of a randomized virtual supermarket experiment. American Journal of Clinical Nutrition 114: 628-637. https://doi.org/10.1093/ajcn/nqab057

Sunstein, C.R., 2014. Why nudge? The politics of libertarian paternalism. Yale University Press, New Haven, CT, USA.

Sunstein, C.R., 2015. The ethics of nudging. Yale Journal on Regulation 32 413-450. https://digitalcommons.law.yale.edu/yjreg/vol32/iss2/6

Sunstein, C.R., 2016. The ethics of influence: government in the age of behavioral science. Cambridge University Press, Cambridge, UK.

Swinburn, B.A., Sacks, G., Hall, K.D., McPherson, K., Finegood, D.T., Moodie, M.L. and Gortmaker, S.L., 2011. The global obesity pandemic: shaped by global drivers and local environments. The Lancet 378: 804-814. https://doi.org/10.1016/S0140-6736(11)60813-1

Thaler, R.H. and Sunstein, C.R., 2008. Nudge: improving decisions about health, wealth, and happiness. Yale University Press, New Haven, CT, USA.

Tørris, C. and Mobekk, H., 2019. Improving cardiovascular health through nudging healthier food choices: a systematic review. Nutrients 11: 2520. https://doi.org/10.3390/nu11102520

Van Kleef, E., Seidell, K., Vingerhoeds, M.H., De Wijk, R.A. and Van Trijp, H.C.M., 2018. The effect of a default-based nudge on the choice of whole wheat bread. Appetite 121: 179-185. https://doi.org/10.1016/j.appet.2017.11.091

Vecchio, R. and Cavallo, C., 2019. Increasing healthy food choices through nudges: a systematic review. Food Quality and Preference 78: 103714. https://doi.org/10.1016/j.foodqual.2019.05.014

Vugts, A., Van den Hoven, M., De Vet, E. and Verweij, M., 2020. How autonomy is understood in discussions on the ethics of nudging. Behavioural Public Policy 4: 108-123. https://doi.org/10.1017/bpp.2018.5

Wachner, J., Adriaanse, M.A. and De Ridder, D.T.D., 2020. And How Would That Make You Feel? How People Expect Nudges to Influence Their Sense of Autonomy. Frontiers in Psychology, 11. https://doi.org/10.3389/fpsyg.2020.607894

Waterlander, W.E., De Mul, A., Schuit, A.J., Seidell, J.C. and Steenhuis, I.H., 2010. Perceptions on the use of pricing strategies to stimulate healthy eating among residents of deprived neighbourhoods: a focus group study. International Journal of Behavioral Nutrition and Physical Activity 7: 44. https://doi.org/10.1186/1479-5868-7-44

Wernaart, B.F.W., 2022. Food ethics. In: Wernaart, B.F.W. and Van der Meulen, B.M.J. (eds) Applied Food Science. Wageningen Academic Publishers, Wageningen, the Netherlands, pp. 45-64.

Wyse, R., Jackson, J.K., Delaney, T., Grady, A., Stacey, F., Wolfenden, L., Barnes, C., Mclaughlin, M. and Yoong, S.L., 2021. The effectiveness of interventions delivered using digital food environments to encourage healthy food choices: a systematic review and meta-analysis. Nutrients 13: 2255. https://doi.org/10.3390/nu13072255

19. Food market analysis

This is what you need to know about market analysis theories

Iris van Hest

School of Business and communication, Fontys University of Applied Sciences, P.O. Box 347, 5600 AH Eindhoven, the Netherlands; iris.vanhest@fontys.nl

Abstract

Business success is sometimes viewed as an internal matter: a company has an extremely innovative business concept, employees are real ambassadors, product quality is excellent and so on. Apart from these internal factors, external forces such as customers, suppliers, industry and macro factors create opportunities and threats that greatly influence a company's success. To track down where and what to improve is the key and a challenge for every enterprise that aims to be in business in the long run and therefore a thorough market analysis is essential. Widely known theories describe red and blue oceans as a metaphor for a business environment. Analysing this environment is a basic step towards gaining insights and is necessary in order to build commercial strategies, but also to swim from a red ocean towards a blue ocean. In this chapter, we discuss the relevant market analysis theories. An external analysis determines opportunities and threats in the marketplace and is therefore the key to innovation and business success.

Key concepts

- ► To determine the attractiveness of a market, we assess the external business environment.
- ► Data-driven decision-making is relevant in today's complex and diverse societies.
- ► Customers play a key role in our external environment because there is no enterprise that can survive without customers.
- ► Suppliers are crucial in adding value to a product in the so-called value chain.

Bart Wernaart and Bernd van der Meulen (eds)
Applied food science
DOI: 10.3920/978-90-8686-933-6_19, © Iris van Hest 2022

- ▶ Besides the power of the suppliers, other industry factors, such as the rivalry among existing competitors, the possibility of new market entrants, the power of customers and the threat of substitute products or services are important to take into account when assessing market attractiveness.
- ▶ Macroeconomic trends offer opportunities and can cause threats. A business strategy is required to anticipate these changes.
- ▶ Our highly complex business environment can be understood by making sense of relevant data. Technology enables us to analyse digital data rapidly, sometimes even in real time. Prototyping can be an appropriate method for creating insights, especially related to (potential) customer needs.

Case 19.1. How Starbucks keeps on swimming towards blue oceans ...

Although it is globally consumed, when it comes to coffee each country seems to have its own preferences. With this in mind, setting up a successful global coffee brand seems like a mission impossible. Moreover, global brands in product sectors where culture plays a key role are sometimes perceived as Trojan horses (Thompson and Arsel, 2004). Trojan horse brands are international enterprises that colonise local cultures (Ritzer, 1993), in reference to the legend of the Greeks during the Trojan War.

Despite that, the marketing success of Starbucks is legendary. Starbucks is the largest coffee chain in the world, has more than 15,000 locations in the US only, and has over 30,000 establishments worldwide (2019) (Bennett, 2020). This raises the question of how Starbuck managed to become so successful.

In 1983, Howard Schultz travelled to Italy and became captivated with Italian coffee bars and the romance that is associated with coffee. It was Howard's vision to bring the Italian coffeehouse culture to the United States. That trip to Italy laid the foundations of the company we now know as Starbucks. This brand seeks to represent an amazing coffee experience through high-quality products. On top of that, there is a warm, friendly, welcoming, and inspiring environment in the Starbucks establishments, inspired by the Italian coffee houses. As Starbucks states on their website: '*We're not just passionate purveyors of coffee, but everything else that goes with a full and rewarding coffeehouse experience. We also offer a selection of premium teas, fine pastries and other delectable treats to please the taste buds. And the music you hear in store is chosen for its artistry and appeal.*' Finally, the community involvement and personal connection is a pillar that Starbucks focuses on. Think of the practice in which employees write the name of the customer on their coffee cup. Starbuck describes this as follow: '*It's not unusual to see people coming to Starbucks to chat, meet up or even work. We're a neighbourhood gathering place, a part of the daily routine – and we couldn't be happier about it. Get to know us and you'll see: we are so much more than what we brew'* (Starbucks, n.d.).

A thorough market analysis is the key to business success, especially in a market where there are many strong players: a red ocean. A red ocean can be defined as a known market space, where industry boundaries are defined and companies try to outperform their rivals to grab a greater share of the existing market. Blue oceans are the unknown market space, unexplored and untainted by competition. These blue oceans are powerful in terms of opportunity and profitable growth. By differentiating your brand from competitors it is possible to outrun your competition (Kotler, 2019). In other words: a company can swim from a red ocean towards a blue ocean if it differentiates itself so that new (sub)markets arise (Kim and Mauborgne, 2014). Starbucks created an uncontested market by turning the simple coffee drinking experience into a way-of-life experience; this was done by drastically redefining the coffee shop environment by adding music, Wi-Fi, relaxed seating, luxurious interiors but also a wide range of indulgent coffee flavours. By doing this, Starbucks is able to safely offer specialty drinks at a higher price (Miller, 2009). When taking a further look at the coffee market, the five forces of Porter (1979) reveal that threat of entry is high in this industry: incumbents must hold down their prices or boost investment to deter new competitors. Therefore, these relatively low entry barriers mean that Starbucks must invest aggressively in innovating their stores and menus (Porter, 2008). Innovation is incorporated in the culture of Starbucks. Customers are able to contribute their ideas through www.ideas.starbucks.com. Many of its products and services are a direct result of suggestions from employees and customers, e.g. the customised CD music collections, the sale of sandwiches, gums, and chocolates. Also, new macroeconomic trends are closely monitored, which leads to continuous adaptions of the Starbucks business strategy, for instance in the field of contributing to a circular economy: *'Starbucks' aspiration is to become resource positive,'* according to Michael Kobori, chief sustainability officer at Starbucks. *'This aspiration, coupled with the insight that our customers are looking for more plant-based choices, has inspired the development of exciting and delicious plant-based beverages and food. Our customers continue to look for new ways to enjoy plant-based options at Starbucks and customize their Starbucks moment, and we are delighted to introduce these new menu items to our customers over the coming years.'* (Starbucks, n.d.).

19.1 The market

The Starbucks case reveals that even with an everyday product it is possible to build an internationally successful company. Although there is a wide variety of interesting products on the market, the number of product failures in the fast-moving-consumer-goods industry (FMCG) is still high. Considering the fact that American families, on average, repeatedly buy the same 150 items, which constitute as much as 85% of their household needs, it is a real challenge to get a foot in the door (Schneider and Hall, 2011). An external analysis to determine opportunities and threats in the marketplace is therefore the key to business success. Such analyses enable businesses to distinguish themselves from their competition, to be more relevant to customers and therefore create blue oceans – a place where a company can safely offer their products at a higher price level – instead of red oceans. In the rare case of a monopoly market this need is less relevant. However, in almost every industry there is at least some level of competition and a need to develop strategies to swim towards blue oceans.

market analysis

Organisations interact with the market, most obviously by interacting with customers, suppliers, competitors, government and banks. The goal of a thorough market analysis is to create insights that support managers in strategic decision-making. In a market analysis several aspects are essential, as demonstrated in Figure 19.1.

- ► Customers: people or enterprises that purchase a product or service. Customers are extremely important stakeholders of a business. After all, without sales there can be no durable business endeavours.
- ► Suppliers: people or enterprises that provide a raw material, (semi-finished) product or service to another entity. The suppliers are also crucial stakeholders because they fulfil (business) needs.
- ► Industry factors: the industry sector(s) in which a company operates forms an important part of the external environment. Within such environmental domains, events and trends arise that affect an enterprise's actions and prospects. Moreover, the stage of the products' life circle influences the opportunities and threats that arise. In addition, the intensity and strength of the competitors play a key role.
- ► Macroeconomic factors: organisations also need to consider trends and changes wider afield than industry markets. The macro and global economic drivers generate additional opportunities, but also threats and challenges that need to be anticipated via strategic decisions.

customers

A first step in an external market analysis is to focus on customers for the simple reason that without customers a company cannot have a functional business model. Not surprisingly, this is why customers are at the top of the triangle (Figure 19.1). Every organisation, including a non-profit organisation, has customers. Since the relationship between an organisation and a customer is crucial, it is important to understand the customers. In markets where supply exceeds the demand, also referred to as overcapacity, a clear focus on the customer is especially critical. The profiling of customers to discover their wishes and needs is a marketing approach that is widely adopted. We are all

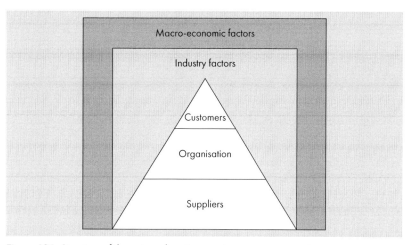

Figure 19.1. Structure of the external environment.

unique creatures with different needs and wishes that can lead to different buying criteria. At the same time, circumstances can heavily influence our customer choices (see also Case 19.2).

Case 19.2. Hurry up, I need food (and fast!).

Imagine you have an extremely busy day, working on a final presentation for your degree or job and you are feeling hungry. You glimpse a stall selling fast food. In my case, on such an evening I would very likely be one of the customers at that fast food stall. In this situation I am in need of – literally – fast food, since my presentation is not yet finished and I am feeling hungry. Two days later, on a Saturday, I might come across the same fast food place but instead pop into the shop next door and buy fresh herbs for a Thai salad. Although I am the same person, the circumstances alter my buying criteria.

suppliers

When considering the key business of a company, producing products and/or services, suppliers are crucial (Cravens and Piercy, 2013). In a so-called supply chain each enterprise adds additional value to a product, which leads to the end product. This chain can change a raw coffee bean into a delicious latte macchiato. Porter (1985) pointed out the concept of a value chain: this value chain describes the full range of activities that add value to a product or service in order to create competitive advantage. These activities are required to bring a product or service from conception, through the different phases of production (involving a combination of physical transformation and the input of various producer services) and delivery to end customers. Last but not least, the final phase of this chain involves the disposal after use of the product or service (Kaplinsky and Morris, 2000).

industry

The external environments are multi-layered, as shown in Figure 19.1. The layer immediately surrounding the organisation is the industry. This comprises the set of enterprises that supply similar products or services in comparable ways. Very simply stated, all enterprises in a sector are direct rivals to an organisation. However, competition encompasses more than just these industry competitors. After all, a specific need of a customer can be satisfied in various ways. It is possible that a product or service in another industry sector satisfies a particular demand equally well, or even better! For example, a customer's need could be to eat a tasty snack, because they are hungry. This particular need can be fulfilled with, for instance, a piece of chocolate cake or a fruit salad. Shiv and Fedorikhin (1999) even found out that customer decision-making is influenced by the complexity of tasks that are presented to us. The experiment revealed that with a higher cognitive load, a more complex task, it is harder to resist the chocolate cake!

macro layer

Beyond the industry level is the seemingly remote macro layer (Figure 19.1). This macro environment may include any force, trend or event with a possible influence on what the organisation is, does or could do (Pitt and Koufopoulos, 2012). A trend is a direction of change in values and needs, which is driven by forces (like digitalisation, urbanisation) and manifests itself already in various ways within certain groups in a society (Dragt, 2017). A too narrow a view on customer's needs can lead to biased viewpoints, creating (false) dogmas that are set in concrete: for instance, the idea what we only consume indulgence products like ice cream because of their taste. The ice cream brand Ben & Jerry's were one of the first in this industry to strongly communicate the idea of making the world a better place, with ice creams. As they state: 'If it's melted, it's ruined. It's true for ice cream, and it's true for the planet' (Ben & Jerry's, n.d.).

19.2 Key element of a market analysis

In this paragraph we take a deep dive into four different elements of the market analysis: customers, suppliers, industry factors and macro factors. First of all, we start with 'customers'. We extensively elaborate on this element, because gaining insights into the customer's mind is at the heart of a thorough market analysis. The reason behind this is obvious: as pointed out before, without sales there can be no durable business endeavours. Suppliers, industry and macro factors are also outlined, because together with the customers they form the basis of a thorough market analysis.

19.2.1 Customers

As stated before, customers are people or enterprises that purchase a product or service. An enterprise can have customers that are very much alike regarding their product wishes and needs, or diverse customers, which leads to different customer segments. A customer can be a single person, or (often seen in business-to-business environments) a decision-making unit – a team of individuals that is responsible for the purchase of certain products and/or services. Also, keep in mind that the customer is not always the (only) user of a product. For example, soft drinks can be bought by parents and their children might consume these products (as well).

customer satisfaction

As Kotler (2003) states: 'Market share is a backward-looking metric and customer satisfaction is a forward-looking metric'. Therefore, it is relevant that enterprises monitor and improve the level of customer satisfaction. Higher customer satisfaction leads (in general) to higher retention rates (percentage of customers an enterprise retains over a given period of time). There is a simple logic to this: if you are satisfied, or maybe even delighted with a certain product, you are very likely to return to the same place where you bought that product.

customer value

wishes and needs

Customer satisfaction is determined by what we 'get' in terms of customer value and what we have to 'give up' in terms of costs. Customer value is driven by individual wishes and needs. A customer or customer segments can have many different wishes and needs in the process of purchasing a product. One person, for example, buys a prepacked sandwich for convenience, the other purchases the same sandwich mainly for its healthy ingredients and yet another chooses the sandwich because of its price. Our wishes and needs lead to what we perceive as value. Customer value can be conceptualised as function, experiential and symbolic value (Smith and Colgate, 2007). Woodruff (1997) suggests function value relates to aspects such as product quality, product characteristics, reliability, efficiency, performance quality and financial benefits. Experiential value relates to the extent to which a product creates appropriate experiences, feeling and emotions for the customer. Retailers often focus on sensory values such as ambiance, aromas, feel and tone. Symbolic value is concerned with the extent to which customers associate psychological meaning with a product. Luxury goods appeal symbolically to consumers' self-concept and self-worth (Smith and Colgate, 2007).

> **Case 19.3. Customer needs or nudging?**
>
> On top of focusing on wishes and needs, in order to create high retention rates or even reach for loyal customers, do not forget that marketers can also rely on nudging activities. Nudging is a way to steer people's behaviour; this concept depends on the idea that a small and apparently insignificant detail can have major impacts on people's behaviour. Just consider what can be achieved if you reduce the plate sizes in hotel restaurants? Kallbekken and Sælen (2013) were able to reduce food waste (in hotels) by as much as 22% with the use of smaller plates, without any change in guest satisfaction. Another remarkable example of nudging is how auditory stimuli strongly influence consumer selection. North *et al.* (1999) revealed that stereotypical music activates related knowledge structures concerning the subject, which causes certain sections in the brain to get primed. This nudging effect even holds true in the digital environment, like a webshop! Damen *et al.* (2021) concluded that stereotypical music, French versus German music, significantly influences the wine selection in an online webshop.

KANO model

must-be elements

satisfier elements

delighters

It is important to take into account that adding customer value does not necessarily mean satisfied customers and repeat purchases. The KANO model (Kano et al., 1984) suggests that customers can perceive three different product elements that lead to customer satisfaction. First of all, must-be elements: when these requirements are not fulfilled, the customer will be extremely dissatisfied. On the other hand, the customer takes these elements for granted and fulfilling them will not increase customer satisfaction. For example, when you purchase a cookbook you expect the recipes to be supported by pictures of the dishes. If there are no pictures, it can be quite frustrating. Secondly, Kano points out satisfier elements. With regard to these elements, customer satisfaction is proportional to the level of fulfilment, which means: the higher the fulfilment, the higher the customer's satisfaction and vice versa. Looking back at the example of a cookbook, a satisfier element can be QR codes that support the recipes with short instruction movies. Finally, delighters are those elements that have the greatest influence on how satisfied a customer will be with a product. Fulfilling delighters leads to more than proportional satisfaction. However, if these elements are not met, there is no feeling of dissatisfaction. A delighter could be, in the case of the cookbook, the option to switch each recipe into a vegetarian version. Often we find that over time a delighter or satisfier element moves towards a must-be element. For example, the amount of visual support in cookbooks has significantly increased over recent decades, whereby this element has moved from a delighter towards a satisfier and now even a must-be element.

customer loyalty

Considerable attention has been given to customer satisfaction as a potential determinant of customer loyalty (Fornell, 1992; Oliver, 1999). Loyal customers are those clients that chose your product, brand or company over the offers of a competitor, even when promotional efforts are made by these competitors (see also Chapter 20; Floto-Stammen, 2022). Customer loyalty leads to increased revenue, reduced customer acquisition costs, and lower costs of serving repeat purchasers, leading to greater profitability (Reichheld, 1993).

Net Promotor Score (NPS)

There are many measurements with which to assess loyalty. The Net Promotor Score (NPS) is a very popular metric that measures customer loyalty with one simple question: 'How likely is it that you would recommend our company to a friend or colleague?' The scale is 11-point, a 0-to-10 rating scale, and groups the respondents into 'promoters' (9-10 rating – extremely likely to recommend), 'passively satisfied' (7-8 rating), and 'detractors' (0-6 rating – extremely unlikely to recommend). In order to calculate the score, one should subtract the percentage of detractors from the percentage of promoters. Although the NPS is used in a wide variety of settings, there is also criticism regarding this single customer metric. One of the main concerns is the so called 'passive' consumer. These customers may be passive in terms of being favourable or unfavourable towards a specific company or brand, but they are probably very willing to shop for better value (Fisher and Kordupleski, 2019).

Traditional market research often uses techniques such as surveys, in-depth interviews or focus groups to gain insights. In today's digital world these options get expanded by data such as online communities and analysing the voice of the customer via online reviews. However, to fully understand customers we may need to reflect on how we approach our customers. Daniel Kahneman (2012) showed in his Nobel Prize winning theory that we have two different systems in our mind. System 1 and system 2: system 1 is fast, automatic and intuitive, while system 2 is relatively slow. System 2 is our analytical mode and it is the place where reason dominates. Asking customers, for example, about their wishes and needs through a survey or via an in-depth interview can trigger system 2. We ask customers about what they need, want or think. Logically system 2 dominates because we ask for reasoning. It seems logical that we know all the answers about our own brain. However, if I were to ask you 'Are you intuitively more attracted to products like wine with a higher price level?', I believe many of us would say no. I would certainly say 'no'! Recent brain research (Schmidt et al., 2017) discloses that our brain is sensitive to cues such as the price with regard to luxury products like wine. These cues can trigger expectations about its taste, quality and thereby modulate the sensory experience. In summary, we can actually perceive the taste of the same wine with a different price differently. With this knowledge in mind, it is important to realise that when investigating the customer, one should not just take verbal feedback into account. Observing customers, by e.g. MRI research, monitoring online behaviour, or (online) A/B testing, can reveal subconscious aspects that play a major role in our decision-making.

On top of gaining insights into the customer, it is also relevant to assess the strength of your customers. After all, customers with a high level of power can capture more value by forcing down prices, demanding better quality or more service (thereby driving up costs) and generally playing industry participants off against each other, all at the expense of industry profitability. A powerful customer has leverage over industry participants and can pressure the industry in order to gain price reductions (Porter, 2008). Finally, boundaries are fading in the real world and sometimes a customer can also become a competitor. An obvious business-to-business example is the fast-moving-consumer-goods industry. Modern-day store brands or private labels are generally owned and marketed by retailers (the customer). Retailers have been a part of the manufacturer (the supplier) network for decades and private labels are not new on the market. Nevertheless, we see that over time, these private label brands have evolved from generic, cheap, low-quality, economy or budget private labels to lower priced than industry-brand but acceptable quality value or standard private labels. Over time, retailers have even extended the value proposition to the customer segment seeking higher quality by offering premium private labels, like Wal-Mart's Sam's Choice (USA), Tesco Finest (UK) and AH Excellent (the Netherlands). This strategy, whereby retailers offer high-quality private labels to the price-sensitive but not quality-sensitive customers, changes the industry. Over the last 40 years (1980-2020), these versions of private labels have witnessed substantial growth around the whole world, though the growth is said to be tapering in recent times (Gielens *et al.*, 2021). This example clearly shows that it is relevant to consider the concept of a 'customer' in a broader sense: customers should not be perceived as merely a buying party. Entities that are today's customers may be tomorrow's competitors.

19.2.2 Suppliers

Although you are a customer to your supplier and this supplier might see you as an essential asset, analysing the supplier and especially the power of the supplier is extremely important. A powerful supplier might take a serious bite out of the value of your end product or service by charging higher prices. Alternatively, these suppliers can limit quality, services and in some cases even shift costs to industry participants. Powerful suppliers, including labour suppliers, can squeeze profitability out of an industry that is unable to pass on cost increases in its own prices (Porter, 2008). A supplier group is stronger when it is more concentrated than the industry to which it sells products. A good example is the cacao industry, where more than 70% of the business in cacao is owned by only a small number of large companies. New businesses, like Tony's Chocolonely, faced the incredible power of suppliers. When there are only a few suppliers – cacao manufacturers – and they are unaware of the conditions in which the cacao is obtained, producing a slave-free chocolate bar can be a real challenge for an enterprise like Tony's Chocolonely (Siebelink, 2018). Besides this, the power of a supplier is also higher when the supplier group does not depend heavily on the industry for its revenues. In other words: if your company is just one of the many clients of your supplier, this supplier has a lot of power over you, simply because the amount of profit you provide for this company is relatively low. If your supplier also produces products for multiple industries, the power of this supplier further increases, because this supplier is less vulnerable towards threats in a certain industry. This power can also be higher when the industry participants are dealing with switching costs when changing suppliers, the products of suppliers are differentiated (for example, in terms of quality) and the products the supplier offers cannot be substituted.

Although suppliers are critical for buyers' success and add value in the value chain, they have historically been viewed merely as the entities that represent costs to the buyers. However, over the last few decades a mentality switch toward suppliers became visible: instead of focusing on cost, customers began to evaluate their suppliers on other dimensions than just costs. Aspects like quality, delivery, production process, R&D capability and service are now also taken into account (Tang, 1999). A very famous example of this is WalMart (the customer), which provides store sales information via Electronic Data Interchange (EDI) to Procter & Gamble (the supplier). After receiving the sales data, P&G is responsible for monitoring and replenishing WalMart's inventory. This collaboration has helped WalMart to reduce inventory and add value in terms of improving customer service (Buzzell and Ortmeyer, 1995).

According to Tang (1999) there are two key factors that play an important role in determining the supplier relationship. These are: strategic importance of the supplied part to the buyer and the buyer's bargaining power (these two factors are essential if the market risk is relatively low). Focusing on the strategic importance of the part to the buyer and the buyer's bargaining power leads to four different types of supplier relationships. A **vendor** tends to be the type of supplier who makes common parts and competes solely on price. Since there are numerous suppliers in the market, the contract term tends to be short and the customer is likely to switch supplier whenever there is a cheaper source of supplier available. The communication level between the customer and supplier is low and the interaction usually only occurs when the customer places an order. A **preferred supplier** tends to provide more complex/unique products or services, in a market where the bargaining power of the customer is still high. The switching cost for the customer is relatively high and the contract

vendor

preferred supplier

exclusive supplier

partner

term tends to be longer. An exclusive supplier tends to provide a product or service that very few other suppliers can provide. In this case, the switching cost would be very high for the customer, and hence, the contract term is usually even longer than that of a preferred supplier. A partner can be classified as a supplier who provides unique products and services and on top of that also commits to revenue and risk sharing with the customer (Tang, 1999).

19.2.3 Industry factors

Taking a first glimpse at an industry, the industry size, growth and the cyclical sensitivity are relevant factors. Cyclical sensitive industries are those whose revenue generation capabilities are tied to seasons or economic cycles. A well-known example is the agriculture industry: agricultural returns tend to be cyclical in nature; a few years of good returns are in general followed by a few years of negative returns. This cycle is inherent to agriculture. In general, large industries with high growth numbers that are relatively insensitive to economic fluctuation are attractive (Porter, 2008).

Keep in mind that an industry is a dynamic playing field and changes over time. Some, but not all, industries undergo a classical evolution. When an industry is new, there are often many companies entering the market. The rate of product innovation is high and market share changes rapidly. Although there is continued market growth, entry subsequently slows down, exit overtakes entry and there is a shakeout in the number of market participants. After this stage the rate of product innovation and the diversity of competing versions of the product declines. This results in a stabilisation of market shares, and the effort that is devoted by the enterprises to improving the production process diminishes. This process is also referred to as the product life cycle (PLC) (Klepper, 1996). In addition, a common mistake is to assume that fast-growing industries are always attractive. Growth does tend to mute rivalry, because an expanding pie offers opportunities for all competitors. However, fast growth can put suppliers in a powerful position and high growth with low entry barriers will draw in entrants. Even without new entrants, a high growth rate will not guarantee profitability if customers are powerful or substitutes are attractive. A narrow focus on growth is one of the major causes of bad strategy decisions (Porter, 2008). With this in mind, it is important to realise that analysing the industry requires deeper investigation. The five forces of Porter are a powerful model revealing why industry profitability is what it is. Often managers define competition too narrowly, as if it occurs only among today's direct competitors. Yet, this model argues that competition for profits goes beyond established industry rivals and includes four other competitive forces as well: customers, suppliers, potential entrants and substitute products (Figure 19.2). Below, we will take a closer look at factors such as potential entrants, substitute products and rivalry among existing competitors. The bargaining power of the customers and the suppliers are covered in Section 19.2 'Customers' and 'Suppliers'.

new organisations New organisations that enter an industry bring new capacity and a desire to gain market share. This puts pressure on the current prices, costs, and the rate of investment necessary to compete. In general, it is relatively easy to enter a new market if products are homogeneous, there is no specific knowhow required, entry costs are low and economies of scale (whereby more units of a good or service can be produced on a larger scale, with per item fewer input costs) are not relevant. In addition, the availability of channels plays an important role. For example, a new food item must displace others from the supermarket shelf via price breaks, promotions, intense selling efforts, or other means. The more limited the wholesale or retail channels are and the more grip existing competitors have on them, the tougher entry into an industry will be for a new player. Taking a closer look at this supermarket industry, we even find that limited freezer space in grocery stores is making it harder for new ice cream makers to gain access to distribution in North America and Europe (Porter, 2008).

substitute A substitute performs the same or a similar function as the industry's product but through different means. Sometimes, the threat of substitution is downstream or even indirect. When the threat of substitutes in a certain industry is high, this may negatively affect the profitability within that industry, because substitute products or services limit an industry's profit potential by

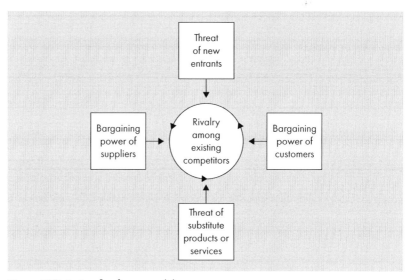

Figure 19.2. Porters five forces model.

placing a ceiling on prices. If a particular industry does not distinguish itself from substitutes through product performance, marketing, or other means, it will harm profitability and growth potential. An example is the evolution of our retail landscape: throughout the 20th century supermarkets have substituted many small grocery stores in Western Europe, creating convenience for shoppers by offering the same products, all in one store.

rivalry among existing competitors Rivalry among existing competitors refers to the intensity, the strength of the existing competitors within an industry and the basis on which they compete with each other. Rivalry among existing competitors is often reflected by price discounting, new product introductions, advertising campaigns and service improvements. Fierce competition limits the profitability of an industry. The intensity of rivalry is greatest if competitors are numerous or are roughly equal in size and power, industry growth is slow and exit barriers are high. On top of that, other aspects can also play a role here, e.g. when the products or services of competitors are nearly identical and there are few switching costs for customers, fixed costs are high and marginal costs are low. This creates intense pressure for competitors to cut prices below their average costs, even close to their marginal costs, to steal incremental customers while still making some contribution to covering fixed costs. Capacity must be expanded in large increments to be efficient. The need for large capacity expansions disrupts the industry's supply-demand balance and often leads to long and recurring periods of overcapacity and price-cutting. Finally, competition is influenced by the perishability of the product, often found in the fast-moving-consumer-goods (FMCG) industry. Perishability creates a strong temptation to cut prices and sell a product while it still has value.

While profiling competitors, it is relevant to analyse to what degree rivals compete on the same dimensions. When all or many competitors aim to meet the same customer's needs or compete on the same attributes, the result is zero-sum competition. Here, one firm's gain is often another's loss, driving down profitability, most of the time across the entire industry. While price competition involves a higher risk than non-price competition to become zero sum, this may not happen if companies segment their markets, e.g. by targeting their low-price offerings to different customers. In contrast, rivalry can also be positive sum and actually increase the average profitability of an industry. In this case, each competitor aims to serve the needs of different customer segments, with different mixes of price, products, services, features, or brand identities. This form of competition can not only support higher average profitability but also expand the industry, as the needs of more customer groups are better met (Porter, 2008). Therefore, a thorough analysis, via market research concerning the strength and weaknesses and unique selling points (USPs) of competitors, is crucial.

19.2.4 Macro factors

Be aware of the fact that drivers outside the industry can also create opportunities and threats for (new) enterprises. There are different frameworks within which to assess macro environmental factors, like PEST (also known as STEP), PESTLE, DESTEP and STEEPLE. The acronyms used represent a particular area of interest that needs to be structured to understand the macro environment of a business. Some frameworks include more factors than others do, which make them supplementary. In this section we use a widely known framework, the PEST analysis, and later we add some extra dimensions that relate to the other frameworks. The PEST analysis stands for political, economic, social and technological drivers. These drivers represent important macroeconomic features that lie at the core of a coherent macro-analysis.

political

Political: these drivers relate to the extent to which a government intervenes in the economy. Government policy is a factor that can hinder or aid (new) entries directly, as well as magnify the other entry barriers. Licences, requirements and restrictions, for example on foreign investment, can limit or even foreclose entry into industries (Porter, 2008). Please note that there are some frameworks that distinguish between political and legal drivers. These frameworks prefer to separately address political drivers, which involve all aspects related to the current political playing field and legal aspects like legislation, rules and regulations. In this section we consider both political and legal aspects as one category, because these drivers are highly correlated. Relevant examples are: government policy, political stability or instability in overseas markets, foreign trade policy, tax policy, labour law, environmental law, trade restrictions and so on. It is important to investigate the political environment in which business operates, so that they are well equipped to respond to and anticipate these factors by adjusting strategy policies accordingly. An interesting example regarding political drivers is the approval in 2021 by the European Commission (Commission Implementing Regulation (EU) 2021/882) of dried yellow mealworms for consumption in the EU. This is the first time an insect has been approved as a food in the European Union (see also Chapter 20; Floto-Stammen, 2022). The authorisation means that within the 27 EU Member States mealworms can be put on the menu as a food (Galán Feced and Moynihan, 2021).

economic

Economic: every organisation is bound by the influence of the economy, in areas such as growth rates, inflation, GDP, interest rates, international exchange rates, international trade, labour costs, consumer disposable income, consumer purchasing power, unemployment rates, availability of credit and raw material costs. Economic forces are beyond the control of an organisation; however, economic data can be input for adjusting company strategies and policies.

Newspapers, trend reports, market analysis papers are filled with data regarding economic trends. Take, for example, the COVID-19 pandemic: while the pandemic is first and foremost a health crisis, it also led to a global economic crisis. When looking at the US market household saving rates almost doubled in 2020 to 12.9%, due to COVID-19. Economists and central bank officials claim these pandemic-driven trends are most likely temporary effects. Despite that, investors and politicians are worried that prices will keep climbing, potentially pressuring the Federal Reserve to raise interest rates sharply. That could slow economic growth and send stock prices into freefall (Smialek, 2021).

social

Social: social drivers are connected closely with demographic aspects. Demographics covers the traits and characteristics of the population. Here, factors such as age range, geography, education levels, gender, and family structures play a major role. These factors can affect our attitudes, opinions, and interests. Therefore, it can influence the sales of products and revenues earned by organisations. The social factors shape who we are as people, and as a result it affects how we behave and what, when and how we buy. Elements that can be explored are – amongst others – lifestyles, buying habits, religion and beliefs, social classes, family size and structure. A good example of a social driver is how people's attitude towards health (power) food is changing. In the United Kingdom for instance, one of the leading reasons for being vegan is the desire to embrace a healthier lifestyle. The number of vegans in the UK has been growing for years and will likely increase further. Similarly, a leading reason for purchasing organic products is that customers perceive them to be healthier than other products. As a result, the organic food and drink market in the UK generated over two billion GBP in 2020 (Wunsch, 2020). For companies that offer unhealthy products, this trend could be perceived as a threat. However, in response to these societal trends, Coca-Cola seized on new opportunities instead (Case 19.3).

Case 19.3. Coke Zero – from a bloke coke to a healthy alternative.

In 2006, a new coke was launched targeting young, male drinkers with a zero-sugar alternative to support the flagship Coke brand. It was Coca-Cola's biggest launch since Diet Coke 22 years ago, backed by an £8m advertising budget, and more than 4 million samples were handed out (Anomymous, 2008). When Coke Zero launched in 2006, it was positioned as a 'bloke Coke': the no-sugar option for boys. Since diet Coke had a predominantly female audience, diet Coke Zero was positioned as a 'gender agnostic' brand, targeting a different segment (O'Reilly, 2016).

Over the years, after a thorough analysis of the market (shifts), Coca-Cola further developed the recipe of Coca-Cola Zero closer to the original Coca-Cola. This was done to enable the Coca-Cola Classic drinkers to give Zero a try. It contains zero calories and zero sugar, while providing the signature Coca-Cola Classic flavour. With this new branding of Coke Zero, customers (male and female) searching for a healthy alternative are targeted.

technological

Technological: these drivers relate to the existence, availability and development of technology. In general, technology can have a profound impact on how societies progress (Mazali, 2018). More specifically, technology can change the way people interact, do business or purchase products. This could include aspects from computational power to engine efficiency. Just consider the cans that are made for all of our conserved vegetables, the computer servers that are used to maintain the webshops we use, or the online firewalls that are in place when we make an online payment. Some general technological drivers that could influence an organisation are: 3D technology, digitalisation, internet connectivity, security in cryptography, automation and wireless charging. The availability of new food technology can have a considerable impact on how food is produced and distributed. Optimising logistic processes may provide customers with fresher foods, new technologies may lead to more efficient or environmentally friendlier production processes, or lead to a larger product quantity per square metre (Gaffney *et al.*, 2019). Besides plant-based food options, technology also opens up new possibilities for those who still want to eat meat. Cellular agriculture uses culturing techniques, which aim to produce animal proteins using fewer animals. Cultured meat involves applying the practices of tissue engineering to the production of muscle in order for it to be consumed as a food (Stephens *et al.*, 2018). These technological trends not only affect the efficiency or productivity of the food business sector, but can also contribute to sustainability (Gaffney *et al.*, 2019). At the same time, the availability of these technologies can be limited by intellectual property, and create advantages for those who have access to the relevant patents (and disadvantages for those who do not).

environment

Although the PEST analysis determines a broad range of macro environmental drivers, there are some drivers that are relevant in today's world that it does not cover. In this section we take a look at these drivers, starting with environment. The sustainability of our planet has become a major issue in enterprises' macro environments, experienced as pressures to reduce energy consumption, avoid pollution and increase recycling of products and materials. Nowadays, the costs of pollution and the benefits of environmental sustainability are increasingly recognised worldwide. This shift is not just occurring in the Western world. China, some of the Gulf States, and India are also investing in

ethical aspects

green energy on a scale that would have been considered improbable even a decade ago (Sneader and Singhal, 2020). In addition to that, ethical aspects can greatly influence an enterprise and should therefore be taken into account when analysing the macro environment (see Chapter 3; Wernaart, 2022). Ethics provides values on the basis of what is right and what is wrong. The ethical ideas of a people in a country or (sub) culture will not change overnight. However, small changes in morality take place over time. An interesting example is the product range of Harrods, one of the UK's most famous department stores. Harrods sold exotic animals for more than 50 years. Customers were able to buy all sorts of exotic animals like lions, alligators, camels and even elephants! It was the place where, if your wallet and your home were large enough, you could buy almost any wild animal as a household pet. Over time, the number of critics rose, with people believing it was wrong to sell exotic pets, especially in a department store. Ethical values change over time and are very much dependent on a culture or subculture. This leads to

culture

another driver, culture, that could be worth further assessment (often covered under the section social-cultural drives). Culture refers to the fundamental, deeply rooted qualities of a nation or society. There are theories that, due to globalisation, cultural differences are disappearing. Van Oudenhoven (2008) claims that we overestimate cultural similarities and we, in general, tend to underestimate cultural differences. One of the most famous, but also criticised, methods of examining culture is Hofstede's cultural dimensions. With only six different dimensions, power distance, individualism, masculinity, uncertainty avoidance, long-term orientation and indulgence a culture is analysed. Please do take into account that even though this is a wide used method, there are many arguments run against Hofstede's work (Jones, 2007; Shaiq *et al.*, 2011). One of the most popular criticism relates to cultural homogeneity: Hofstede's theory assumes the domestic population is homogenous. However, most nations are groups of subcultures and ethnic units (Nasif *et al.*, 1991; Redpath and Nielsen, 1997).

19.3 Looking ahead

Analysing our environment has been an important factor thorough history. As cavemen, we used to scan our environment to identify the potential threats from wild animals, or an opportunity to hunt for food. We believe that a (market)scan of the environment is still relevant and will be even more so in the future. Today, there are some differences compared to 100 or 1000 or more years ago. First, and maybe most importantly, we live in a highly connected, hybrid world where boundaries are fading. Rapid changes in industrial structure and global competition have occurred in recent decades. As mentioned before, environmental diversity makes it difficult to link customers and the goods and services that meet customers' needs and wants in the

digitalisation

marketplace (Cravens and Piercy, 2013). This means that long-term market strategies of 5 to 10 years are no longer realistic in some industries because of the dynamics of the environment. Fast and maybe even continuous insights that drive innovation in the market are key in a world that is capricious. Digitalisation delivers us insights into data, e.g. (online) customer behaviour that was not conceivable before this digital era. Just consider a loyalty card for our preferred supermarket. Retailers are able to analyse what you buy, when you buy, if you are sensitive to sales promotions, and so on. When exploring customer behaviour there is a tendency to shift from verbal techniques such as surveys, interviews, focus groups toward (digital) observation methods. Why ask people about their buying behaviour when we are able to entirely monitor customer behaviour (via cookie tracking)? A next step that some enterprises have already implemented perfectly is building models (algorithms) based on these data in order to make predictions about future customer behaviour. Finally, with the presence of large social platforms such as Instagram and Facebook, network analyses have become more accessible due to the availability of big data.

prototyping

design thinking

Apart from the enormous possibilities big data offers, another strategy that can be implemented is prototyping. Due to the fact that our environment is becoming increasingly complex, experimenting with prototypes can be a solution in various industries. In some cases, it will be useful to combine a market analysis with prototyping and experimenting with this prototype to find out whether an innovation like a new product, service, solution or idea will be successful. Prototyping is not necessarily a new thing. Consider great masters like Antoni Gaudí, who made many prototypes like the Sagrada Familia (still visible in the Sagrada Familia museum). Prototyping is a well-known technique in, for example, creative industries and has recently entered the field of economics via design thinking. Design thinking is an iterative process where the focus is not only on analysing the context, but also on investigating the solution (Glen *et al.,* 2014). By also focusing on the solution and quick, iterative research cycles, enterprises become more flexible when analysing the market.

co-creation

In this chapter, we discussed the market, where we addressed customers, suppliers, and industry- and macro factors as separate issues. We believe that in the near future the boundaries between these issues will further fade. Collaborations within the industry or outside the industry will increase in our highly connected world. On top of that, it seems more likely that enterprises will team up with suppliers in order to serve customers better or to gain market share. Finally, enterprises incorporate (the voice of) the customer more and more while innovating and, in some cases, apply co-creation. This can be executed in various ways, but online brand communities form an interesting base. These communities encourage customers to share their ideas, discuss and evaluate other customers ideas and develop ideas into better ones through interaction (Gangi and Wasko, 2009). So, encouraging customers to interact with other customers and extracting the useful 'voice of the customer', critical for brand communities, can help organisations innovate (Lee *et al.,* 2014).

19.4 Conclusions

Traditionally, a market is a place where customers and sellers meet. In today's complex world, this market involves the customer and the seller, but suppliers, industry factors and macro-environmental drivers also play an important role.

The relationship an organisation has with a customer is always crucial: if there are no customers, the business organisation loses its reason to exist. Getting to know this customer or customer segments, often via market research, is therefore crucial. Today's digital world offers us rich possibilities to analyse the voice of the customer in online communities, track down customer reviews or follow customers throughout the entire online customer journey via cookie tracking. The five forces model of Porter investigates the power of the customers, as well as the power of the suppliers. Besides, it takes a deep dive into the threat of new entrants and the threat of substitute products in a certain industry. Finally, we are rarely the only seller in a particular industry, which means competitors also play an important role. A thorough analysis of the strength and weaknesses of your most crucial competitors is therefore essential. When moving from an industry perspective towards a macro perspective, other forces can be relevant for enterprises. The PEST analysis offers a framework by which political, economic, social and technological opportunities and threats can be detected. Besides these forces one should take a deep dive into forces related to environmental, legal, ethical and cultural aspects. For an organisation it is relevant to be aware of these forces, in order to anticipate on these trends with a suitable business strategy.

To remain in business all enterprises are (actively) searching for blue oceans. Insights and innovation are key to success, in our fast-changing environment. By analysing the market companies can develop strategies that implement relevant innovations, based on the needs of a customer, industry opportunities or threats and macro drivers. As the case of Starbucks pointed out: it is innovation and insights that can make companies swim from red oceans towards blue oceans!

References

Anonymous, 2008, 7 August. From Zero to hero. Marketing Week. Available at: https://www.marketingweek.com/from-zero-to-hero/

Ben & Jerry's, nd. We believe that ice cream can change the world. Available at: https://www.benjerry.com/values

Bennett, P., 2020, 29 September. 10 of the best coffee chains in the US. Available at: https://www.insider.com/best-fast-food-restaurants-that-have-coffee-in-the-us#starbucks-is-the-biggest-coffee-chain-in-the-world-1

Buzzell, R.D. and Ortmeyer, G., 1995. Channel partnerships streamline distribution. MIT Sloan Management Review 36: 85.

Cravens, D. and Piercy, N., 2013. Strategic marketing. McGraw-Hill, Columbus, OH, USA.

Damen, M., Van Hest, I. and Wernaart, B., 2021. The effect of stereotypical music on the customer selection of wine in an online environment. Journal of Innovations in Digital Marketing 2: 29-37. https://doi.org/10.51300/jidm-2021-35

Dragt, E., 2017. How to research trends: move beyond trend watching to kickstart innovation. BIS Publishers, Amsterdam, the Netherlands.

European Commission (EC), 2021. Commission Implementing Regulation (EU) 2021/882 of 1 June 2021 authorising the placing on the market of dried *Tenebrio molitor* larva as a novel food under Regulation (EU) 2015/2283 of the European Parliament and of the Council, and amending Commission Implementing Regulation (EU) 2017/2470. Official Journal of the European Union L 194, 2.6.2021: 16-20.

Fisher, N. and Kordupleski, R., 2019. Good and bad market research: a critical review of net promoter score. Applied Stochastic Models in Business and Industry 35: 138-151. https://doi.org/10.1002/asmb.2417

Floto-Stammen, S., 2022. Food marketing. In: Wernaart, B.F.W. and Van der Meulen, B.M.J. (eds) Applied Food Science. Wageningen Academic Publishers, Wageningen, the Netherlands, pp. 453-479.

Fornell, C., 1992. A national customer satisfaction barometer: the Swedish experience? Journal of Marketing 56: 6-21. https://doi.org/10.1177/002224299205600103

Gaffney, J., Bing, J., Byrne, P.F., Cassman, K.G., Ciampitti, I., Delmer, D., Habben, J., Lafitte, H.R., Lidstrom, U.E., Porter, D.O., Sawyer, J.E., Schussler, J., Setter, T., Sharp, R.E., Vyn, T.J. and Warner, D., 2019. Science-based intensive agriculture: sustainability, food security, and the role of technology. Global Food Security 23: 236-244. https://doi.org/10.1016/j.gfs.2019.08.003

Galán Feced, C. and Moynihan, Q., 2021, 6 May. The European Commission has authorized the consumption of yellow mealworms in the EU. Available at: https://www.businessinsider.com/insects-bugs-edible-environment-sustainability-eco-friendly-health-sustainable-farming-2021-5?international=true&r=US&IR=T

Gangi, P. and Wasko, M., 2009. Steal my idea! Organizational adoption of user innovations from a user innovation community: a case study of Dell IdeaStorm. Decision Support Systems 48: 303-312. https://doi.org/10.1016/j.dss.2009.04.004

Gielens, K., Ma, Y., Namin, A., Sethuraman, R., Smith, R.J., Bachtel, R.C. and Jervis, S., 2021. The future of private labels: towards a smart private label strategy. Journal of Retailing 97: 99-115. https://doi.org/10.1016/j.jretai.2020.10.007

Glen, R., Suciu, C. and Baughn, C., 2014. The need for design thinking in business schools. Academy of Management Learning and Education 13: 653-667. https://doi.org/10.5465/amle.2012.0308

Jones, M.L., 2007. Hofstede – culturally questionable? Oxford Business & Economics Conference, Oxford, UK, 24-26 June, 2007.

Kahneman, D., 2012. Thinking, fast and slow. Penguin Books Ltd, New York, NY, USA.

Kallbekken, S. and Sælen, H., 2013. 'Nudging' hotel guests to reduce food waste as a win-win environmental measure. Economics Letters 119: 325-327. https://doi.org/10.1016/j.econlet.2013.03.019

Kano, N., Seraku, N., Takahashi, F. and Tsuji, S., 1984. Attractive quality and must-be quality. Journal of the Japanese Society for Quality Control 41: 39-48. https://doi.org/10.20684/quality.14.2_147

Kaplinsky, R. and Morris, M., 2000. A handbook for value chain research. University of Sussex, Institute of Development Studies, Brighton, UK.

Kim, W.C. and Mauborgne, R., 2014. Blue ocean strategy, expanded edition: how to create uncontested market space and make the competition irrelevant. Harvard Business Review Press, Brighton, MA, USA.

Klepper, S., 1996. Entry, exit, growth, and innovation over the product life cycle. American Economic Review 86: 562-583. http://www.jstor.org/stable/2118212

Kotler, P., 1999. Marketing Management. Prentice Hall, Singapore, Singapore.

Kotler, P., 2003. Marketing insights from A to Z: 80 concepts every manager needs to know. John Wiley & Sons, Hoboken, NJ, USA.

Lee, H., Han, J. and Suh, Y., 2014. Gift or threat? An examination of voice of the customer: the case of MyStarbucksIdea.com. Electronic Commerce Research and Applications 13: 205-219. https://doi.org/10.1016/j.elerap.2014.02.001

Mazali, T., 2018. From industry 4.0 to society 4.0, there and back. AI and Society 33: 405-411. https://doi.org/10.1007/s00146-017-0792-6

Miller, C., 2009, August 20. Will the hard-core Starbucks customer pay more? The chain plans to find out. New York Times. Available at: https://www.nytimes.com/2009/08/21/business/21sbux.html

Nasif, E.G., Al-Daeaj, H., Ebrahimi, B. and Thibodeaux, M.S., 1991. Methodological problems in cross-cultural research: an updated review. MIR: Management International Review 31: 79-91. http://www.jstor.org/stable/40228333

North, A.C., Hargreaves, D.J. and McKendrick, J., 1999. The influence of in-store music on wine selections. Journal of Applied Psychology 84: 271-276. https://doi.org/10.1037/0021-9010.84.2.271

Oliver, R., 1999. 'Whence consumer loyalty?' Journal of Marketing 63: 33-44. https://doi.org/10.1177/00222429990634s105

O'Reilly, L., 2016, 8 July. Coca-Cola explains its Coke Zero rebrand: 'Coke and Coke Zero Sugar are like ham and egg'. Business Insider. Available at: https://www.businessinsider.com/coca-cola-explains-its-coke-zero-rebrand-coke-and-coke-zero-sugar-are-like-ham-and-egg-2016-7?international=true&r=US&IR=T

Pitt, M. and Koufopoulos, D., 2012. Essentials of strategic management. Sage, London, UK.

Porter M.E., 1985. Competitive advantage: creating and sustaining superior performance. Free Press, New York, NY, USA.

Porter, M.E., 1979. How competitive forces shape strategy. Harvard Business Review 57: 137-145.

Porter, M.E., 2008. The five competitive forces that shape strategy. Harvard Business Review 86: 78.

Redpath, L. and Nielsen, M.O., 1997. A comparison of native culture, non-native culture and new management ideology. Canadian Journal of Administrative Sciences/Revue Canadienne des Sciences de l'Administration 14: 327-339. https://doi.org/10.1111/j.1936-4490.1997.tb00139.x

Reichheld, E.E., 1993. Loyalty-based management? Harvard Business Review 71: 64-73.

Ritzer, G., 1993. The McDonaldization of society. Sage, London, UK.

Schmidt, L., Skvortsova, V., Kullen, C., Weber, B. and Plassmann, H., 2017. How context alters value: the brain's valuation and affective regulation system link price cues to experienced taste pleasantness. Scientific Reports 7: 8098. https://doi.org/10.1038/s41598-017-08080-0

Schneider, H. and Hall, J., 2011. Why most product launches fail. Harvard Business Review 89(4): 21-24.

Shaiq, H.M.A., Khalid, H.M.S., Akram, A. and Ali, B., 2011. Why not everybody loves Hofstede? What are the alternative approaches to study of culture? European Journal of Business and Management 3: 101-111.

Shiv, B. and Fedorikhin, A., 1999. Heart and mind in conflict: the interplay of affect and cognition in customer decision making. Journal of Customer Research 26: 278-292. https://doi.org/10.1086/209563

Siebelink, J., 2018. Het wereldschokkende en onweerstaanbaar lekkere verhaal van Tony's Chocolonely. Thomas Rap, Amsterdam., the Netherlands.

Smialek, J., 2021, 12 May. Jump in customer prices raises stakes in inflation debate. New York Times. Available at: https://www.nytimes.com/2021/05/12/business/inflation-customer-price-index-april.html

Smith, J.B. and Colgate, M., 2007. Customer value creation: a practical framework. Journal of Marketing: Theory and Practice 15: 7-23. https://doi.org/10.2753/MTP1069-6679150101

Sneader, K. and Singhal, S., 2020, 4 January. The next normal arrives: trends that will define 2021 – and beyond. Available at: https://www.mckinsey.com/featured-insights/leadership/the-next-normal-arrives-trends-that-will-define-2021-and-beyond#

Starbucks, nd. Onze geschiedenis. Available at: https://www.starbucks.nl/about-us/our-heritage/

Stephens, N., Di Silvio, L., Dunsford, I., Ellis, M., Glencross, A., and Sexton, A., 2018. Bringing cultured meat to market: technical, socio-political, and regulatory challenges in cellular agriculture. Trends in Food Science and Technology 78: 155-166. https://doi.org/10.1016/j.tifs.2018.04.010

Tang, C.S., 1999. Supplier relationship map. International Journal of Logistics: Research and Applications 2: 39-56. https://doi.org/10.1080/13675569908901571

Thompson, C.J. and Arsel, Z., 2004. The starbucks brandscape and customers (anticorporate) experiences of glocalization. Journal of Customer Research 31: 631-642. https://doi.org/10.1086/425098

Van Oudenhoven, J.P., 2008. Crossculturele psychologie: de zoektocht naar de verschillen en overeenkomsten tussen culturen. Uitgeverij Coutinho, Bussum, the Netherlands.

Wernaart, B.F.W., 2022. Food ethics. In: Wernaart, B.F.W. and Van der Meulen, B.M.J. (eds) Applied Food Science. Wageningen Academic Publishers, Wageningen, the Netherlands, pp. 45-64.

Woodruff, R.B., 1997. Customer value: the next source for competitive advantage. Journal of the Academy of Marketing Science 25: 139-153. https://doi.org/10.1007/BF02894350

Wunsch, N.G., 2020, 19 August. Health and wellness food trends in the United Kingdom – statistics & facts. Statista. Available at: https://www.statista.com/topics/6843/health-and-wellness-food-trends-in-the-uk/#dossierSummary

20. Food marketing

This is what you need to know on how marketing can support the agri-food transition in times of change

Sonja Floto-Stammen

Research Group Business Innovation, Fontys University of applied sciences, 5900 BC Venlo, the Netherlands; s.flotostammen@fontys.nl

Abstract

You can sell anything if you just use the right tools. That is certainly more true today than ever. With the knowledge of psychological and neurological processes of the consumer, one can predictably manipulate behaviour in many ways. No company can do without these methods, because nobody is alone in the market and everyone has to stand out in a flood of loud and colourful competitors. Marketing is about the strategy of successfully staging something. Whereby success depends on the goal that the company strives for. The bottom line is that it is always about economic goals, thus about profit maximisation. But success can also be defined in terms of maximum resource conservation, maximum utilisation of leftovers and maximum customer satisfaction. Talking about customer satisfaction requires knowledge of the customer. Who is she or he exactly? Where can I contact her or him? What does she or he want? Not even the best product from an ecological, health or simply pleasure point of view comes onto the market without being equipped with the messages that the customer needs in order to make the decision to buy. In this chapter, food marketing is discussed against the background of current challenges and thus new and urgent social changes in production and consumption habits. Classic marketing methods are briefly explained. To show possibilities for a paradigm shift, the new food category of edible insects is presented as an example.

Bart Wernaart and Bernd van der Meulen (eds)
Applied food science
DOI: 10.3920/978-90-8686-933-6_20, © Sonja Floto-Stammen 2022

Key concepts

- ► Marketing is the activity, set of institutions, and processes for creating, communicating, delivering, and exchanging offerings that have value for customers, clients, partners and society at large.
- ► The Marketing strategy is based on defined goals which depend on the societal context.
- ► The Marketing Mix is the toolbox for developing the marketing concept and is based on the 4 Ps:
  - ► the Product – and its unique selling point;
  - ► the Price – and why the value is in the eye of the betrayer;
  - ► the Place – where the consumer expects it to be;
  - ► the Promotion – to catch the attention of the consumer.

20.1 Food marketing

Compared to other sciences, marketing is a young discipline that came to Europe from the USA at the end of the 1960s. Marketing in general and thus also food marketing replaced the previously prevalent sales theory consisting of procurement, production and sales. While the aim of sales theory was to sell a company's food products to consumers for an adequate price, the aim of marketing was to design and produce food products tailored to consumers' needs and desires. A supply-oriented perspective (sales theory) became a demand-side oriented perspective (marketing). At the same time, other market participants were also being considered because the number of competitors was growing steadily. In addition to their own lemonade, suddenly there was also one from the competition. A third and fourth carbonated refreshment entered the market, so it was necessary to analyse what the other soft drinks had to offer compared to their own product. The company's own strengths and weaknesses came into focus on the management floors of the companies as well. Entrepreneurs thought more about reliability, trustworthiness, flexibility and quality. Was their product easy to copy or did the recipe include inimitable secrets? These factors helped the new discipline of marketing gain significance as a corporate function. The marketing departments became overarching corporate functions or leading philosophies in many corporations (Gelbrich *et al.*, 2018).

Within contemporary marketing literature, marketing is considered to be an entrepreneurial mindset. It is concretised in the analysis, planning, implementation and control of all internal and external corporate activities, which aim to achieve sales-oriented corporate goals by aligning corporate services with customer benefit in the sense of consistent customer orientation (Bruhn, 2019).

The American Marketing Association currently proposes the following definition: 'Marketing is the activity, set of institutions, and processes for creating, communicating, delivering, and exchanging offerings that have value for customers, clients, partners and society at large' (AMA, 2017).

paradigm shifts

Substantial changes in market conditions over the past few decades have also triggered a change in the definition of the role of marketing. Three major paradigm shifts mark the history of the development of marketing. First, from a society in shortage to a consumer society, and thus, the shift from the above-explained sales economy to the marketing idea. It is the road from offering products which satisfy basic needs to a competitive market where it matters to fulfil the consumer's needs. Second, from the consumer to the affluent society, in the sense of a society in which the material benefits of prosperity are widely available, and thus, from heuristic to analytical marketing. The sheer volume of similar products addressing the consumer's supposed needs fuelled research interest in consumer behaviour. Numerous concepts were developed for proving efficiency of marketing measurements, especially in the field of psychology and perception. The third paradigm shift was from the affluent society to the co-consumer society, and thus, from analogue to digital marketing (Gelbrich *et al.*, 2018). The revolution of the Internet triggered changes in buying behaviour (e.g. e-commerce), information practices (e.g. influencer) and communication activities (e.g. social media). In the currently developing, so-called *co-operative consumer society*, consumption becomes a shared experience through increasing digital interconnectedness. Every consumer is just one click away from the corporation and micro-targeting audiences for communication purposes is increasingly the norm.

The definition of marketing shows clearly that next to the *how* (*marketing activities are done*), the *why* adds other perspectives to the role that marketing plays today. Talking about food allows us to focus on the vast challenges our agri-food system is facing. E.g. the need to feed the growing world population while minimising environmental burden. The apparent gap between global production potential and future food demand forces a system change. It might thus be time for a fourth paradigm shift for the role of marketing. This time it is an urgent change from the co-consumer society to the rational society. We are not referring here to a rationality à la Homo Oeconomicus, which is a model that has attracted a lot of criticism because of its serious flaws in depicting human nature. Rather, a rational society means one with a common sense of survival. From this perspective, food marketing can take on the task of educating and convincing consumers to make social and ecological decisions.

20.2 Methods in food marketing

In the following, we will stick to the definition of marketing as market-oriented corporate management. This self-image assumes that marketing is mainly responsible for the development, expansion, penetration and cultivation of markets. The classic product-market matrix by Ansoff (1966) is typically used for this approach. A visualisation of the matrix can be found in Figure 20.1. The goal is to identify the overarching strategy and tactics that should be used in the marketing activities.

Ansoff Matrix

The Ansoff Matrix distinguishes between existing and new products, and between existing and new markets. This results in four growth strategies: market penetration, market development, product development, or diversification. In the sweet snack market from Ferrero, for example, all strategies are used in the *Kinder* product range for further growth. In addition to children, other suitable target groups/user groups are addressed for market penetration, and the white spots on the map are systematically processed for market development, i.e. geographical gaps are eliminated. In product development, you will find various innovations from individually wrapped chocolate bars to bonbon pralines to a kind of snack bar, and in terms of diversification, the Kinder brand can now also be found in the ice cream segment.

20.2.1 Marketing strategy

What is a promising strategy for a start-up like Mybugbar (Case 20.1)? With a multiplicity of analytical practices, the appropriate marketing strategies can be chosen and evaluated. The most widely known tools for analysis are, for example, the SWOT analysis, the portfolio model, PIMS (Buzzell and Gale, 1987) or the gap analysis (Grand View Research, 2019). All of these analyses

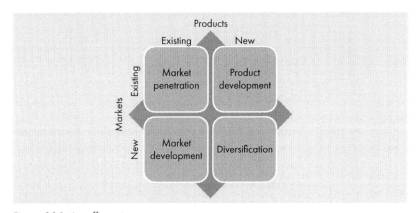

Figure 20.1. Ansoff matrix.

can determine the starting position of a company. They show both strengths and weaknesses and highlight risks and opportunities. They mark the position of the competitors and take trends and developments into account. To get back to the buffalo worm, it might be good to know how long it will take for competitors to copy successful products. In addition, it should be taken into account which growth is possible and feasible and which target market would be suitable in view of a product that requires explanation. The strategy is closely interwoven with the company's goals. This makes it necessary to make decisions, for example whether sustainable production or rapid growth is the priority (Case 20.2).

Case 20.1. We bring insects to Europe's dining tables.

'Insects are the ideal compromise between functionality and sustainability. In Europe, however, they are only vaguely known as food. We want to change that now!' (Entorganics GmbH, 2019)

The start-up 'Entorganics GmbH' was founded in 2019 by Kai Funada Classen and Finn Bußberg. After graduating from high school, the two friends travelled through Asia and got to know *edible insects* for the first time. Back in Germany, they transformed their enthusiasm for edible insects into a company. With the help of the xStarters Accelerator, a start-up programme of Volkswagen AG, the idea was realised, and the young company was able to produce its first products at the beginning of 2020. Under the brand name *Mybugbar*, protein powder was sold in the company's online shop and local supermarkets. Since then, the young start-up has grown continuously and has expanded its product portfolio with insect porridge and flour. The processed buffalo worms come from a modern farm in the Netherlands. The other ingredients are bought in Germany. Entorganics aims to be a sustainable company. To that end, they use only natural ingredients. They prefer short transport routes and environmentally friendly materials. Furthermore, they try to avoid unnecessary packaging, plastic and advertising measures (Entorganics GmbH, 2021).

Is the European market ready for insects as food?

Entorganics GmbH is one of the numerous companies following the idea of bringing sustainable insect proteins onto the EU market. Out of this context, some questions arise for food marketing. How can food marketing support the success of a novel food? What is the role of food marketing? What are the tools and options? Also, where are the limits of feasibility in the marketing framework?

This case will guide us through this chapter. We will analyse how marketing, as defined by the state of the art, is best applied.

Case 20.2. Strategies for sustainable food.

Market analysis carried out by big commercial market research institutes, such as Grand View Research (2019), revealed that the insect protein market is projected to reach USD 1,336 million by 2025, from USD 144 million in 2019, at a compound annual growth rate (CAGR) of 45.0% during the forecast period. The market is driven by factors such as the shift in consumer preference from protein from warm-blooded animals (and fish) to alternative protein. Additional drivers are the increased concerns over future sustainability and the increased public and private support for new insect protein development projects in both developed and developing economies.

According to this forecast, the market seems attractive. For a European insect company like Protifarm the strategy might read like this: 'We offer a product innovation in a new market with the aim of rapid market penetration for Europe'.

The drivers behind the strategy are stated on Protifarm's website. 'Protifarm is a response to our urgent need to eat and drink more effectively. With the help of the very best food scientists, engineers, and experts from universities and research institutes, our passionate team is offering planet-friendly food solutions for a changing and more challenging planet.

With our functional, nutritious, and sustainable ingredient line, AdalbaPro, Protifarm is feeding the world today and nourishing the world of tomorrow. In the years to come, eating insects will be the new normal for a healthy lifestyle. And we believe that the future begins today.'

The company was founded in 2015 and has created the world's first and largest vertical farm for breeding the *Alphitobius diaperinus*, better known as the buffalo beetle. The facility in Ermelo (the Netherlands) is fully operational and can deliver sufficient protein to feed a small city, all with the footprint of less than a parking lot. This first facility offers a template for cultivating nutrient-dense buffalo larva anywhere in the world, delivering a healthier planet with healthier inhabitants.

In 2021 the insect production company Ynsect acquired Protifarm. Together they strengthen their leading market position in mealworm protein production. Next to human consumption, the segment of pet food, livestock feed and other purposes are covered by the fusion.

20.2.2 The marketing mix – ingredients for good marketing

Since Neil Borden (1964), the Marketing Mix term has been used to describe a conscious selection of marketing instruments or marketing policies and their coordinated use. The four components of the marketing mix are called the *four Ps* for product, price, place and promotion (Figure 20.2). Combining these factors leads to a successful marketing concept or, the right mix. The starting point for the right mix are the company's defined goals and strategies.

The money that manufacturers invest in developing, pricing, promotion, and placing their products helps differentiate a food product based on quality and brand-name recognition. Overall, the marketing mix is meant to add value to a food organisation's product.

Product – and why it needs a unique selling point

unique selling point A key success factor in marketing is the added value for the customer (clearly) superior to competing offers. The underlying idea is the unique selling proposition or unique selling point (USP). This specific value is the trigger for consumers to buy the product. It is the differentiator that provides an advantage that competing products do not have. Often the brand in itself is an added value. Many food products are identical in recipe, taste and look, and differ only in design and communication on and around the packaging. The added value is, so to speak, in the eye of the beholder. Only the expectation, the belief in eating a special product of superior quality, changes our perception. This mechanism is often the strategy for the success of expensive branded products, which exist as identical recipes to cheaper no-name versions. There are various experiments in which product tests have shown that blind tasting delivers completely different preference results from those where participants know

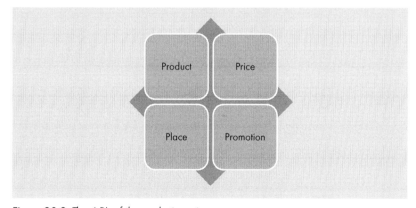

Figure 20.2. The 4 P's of the marketing mix.

what they are trying (see also Case 20.3). One of the most famous is the Coke/Pepsi experiment in which Coke is superior to Pepsi in open tasting, while it does not achieve higher values or is even inferior in blind tasting (Wei and Green, 2013).

Case 20.3. When does a grasshopper taste delicious?

People prefer foods and beverages that taste the way they expect them to taste. They do not like surprises, especially when it comes to the stimuli that enter the mouth, and hence have the potential to poison them (Spence, 2015). Red colour triggers the expectation of a strawberry or cherry taste. This is especially the case when combined with ice cream texture or a beverage. Thus, in an experiment carried out by Yeomans *et al.* (2008) with pink-coloured ice cream, participants were disappointed with a salmon savoury taste when not informed what to expect. The same salmon sample has obtained high grades of acceptance when promoted as an innovative savoury ice cream. For a new product category, it is a challenge to set expectation standards. What does a grasshopper taste like (Figure.20.3)? Most people do not expect anything nice. This is why food development for a new category has to think of ways to trigger a positive expectation. This is done with the most favourable and familiar tastes like vanilla or chocolate in the sweet range and pepper, onion, curry in the savoury range. Also, the texture described as crisp, crunchy or melting helps to stimulate a positive expectation. Insects struggle with a negative taste expectation of the consumers. Disgust is the main barrier to trying an insect food. Colour, which automatically signals a taste and description of familiar and preferable tastes, can be very helpful in overcoming the challenges.

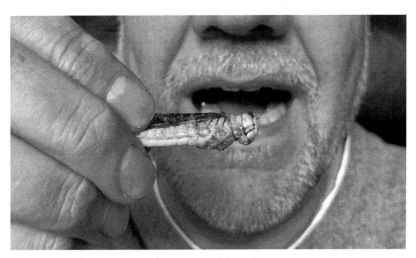

Figure 20.3. When does a grasshopper taste delicious?

product innovation

Real USPs that differentiate a food product from their competitors in a saturated market are rare. A new flavour like, for instance, salted caramel provides a clear but limited USP. Product innovation can be the solution for a specific problem or the answer to a particular need. It provides a great incentive if the favourite product is now also available in a healthier version. Extra points can be won if it additionally helps to save the planet and ensures animal welfare. As *value* is contextual, it is essential to consider that the healthy, environmentally friendly product is one that consumers (and potentially their friends and family) also like to eat, know how to cook and are able to buy everywhere. Otherwise, for instance, light cheese made of pea-protein with additional calcium would be a hit, as this product covers several USPs. The cheese contains less saturated fat and is thereby healthier. Made from pea protein, it supports plant-based diets. The additional calcium is beneficial for bone health thus preventing osteoporosis, especially for people who avoid dairy which is a main source of calcium. Yet, such innovations are not conquering the food market by themselves. Mobilising a group of people open to trying this product to function as trendsetters can help in the process of successfully placing a novel food on the market.

brand loyalty

Do you have a favourite chocolate, coffee or ice cream brand? Manufacturers that have invested a great deal of money in brands may have developed a certain level of consumer brand loyalty that is, a tendency for consumers to continue to buy a preferred brand even when an attractive offer is made by competitors. For loyalty to be present, it is not enough to merely observe that the consumer buys the same brand consistently. The consumer, to be brand loyal, must be able to actively resist promotional efforts by competitors. A brand loyal consumer will continue to buy the preferred brand even though a competing product is improved, offers a price promotion or premium, or receives preferential display space. Some consumers have multi-brand loyalty. Here, a consumer switches between a few preferred brands. The consumer may either alternate for variety or may, as a rule of thumb, buy whichever one of the preferred brands is on sale. This consumer, however, would not switch to other brands on sale. Brand loyalty is, of course, a matter of degree. Some consumers will not switch for a moderate discount but would switch for a large one or will occasionally buy another brand for convenience or variety. Growing up with Nutella, which has existed since 1964 on the North-European market, might be a reason for not even knowing that there are some competitive products out there (author´s note: since Nutella was the first spread of its kind and has a dominant position today, the name Nutella has become a generic name for every type of nut chocolate spread).

new products

Consumers like variety and are tempted by the visual stimulus 'new'. In deciding what type of new products a consumer would prefer, a manufacturer can either develop a new food product or modify or extend an existing one. Let's take the example of a protein bar. A new flavour such as salted caramel would be an *extension* of an existing product in the flavours of chocolate, vanilla and strawberry. A crunchy insect protein bar instead would be a newly *developed* product (Case 20.4). There are three steps to both extending and developing: (1) generate ideas, (2) screen ideas for feasibility, and (3) test ideas for appeal. Only after these steps will a food product make it to market. It is usual that of one hundred new product launches only 5-30 will survive on the supermarket shelf. Some are economically not feasible and some do not meet the needs or taste of the consumer.

Case 20.4. Is a buffalo worm food?

Whether a food belongs to the new category or whether it is an extension of an existing food, may be in the eye of the beholder. With *insects*, there is no doubt about the novelty, at least not in Europe.

When start-ups like Mybugbar started offering an insect protein bar in 2020, it was not yet clear whether the buffalo worm was a legal food ingredient.

There are strict rules for novel foods in the EU. The Novel Food Regulation helps food businesses bringing innovative foods to the EU market (EC, 2015). It applies to any food which was not consumed in the EU to a significant degree before 15 May 1997. To get approval, the manufacturer has to submit a dossier with detailed insights into the characteristics of the product during the production process (and in the case of insects also the rearing process). An approval by the European Commission after consulting the European Food Safety Authority (EFSA) guarantees the safety of the product, and leads to inclusion in the list of approved novel foods for the EU.

Insects do not have a history of consumption in the EU. With such a degree of innovation, several questions arise: What is the unique selling point of the product? Do customers accept it? Who is the target group?

Worldwide, two billion people regularly consume insects, such as crickets and mealworms. The question of whether it is food has long been answered by this fact. Nevertheless, the novel food authorisation process in the EU contributes to one of the safest food markets in the world. In January 2021 the first insect product received official approval (EC, 2021), which ended a period of uncertainty for insects on the

---

EU food market. The approval has been granted to dried yellow mealworm *Tenebrio molitor*. The Buffalo worm in the recipe of Mybugbar's protein bar is a different species and still awaiting approval from the EU.

In-depth market research (see also Chapter 19; Van Hest, 2022), including the consumer's opinion, is immensely important. However, it seems quite difficult to get valid results. An interesting example can be examined with the leading brand in the energy drink market, Red Bull. An extensive market study came to the conclusion that the caffeine and taurine-containing lemonade was not promising for European tastes. Red Bull founder Dietrich Mateschitz nevertheless opted for the product, combined with a revolutionary marketing strategy. Instead of a delicious taste, he put adrenaline-fueled action at the centre of his campaign. Adventure and risk were the attributes of the wing-giving drink. Whether or not a customer would want this from their lemonade is certainly a difficult question. The answer – an overwhelming success – is now well known.

strategic marketing

Is now a good moment to enter the market with a food innovation? Should a food company dare take the step and come up with a revolutionary product idea? Can a start-up successfully enter the market with an unknown novelty? All of these questions require a lot of strategic marketing considerations. The market analysis is an indispensable part of marketing decisions. The goals attached to the strategy need to be clear. Targets can, for instance, relate to the level of market share or the spread of awareness. The strategy guides the way to reach the goals.

Contemporary food innovations place a strong focus on health and environmentally friendly products (see also Case 20.5). Innovative concepts mainly revolve around solutions for more effective and efficient production with fewer emissions and/or health benefits related to obesity, malnutrition or food intolerance. These trends are visible in many types of food categories and concepts. Consumers are becoming aware of their role in responsible consumption and, at the same time, generations Y and Z are delivering convincing ideas for responsible and fair product concepts.

Case 20.5. What is my benefit from eating insects?

When it comes to product policy in the marketing mix, everything starts with highlighting the properties of the product. What are the benefits of the product? And how important are the benefits for a specific target group?

Insects are considered highly nutritional; most of them are rich in protein, healthy fats, iron, calcium, and low in carbohydrates. Research by, for example, Schmidt *et al.* (2019) suggested even a natural occurrence of vitamin B12. In fact, the authors of an FAO report (2013) claim that insects are just as – if not more – nutritious than commonly consumed meats, such as beef. Another aspect is the conservation of the environment and resources. Whether environmental friendliness is a personal advantage may be viewed differently. But perhaps there is consensus on the personal threat posed by the consequences of wasteful production methods. Insect farming and processing produce significantly lower greenhouse gas emissions. Not only do insects produce less waste, their excrement, called frass, is an excellent fertiliser and soil amender. This by-product has beneficial amounts of nitrogen, potassium, and phosphorus and is easy to work into soil. In addition, insects can be grown where the protein is needed. In large cities, for example, they can be reared in vertical farms and fed with waste residues.

Protein bars containing insect powder convey two main messages: (1) the functionality of the protein is superior (to competitive products) and (2) the ecological footprint is much lower (than that of competitive products). not surprisingly, the communication around the protein bars highlights precisely these two aspects. It has proven to be an attractive product for a certain group of people seeking products pertaining to health and/or responsibility. A variety of products are entering the market with the same concept. It works for bars in the same way as for protein shakes and powders and is tailored to the target group of athletes, with the idea of shaping their bodies with functional proteins. The kick of the novelty factor of consuming insects might play another role. But does this concept work for a broader audience? The insect industry is experimenting with product concepts for different people. The USP of insect-based foods should appeal to many more people, if only because entomophagy is one of the promising solutions for food security. However, tailor-made value propositions must be convincing for all consumers, including children, families and senior citizens. You can already buy insect-based healthy, fun snacks such as crackers, inexpensive, balanced cooking ingredients such as pasta or functional ready-made mixes such as mueslis. However, they are mostly available online rather than in the supermarket.

For a successful concept, the USP must be convincingly communicated. In the case of insects, there is an extra difficulty to overcome. The insect is the strong and the weak unique selling point at the same time. Communicating and visualising the fact of eating an insect often leads to rejection due to disgust and food neophobia. However, the insect is the argument for being healthier and environmentally friendly. Marketing faces the challenge of rendering the benefit of insects visible without provoking disgust.

Next to the protein-trend market the alternative meat segment seems to be a promising target market for insects. In general it is the nutritional value, the taste, satiety, tradition and food culture that make a piece of meat so appealing. It is difficult for an insect to compete with all of the aspects. While the nutritional value and saturation are equal if not superior, such tradition and food culture is completely missing in Europe. There are no role models yet, no occasions, no recipes, no holiday or snack preferences related to eating habits for insects. Additionally, the taste constitutes a serious obstacle for the success of the new food category. People do not yet know what to expect from the taste of insects. Should it taste like meat? Like beef or chicken? And in the case of protein bars should it even taste of anything?

A lot of research is being done on setting standards for taste categories of insects. The researchers want to establish taste definitions for mealworms and crickets. In this way, it will be easier for food developers to choose the most suitable ingredient for a pasta, a shake or a burger. For now, it seems to be a good strategy to promote the product with a familiar taste, which could be vanilla, chocolate or onion-pepper. These tastes are well known and generally preferred and make it easier for the target customer to know what to expect.

In case these ambitious properties meet the expectations of the customer and the product becomes successful, it is common for larger companies to take a closer look. A trend-oriented product innovation that has already proven itself in initial trials on the market attracts companies to refresh their portfolios. In this way, sooner or later a successful start-up usually ends up in the corporate family of a multinational corporation. The focus then inevitably shifts towards economic goals. Market penetration, market leadership and cost minimisation or cost leadership are becoming new strategic guidelines. The original goal of making a truly better product for the planet and people may be compromised by a company's economic pressures. What follows is a cost-optimised version of a green product. That doesn't have to mean that resource conservation and fair trade fall by the wayside. Many multinational corporations are working flat out on sustainability concepts. For example, METRO, a leading international specialist in food wholesale, is currently defining sustainability criteria for a range of several thousands of products in order to make them 55% more sustainable by 2030 (Metro AG, 2021).

target audience Innovations in the food market are convenient, healthy, sustainable or tasty. They are rarely revolutionary. They address the desire for variety, cooking experience, taste experience, convenience, health, fitness, responsibility for planet, people and animals. Targeting an audience open to the advantage of the product drastically increases the chance of being successful.

Following this, two approaches are possible. Either find a product for the target group or find a target group for the product. Though in the literature, there might be a preference for approach number one, in reality, it often seems to be the second approach that is taken. Innovative concepts like Nespresso's coffee cups are designed for convenience and for the luxury-seeking double income, no kids (DINK's). The jury is still out on whether the buffalo worm peanut spread from the start-up Mybugbar will be liked more by children or students. Yet, it is crucial to come to a decision because the claim could vary from 'funny bugs' to 'for better brains'. Furthermore, packaging, communication channel and USP depend heavily on this choice of target group.

Not everyone will be ready to swap their currywurst for a mealworm croquette. It takes a dose of altruism to give up cherished habits and thereby reduce the pressure on the planet's dwindling resources. But a trend would not be a trend if it were not for a group of innovators setting sail for new shores and habits, trying the new things for us (see also Case 20.6). After a while, the early adopters follow, a group of people willing to experiment and open to new ideas and ultimately not wanting to miss any trends. Who are these people? This theory is used in marketing under the name Diffusion of Innovation (Rogers, 2003) and helps to sharpen the focus on relevant target groups (Figure 20.4). Once a critical mass of trendsetters has been reached, it is easier for others to follow. The task of marketing is to reach out to a suitable group of people. The higher the level of innovation the product displays, the stronger the need to reach out to the innovators. Who are they? Where are they? What appeals to them most?

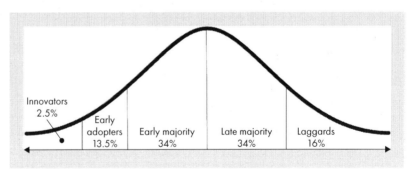

Figure 20.4. Diffusion of innovation theory (Rogers, 2003).

Case 20.6. Athletes, flexigans and entovegans.

Athletes on all levels of performance, which includes sporty people, were identified as the first promising target group for insect protein. The well-established trend of protein supplements mixed with the slowly but steadily growing trend of environmental consciousness supported the idea. The focus is on the functionality of the protein for recovery and building of muscles after and during sport. Environmental friendliness is the additional benefit of insect protein compared to whey protein from milk.

The success of this strategy remains to be seen. Because although some small businesses have launched protein bars, shakes and spreads, and have even been listed in supermarkets and speciality shops alongside their online sales, they remain invisible to the majority of consumers.

Departing from the two main values of insect as food the search continues for the environmentally and/or health-conscious consumers. At the same time, there must be the openness to try novel food. The growing number of vegetarians are by definition excluded from the target customer. Insects are animals and thus not a choice for vegetarians and vegans. A promising target group seems to be the so-called flexitarians (flexible vegetarians). They are heading towards a plant-based diet because of animal welfare and planetary boundaries without being dogmatic about it. This group seems to be highly suitable for the consumption of insects. To understand the developments better, a whole series of scientific studies were conducted in the previous decade (Naranjo-Guevara *et al.*, 2020). Numerous studies conclude that the previously well-defined group of flexitarians can be further subdivided. One subdivision is emerging under the name flexigans. This group is characterised by a desire for both healthy and environmentally friendly products. The rejection of animal products consists of a rejection of environmentally harmful methods and, at the same time, quality deficits caused by cheap production. Flexigans could represent a promising target group for alternative plant-based/insect-based products in this regard (Figure 20.5.)

The label *ento-vegan* is entering the market to establish a new category for a group of health/animal and environmental conscious consumers.

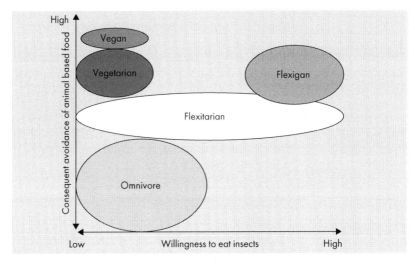

Figure 20.5. Flexigans as a new sub-group of vegetarians.

Defining a suitable target group is an important first step in all marketing activities. The analysis of trends, literature and market research, can lead to a more detailed profile. Such a profile can be translated into *persona*. Personas – also buyer, user or customer persona – were informally developed by Alan Cooper in the early '80s as a way to sympathise with and internalise the mindset of people who would eventually use the software he was designing.

Having a real person in mind (like Selma in Figure 20.6) makes it much easier to think of what he/she really likes. More than this, it is necessary to better understand (eating) habits, (buying) preferences, (cooking) skills, social media usage, opinions, values and emotions. The buyer's persona makes explicit assumptions about the target audience. By creating a character, it is possible to engender more interest and empathy than with a faceless crowd (Adlin and Pruitt, 2010).

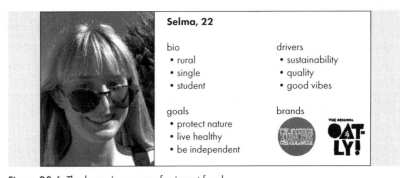

Figure 20.6. The buyer's persona for insect food.

Price

How much would you pay for a cricket snack? That is a rather abstract question as it is difficult to find comparisons in terms of prices. Yet, willingness to pay is a common question in consumer surveys. The results are dubious. With such direct questions, willingness to pay is often underestimated, as the respondent is biased to give socially acceptable responses. They want to present themselves as a smart buyer and therefore are more likely to indicate a relatively low willingness to pay within the survey. Indirect questions are valuable to gain more realistic estimates on willingness to pay. For this, the *Van Westendorp Pricing Model* (Van Westendorp, 1976) can help create an estimated price range accepted by a possibly broad range of customers. In this method, respondents answer four standardised questions, which are averaged to a value from which an acceptable price corridor can be derived. The price corridor indicates the price range between the calculated maximum and minimum (Figure 20.7).

In general, when an offer cannot be easily assessed by a customer, she is drawn to deduce the 'hidden characteristic' of product quality from the price. A high price becomes an indicator for high quality and a low price becomes an indicator for low quality. The price-quality estimates heavily influence the processing of information, expectations as well as the final customer behaviour (Case 20.7).

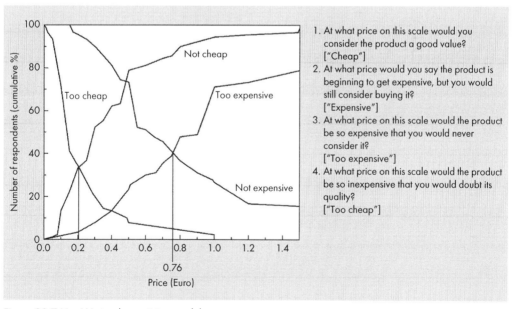

Figure 20.7. Van Westendorp pricing model.

> **Case 20.7. What is the right price for cricket crisps?**
>
> New products lack comparability for consumers. Is he buying the crunchy cricket crisps instead of potato crisps? Or is he serving them to his guests as an appetiser? The event often justifies extreme differences in willingness to pay. He would spend a maximum amount of three euros for his potato crisp alternative, yet up to twelve euros for the stylish appetiser. The unbelievable price of around 400 euros per kilogram of cricket (6/2021) comes about because the production costs are still high in the introductory phase of product innovation. The production processes for larger quantities of this product still have to be standardised. Once insects have established themselves as a food, the goal is to produce a more sustainable and cheaper protein. It can be difficult in the introductory phase to find potential consumers for food innovations, as these often cannot compete (in terms of price) with more established alternatives. Pinaks' strategy is to offer insect crackers with 7.5% buffalo flour, an amount large enough to provide nutrients to the cracker but low enough to keep costs in check.

In the food sector, the attributes of being novel, being hard to compare as well as being made by an unknown producer can lead to increased customer mistrust in the face of low prices. Additionally, the assumption that a low price comes at the expense of the environment, animal welfare and fairness, is playing an increasing role in consumer choices regarding food products.

When pricing food products, the manufacturer must bear in mind that the retailer will add a particular percentage to the price on the wholesale product. This percentage amount differs from food category and country and might easily amount to 50%. The percentage is used to pay for the cost of shipping, storing and selling the food product.

The 'Place' to be *visible and available*

A crucial question in the marketing mix is how the product reaches the customer. There are a multitude of possible channels from supermarkets to specialist retailers, your own business, to franchise companies and the Internet. In an omnichannel strategy the product is available where the target audience is present. For a new product, visibility is a crucial success factor.

The choice of distribution channel is a question of cost, as listing in an established distribution channel such as a supermarket is expensive. In direct sales with an internal sales force, customer contact is a great advantage, but it also incurs high costs. The right sales channel ultimately depends on the target group and the marketing budget.

The big challenge for a food start-up is to get one of the limited spaces on the supermarket shelf. Most of them are already occupied by large food companies who spare no expense or effort to put the products in the best (shelf) light. There is a whole domain of scientists concerned with the phenomenon of where and how a product is best perceived. From the placement at eye level, the display that stands in the way and looks like an offer (without being one), the direction and speed of walking through the store, the light and the atmosphere, nothing is left to chance. Whether store design, packaging design and product placement are planned specifically or left to chance is decisive for success.

It has never been so easy for food start-ups to get onto the supermarket shelf. More and more retailers are offering special programmes, competitions or platforms that support and promote founders in this area. In the German retail sector, around 5% of sales are currently attributable to products from start-ups, with a clear upward trend. This is because small companies often pick up on new trends, such as vegan, organic and fair trade first. With their products, the big chains can differentiate themselves from the competition.

Let's trace a classic route to the supermarket shelf. The manufacturer convinces the retailer's purchasing department to look at the product. There is a presentation with tasting, then quality management checks whether the novelty delivers what it promises. It is then listed in selected markets for a test phase. Depending on the sales success, it comes to all supermarkets or flies out again as quickly as it got in. An additional threat is the me-too strategy of supermarkets copying successful products with their own branding. As soon as the customer has adapted his buying habits to the supermarket brand product with the significantly lower price, the original is listed out.

The economic risk of developing, producing and distributing a new type of product is and will remain high.

online food retailing Online food retailing has really picked up speed since the corona pandemic broke out at the beginning of the year 2020. The supermarket chains are experimenting with numerous delivery concepts. So far, discounters have hardly had any logistics solutions that fit the concept of price leadership. Concepts such as the Dutch *Picnic* or the German *Frischepost* are experimenting with maximum delivery flexibility, with little chance of profitability so far.

Gorillas and *Flink* are examples that developed an infrastructure for fast last-mile-delivery of products for basic human needs. These on-demand delivery start-ups promise to bring the ordered product to the desired place within 10 minutes. Users of the apps can access more than 2,000 basic products at retail price. By breaking with the purchase and supply chain concept of the traditional retail trade, Gorillas and Flink are shaping a new consumer behaviour in the food sector (need-order-get).

While the online sale of wine and delicacies is already established, fruits, vegetables and everyday products have yet to prove themselves. How can tomatoes and avocados be made attractive in the online shop? Many of the factors on which to base a decision are no longer available, such as haptic perception, smell, or inspecting the packaging. Other stimuli are particularly important, such as meaningful product images, detailed product descriptions, videos of the product, reviews of other customers or the emotional approach and last but not least, trust (see also Case 20.8).

Case 20.8. Would you buy insects online?

Offering food online is a challenge. The customer must be convinced both of the safety and the taste of the purely virtual product. A new product can convince with good pictures and descriptions. It is much more difficult for a completely unknown product. Therefore, the conditions to successfully sell insect products in the online shop are extremely unfavourable. There are several factors that make it difficult to convince customers. Although, according to the law, all food that is offered in Europe needs to pass strict safety controls, the Internet still offers a certain anonymity. In fact, the first snack insects were grown in backyards and garages on homegrown organic waste, packed in a cellophane bag, labelled and sent to buyers willing to experiment. The passionate breeders were certainly not always aware that this type of production did not meet the safety requirements of the strict EU guidelines. Food scandals have repeatedly tarnished confidence in the food industry. The cases of horse meat in beef lasagne, fipronil in eggs, deadly germs in fenugreek or antibiotics in poultry meat are evidence that consumers are right to be suspicious. Yet, food products in Europe are among the safest in the world and it can generally be assumed that the safety standards have been met. Online buyers who cannot smell or touch their food, as would be the case in a retail shop, need to trust the product they buy. A well-known brand name or shop, functional and stylish packaging and well-known certifications can make a major contribution to building trust. In the example of a brand-innovative insect product, only some of these success factors can be applied. Pictures, description, packaging, use and reliability of the shop are therefore particularly important. In addition to the online offer, the success of an insect product should also largely depend on whether potential customers can be tempted to 'try it out'. Numerous tests have shown that consumers have negative expectations of the taste and also fear the sight of insect parts in the product. One way to increase acceptance is clearly by trying it. For instance, street festivals or music events provide good opportunities to get people to try novel foods. Influenced by the relaxed atmosphere and the company of other people, potential customers are more likely to try out new things that one would not pay attention to on the supermarket shelf or online shop. The aim of these campaigns is to generate a high level of acceptance through the first purchase. The customer needs to be convinced on the spot to be willing to buy it elsewhere (online!?) afterwards.

It can be costly to launch an online shop. A professional set-up and online marketing are preconditions for success. Another option is to use already established platforms, such as Amazon or market places, such as Shopify, that simplify access for customers. This latter option, however, lowers customer loyalty.

Promotion

There are around 150 different insect food products available in the North-Western EU market (in spring 2021). It is likely that you have not seen any of them yet. The link between the product (development) and the consumer (awareness) is referred to as promotion. And promotion needs to find the target customer through a multitude of possible channels.

Promotion in the marketing mix is defined as the actions used to communicate a (food) product's features and benefits; therefore, persuading the consumer to purchase the product (see also Case 20.9). There are several instruments in a communication policy such as advertising, sales promotion, personal sales, sponsorship, trade fairs, events and public relations. It is a permanent process of communication between the company and existing and potential customers. All promotion activities convey information and messages for the purpose of influencing knowledge, opinion, expectation and behaviour of the target group. The communication theory of the sender-receiver model was defined by Shannon and Weaver in the 1940s and is seen as the basis of communication policy (Shannon and Weaver, 1949). According to the model, a recipient's attention needs to be captured through a hand-tailored message. Contemporary consumers find themselves in constant visual floods. The tailored message needs to contain clues that make consumers recognise it as relevant. The science around advertisement works with all kinds of sensory triggers. Many of them subconsciously influence our behaviour. For the company, it is essential to know how and where to reach the target customer to be able to position the messages where they have a good chance to be acknowledged. The company has to know the desires and needs, the motivation and the drivers of the target customer to design the clues in a highly specific manner. It is important to communicate in the language of the target audience and to attract enough attention by sending a message for which the recipient has the appropriate antenna. The buying behaviour, referred to as consumer behaviour science, encompasses neuroscience and psychology and enables detailed insights into consciousness, emotion, perception, expectation and how these aspects lead to the actual goal, the purchase of the product (Percy and Rosenbaum-Elliott, 2016).

The product design, wording, colour, messages that make the product desirable are entirely subjective. Whether a product is perceived as attractive or not depends on one or some of the sociodemographic aspects like age, gender, education, cultural background and income. Age, for example, allows a brand to market its food towards children, youngsters, parents or the elderly according to their needs. Education comes into play because it is often associated with people buying better food. Income takes into account whether people have more disposable income to spend on more expensive products. Gender further allows food marketers to target women respectively men (archetypes) according to their shopping behaviours.

Furthermore, psychological aspects like values, opinions, desires and lifestyle choices, referred to as psychographics, are important. A person who cares about the environment might be open to meat alternatives such as insects, whereas a person with a strong passion for animals prefers plant-based alternatives.

When choosing the advertising concept, it depends on which target group is addressed. For example, influencer marketing focuses on leveraging people who influence potential buyers. The influencing takes place in a role model function whereby the influencer does or buys something and asks others to imitate it. A celebrity, an influential person or even just a sympathetic peer person with a high reach through many followers on social media channels are examples. Look at this nice guy and how he prepares his mealworm snack – and he has such an athletic body. That will convince many consumers. The suitable (social media) channel varies depending on the age and country of Facebook, Instagram, Pinterest or TikTok. The best choice is changing at a rapid pace and requires some knowledge and money to be successfully found and to gain market reach. Still, for new products this way is inevitable. The same applies for relationship marketing which refers to strategies and tactics for segmenting customers to build loyalty. It uses database marketing, behavioural advertising and analytics to target consumers and create customised offers and loyalty programs. The term viral marketing refers to the phenomenon of people passing on a marketing message, just as they would a virus or disease, from one person to another. News goes viral on social media channels by sharing the post or video at lightning speed. Guerrilla marketing is also a nice strategy for spreading a message very cheaply. It describes an unconventional and creative marketing strategy that aims to get maximum results with minimal resources. It asks explicitly about unconventional, perhaps unknown ways of generating attention. It plays with surprises and unexpected results.

Applied food science

Case 20.9. Promoting insects.

Offering insects with the slogan: 'Be brave and try the monster', might attract the attention of adventurous youngsters. At the same time, it might be the no-go message for the health-conscious parent.

Unfortunately, the target group of a brave jungle camp youth that would buy expensive adventure food such as insects seems to be rather small. The crispy whole insects like grasshoppers, crickets and mealworms are currently the most offered products in the insect as food market. And still they have not made a breakthrough in the past decade.

The diffusion of innovation model suggests addressing the innovators first. For the insect market, the innovators are suggested to be the more educated, younger and open minded. Health conscious athletes believing in superiority of functional products are a promising match. The advertisement for this group is clean, technical and informative. The message is: 'build your body – be smart and use innovation'. The bodies of the role models are shaped like machines. These influential pictures trigger the innovators of this market.

After reaching out to this group via fitness studios, Instagram and some PR campaigns, who will be next? The protein shake or bar might also be attractive for sporty women or just as a snack and for wellness reasons. But do the muscle-shaped innovators have enough influence on the moderate sport group? How important is taste in this phase? Mybugbar is experimenting with peanut spreads and different tastes to attract sports men and women. Other companies such as Beneto Foods (Germany), Plumento (Germany), Gaia foods (the Netherlands), Essento foods (Switzerland) and many others are eagerly working on product concepts for ento-vegans or flexigans. The need for specific micronutrients like vitamin B12 or calcium are especially taken into account because the insect product can be a natural source of them. The innovators for a product range of insect products with the potential for a broader target market are likely to be found among the flexigans. The motivation of this group fits well with the USP of the insect market in general. They might be an alternative for meat such as burger and sausages. They can also be seen in healthy snack, bread and side dishes. Crackers, granola and pasta are entering the market in a seemingly unspecific way of promotion at the moment. There might be potential in targeting kids with funny and educational messages about the planet. Parents as target group to actually buy the product might be pleased by the idea of supporting their children in healthy and conscious eating habits.

Besides that, the introduction of a new food category via the gastronomy channel is likely to be a promising path. It solves some of the main problems that novelties have: (1) chefs can prepare the product in the most attractive and tasty way, (2) consumers trust what the restaurant has to offer, (3) consumers do not need to learn how to prepare/cook the product, (4) consumers and chefs can enjoy the gastronomic tradition of other countries where insects are considered normal in the human diet. So far there are only a few offers in European dining rooms – but given the advantages, this route has great potential.

20.3 Conclusions

The agri and food industry is in a transition phase that is necessary if we are to feed a growing world population within our planetary boundaries. There is a movement favouring the idea of growing more local, seasonal, small scale, organic and in accordance with the conditions of the environment. Urban gardening, renting a cow or hen, being part of a cooperative are examples. And there is a party strongly in favour of technological progress in agriculture and processing, counting on the inventions of laboratory grown meat, fish and dairy, vertically grown fruit, vegetable and legumes and more efficient nutrient concepts on a personal basis. Smart and efficient farming, genetic engineering, synthetic biology are part of the innovations. The solutions might lie in a combination of both ideologies. It is not one or the other but all of them. What is important now is serious pressure to speed up the development process. Amongst other things, alternative proteins are a strategy for more efficiency. Our planetary boundaries do not allow for more food production in the current traditional way. Producing more protein with less land, water, emissions, energy must be our compass. Insects are an example of a food segment that supports the strategy. For a food company, from a purely economic point of view, it is a trend that goes well with current developments in meat alternatives. From an ecological perspective, it is the increasing pressure on the environment that requires a transformation in today's agricultural production.

What does this have to do with marketing? The modern tools of marketing communication can be very persuasive and they have great potential for the establishment of desirable products. This of course raises opportunities as well as risks. There is a German marketing saying that demonstrates the dilemma quite well: 'The worm must appeal to the fish, not to the angler'.

It can be interpreted as such: the fisherman (marketing) is a fraudulent entrepreneur who will use any means! The worm (the advertisement) is the delicious bait to attract victims in a targeted manner! The fish is a harmless and ignorant customer who is gutted under false pretences! Viewed from another perspective it can also mean that the marketing instruments of today are influential and powerful. The opportunity lies in supporting socially desirable goals (Figure 20.8). An agricultural system that will still be able to feed us

Figure 20.8. Paradigm shift in marketing: the worm must appeal to the fish not to the angler.

tomorrow is increasingly demanded by society. The associated establishment of new eating habits is a decisive success factor and at the same time the greatest hurdle. Employing compelling marketing strategies seems helpful for a higher purpose. Last but not least – the new generation of consumers are sceptical about 'worms' and will reward honesty. In this sense food marketing can play an essential role in a faster transition of the food system to one that is fair, sustainable and healthy.

References

Adlin, T. and Pruitt, J., 2010. The essential persona lifecycle: your guide to building and using personas. Elsevier, Amsterdam, the Netherlands.

American Marketing Association (AMA), 2017. Definitions of marketing. Available at: https://www.ama.org/the-definition-of-marketing-what-is-marketing/.

Ansoff, H.I., 1966. Management strategie. Moderne Industrie, München, Germany.

Borden, N.H., 1964. Concept of the marketing mix. In: Schwartz, G. (ed.) Science in marketing. John Wiley, New York, NY, USA, pp. 7-12.

Bruhn, M., 2019. Marketing: Grundlagen für Studium und Praxis. Springer, Wiesbaden, Germany.

Buzzell, R. and Gale, B., 1987. The PIMS principles: linking strategy to performance. Free Press, New York, NY, USA.

Entorganics GmbH, 2021. From food out of insects – superfood von morgen. Available at: www.entorganics.de.

European Commission (EC), 2015. Regulation (EU) 2015/2283 of the European Parliament and of the Council of 25 November 2015 on novel foods, amending Regulation (EU) No 1169/2011 of the European Parliament and of the Council and repealing Regulation (EC) No 258/97 of the European Parliament and of the Council and Commission Regulation (EC) No 1852/2001. Official Journal of the European Union L 327, 11.12.2015: 1-22.

European Commission (EC), 2021. Commission Implementing Regulation (EU) 2021/882 of 1 June 2021 authorising the placing on the market of dried *Tenebrio molitor* larva as a novel food under Regulation (EU) 2015/2283 of the European Parliament and of the Council, and amending Commission Implementing Regulation (EU) 2017/2470. Official Journal of the European Union L 194, 2.6.2021: 16-20.

Food and Agriculture Organization of the United Nations (FAO), 2013. Edible insects: future prospects for food and feed security. Wageningen UR, Wageningen, the Netherlands.

Gelbrich, K., Wünschmann, S. and Müller, S., 2018. Erfolgsfaktoren des Marketing, 2nd edition. Vahlen, München, Germany.

Grand View Research, 2019. Edible insects market size, share & trends analysis report. Edible insects market size | industry growth analysis report, 2025. Available at: https://www.grandviewresearch.com/industry-analysis/edible-insects-market.

Metro AG, 2021. Website. Available at: https://www.metroag.de/en/company.

Naranjo-Guevara, N., Fanter, M., Conconi, A.M. and Floto-Stammen, S., 2020. Consumer acceptance among Dutch and German students of insects in feed and food. Food Science and Nutrition 9: 414-428. https://doi.org/10.1002/fsn3.2006

Percy, L. and Rosenbaum-Elliott, R., 2016. Strategic adverting management, 5th ed. Oxford University Press, Oxford, UK.

Rogers E., 2003. Diffusion of innovations, 5th edition. Simon and Schuster, New York, NY, USA.

Schmidt, A., Call, L.M., Macheiner, L. and Mayer, H.K., 2019. Determination of vitamin B12 in four edible insect species by immunoaffinity and ultra-high performance liquid chromatography. Food Chemistry 281: 124-129. https://doi.org/10.1016/j.foodchem.2018.12.039

Shannon, C.E. and Warren Weaver, W., 1949. The mathematical theory of communication. University of Illinois Press, Champaign, IL, USA.

Spence, C., 2015. On the psychological impact of food colour. Flavour 4: 21. https://doi.org/10.1186/s13411-015-0031-3

Van Westendorp, P.H., 1976. NSS price sensitivity meter. a new approach to the study of consumer perception of price. Proceedings of the 29th ESOMAR Congress, Amsterdam, the Netherlands, pp. 139-167.

Wei, D. and Green, W., 2013, 2 May. We recreated the Pepsi Challenge to see what people really like. Business Insider. Available at: https://www.businessinsider.com/pepsi-challenge-business-insider-2013-5?international=true&r=US&IR=T.

Yeomans, M., Chambers, L., Blumenthal, H. and Blake, A., 2008. The role of expectancy in sensory and hedonic evaluation: the case of smoked salmon ice-cream. Food Quality and Preference 19: 565. https://doi.org/10.1016/j.foodqual.2008.02.009

About the editors

Wernaart
Mr dr Bart Wernaart (www.drwernaart.com) is professor Moral Design Strategy at Fontys University of Applied Sciences, the Netherlands, with specialism in human rights law and ethics. He is also a professional drummer, conductor and composer.

Van der Meulen
Prof. dr B.M.J. van der Meulen (www.BerndvanderMeulen.eu) is affiliated with the University of Copenhagen. He is food legal consultant and director of the European Institute for Food Law (www.food-law.nl).

Index

Printed in the United States
by Baker & Taylor Publisher Services